Machiavellian Intelligen

MACHIAVELLIAN INTELLIGENCE

Social Expertise and the Evolution of Intellect in Monkeys, Apes, and Humans

Edited by

Richard W. Byrne and Andrew Whiten
Psychological Laboratory
University of St Andrews

CLARENDON PRESS · OXFORD
1988

Oxford University Press, Walton Street, Oxford OX2 6DP
Oxford New York Toronto
Delhi Bombay Calcutta Madras Karachi
Petaling Jaya Singapore Hong Kong Tokyo
Nairobi Dar es Salaam Cape Town
Melbourne Auckland
and associated companies in
Berlin Ibadan

Oxford is a trade mark of Oxford University Press

Published in the United States
by Oxford University Press, New York

© Richard Byrne, Andrew Whiten, and the contributors
listed on pp. xi–xv, 1988

All rights reserved. No part of this publication may be reproduced, stored in a retrieval system, or transmitted, in any form or by any means, electronic, mechanical, photocopying, recording, or otherwise, without the prior permission of Oxford University Press

This book is sold subject to the condition that it shall not, by way of trade or otherwise, be lent, re-sold, hired out, or otherwise circulated without the publisher's prior consent in any form of binding or cover other than that in which it is published and without a similar condition including this condition being imposed on the subsequent purchaser

British Library Cataloguing in Publication Data
Machiavellian intelligence: Social expertise and
the evolution of intellect in monkeys, apes and
humans
1.   Primates—Evolution
I.   Byrne, Richard,   II. Whiten, Andrew
599.8'0438   GL737.P9
ISBN 0–19–852179–0
ISBN 0–19–852175–8 (pbk)

Library of Congress Cataloging in Publication Data
Machiavellian intelligence: social expertise and the evolution of
intellect in monkeys, apes, and humans/edited by Richard Byrne and Andrew Whiten.
p. cm. Bibliography: p. Includes index.
1. Primates—Behavior.   2. Animal intelligence.   3. Intellect.
4. Machiavellianism (Psychology)   5. Social behavior in animals.
6. Social psychology.   7. Primates—Evolution.   8. Human evolution.
I. Byrne, Richard (Richard W.)   II. Whiten, Andrew.
QL737.P9M273 1988 599.8'0451—dc19
ISBN 0–19–852179–0
ISBN 0–19–852175–8 (pbk.)   87–34794   CIP

Typeset by Quorum Technical Services, Cheltenham

Printed in Great Britain
at the University Printing House, Oxford
by David Stanford
Printer to the University

Preface

How did this book come to be written? It is not just a collection of conference papers, but was planned as a whole before we got in touch with any of the writers of individual chapters. To understand how we arrived at a plan it is necessary to go back to Nick Humphrey's theoretical paper of 1976. His idea—that the intelligence of primates, including humans, is adapted to deal with a complex social environment not a technological one—seemed to us to be a very good one. It soon became widely cited, but at first remained unrelated to any stream of empirical research. Then, a couple of years ago, it dawned on us that this was no longer so, and that in fact a number of exciting lines of investigation converged upon a 'social' hypothesis for the origin of intellect. Indeed, work in the area was growing exponentially. In a few more years' time, we reckoned, it would no longer be possible to put together in a single volume all the most important work; soon, topics like 'deception', 'intentionality', or 'tripartite relations' would each deserve a separate book. Pleasing as this will be, it will obscure the common ground between each topic: that of the social expertise of the animals involved. Thus, we set out to plan a book encompassing the early history, development, and state of the art on the whole area of social expertise and its relation to intelligence.

How did we design the book? Like most really good ideas, Humphrey's had been partly thought out before, by others. Two earlier papers anticipated his insight and by including them we aim to make the complete scholarly development of the subject accessible. For the rest, we identified a series of topics which we felt really must be included to give a thoroughly broad view of the diversity and interest of primate social expertise. Then we picked the people who we thought would write the most informed and exciting chapters on each topic, and asked them to contribute. And at this stage we would like to record our thanks to our authors, who not only accepted an often heavy editorial hand, but were also notably efficient in keeping to deadlines. In a few cases we did not even have to go that far, because the job had already been done in recent published literature, and there we simply sought permission to reprint the key extracts. In order that students could properly evaluate the strength of the Machiavellian intelligence hypothesis, we also solicited two chapters which looked at the main alternative hypotheses for the evolution of ancestral human intelligence. These are that intelligence evolved in response to an increasing technological sophistication of tools, or that it evolved to deal with the spatial memory problem of finding widely dispersed, ephemeral, but predictable food resources. In addition, to help the reader to relate the various approaches into a coherent whole, we contributed two editorial chapters. One distinguishes the different strands of the Machiavellian intelligence hypotheses; the other represents an introduction to all the research papers which constitute the rest of the book, for it analyses some of the many ways in which humans or animals might be considered socially intelligent. Finally, we asked Alison Jolly, one of the originators of the hypothesis, to undertake the difficult task of reading all the other chapters and making an audit of what has been achieved.

Why did we call the book 'Machiavellian Intelligence'? We expect to be criticized for appearing to emphasize the nastier side of primate social behaviour by the use of Machiavelli's name, which conjures up the use of superior knowledge and skill to deliberately manipulate, exploit, and deceive social companions. We started off, in fact, with a much more neutral title, but the fact is that the strong thrust in the data available at present is Machiavellian: in most cases where uses of social expertise are apparent they are precisely what Machiavelli would have advised! Co-operation is a notable feature of primate society, but its usual function is to out-compete other rivals for personal gain. Having said that, we do not want anyone to take us as arguing that primate uses of intelligence, let alone human ones, are limited to purely exploitative ones. In particular, it seems likely that the later course of human evolution has been characterized by a much greater emphasis on altruistic uses of intelligence. However, the force of the core idea in this book is that the high intelligence on which we rely originally evolved in response to a need for social manipulation, which we have highlighted by talking of 'the Machiavellian intelligence hypothesis'. Remember, it was not a baboon giving advice to another who said:

[The prince] must be a great simulator and dissimulator. So simple-minded are men and so controlled by immediate necessities that a prince who deceives always finds men who let themselves be deceived. For a prince, then, it is not necessary to have all the [virtuous] qualities, but it is very necessary to appear to have them . . . [It] is useful, for example, to appear merciful, trustworthy, humane, blameless, religious —and to be so—yet to be in such measure prepared in mind that if you need to be not so, you can and do change to the contrary.

St. Andrews R. W. B.
June 1987 A. W.

Editors' note

The planning and execution of this volume was at all stages a genuinely joint enterprise; therefore, there is no 'natural' order for our names on the editorial chapters or on the book itself. While varying the order between these two components seemed to us to be roughly fair, as truly Machiavellian primates we were each left with the worry that—whichever way round we settled for—we might have been outwitted! We took the only course left to primates of superior social expertise: we tossed a coin.

Acknowledgements

We would like to express our gratitude to the following for kind permission to reproduce sections of this book: Cambridge University Press (Chapter 2); *Science* (Chapter 3); University of Chicago Press (Chapter 9); Jonathan Cape (Chapter 10); Academic Press (Chapter 12); *Behavioral and Brain Sciences* (Chapter 14); *Animal Behaviour* (Chapter 15); *Philosophical Transactions of the Royal Society of London* (Chapter 19); and *American Anthropologist* (Chapter 21).

Contents

	List of contributors of new material	xi
	List of contributors of previously published material	xv
1	**The Machiavellian intelligence hypotheses: editorial** *Andrew Whiten and Richard W. Byrne*	1
	The origins of the idea	11
2	The social function of intellect *Nicholas K. Humphrey*	13
3	Lemur social behaviour and primate intelligence *Alison Jolly*	27
4	Social behaviour and primate evolution *Michael R. A. Chance and Allan P. Mead*	34
5	**Taking (Machiavellian) intelligence apart: editorial** *Andrew Whiten and Richard W. Byrne*	50
	What primates know about social relationships	67
6	Do monkeys understand their relations? *Robert M. Seyfarth and Dorothy L. Cheney*	69
7	Mapping social concepts in monkeys *Verena Dasser*	85
8	The cognitive demands of children's social interactions with peers *Peter K. Smith*	94
	Social complexity: the effect of a third party	111
9	Tripartite relations in hamadryas baboons *Hans Kummer*	113
10	Chimpanzee politics *Frans de Waal*	122
11	Alliances in contests and social intelligence *Alexander H. Harcourt*	132
	Are primates mind-readers?	153
12	A group of young chimpanzees in a one-acre field: leadership and communication *Emil W. Menzel*	155
13	'Does the chimpanzee have a theory of mind?' revisited *David Premack*	160
14	The intentional stance in theory and practice *Daniel C. Dennett*	180

Deception 203

15 Tactical deception of familiar individuals in baboons
   Richard W. Byrne and Andrew Whiten 205

16 The manipulation of attention in primate tactical
   deception   *Andrew Whiten and Richard W. Byrne* 211

17 Deception and social manipulation in symbol-using apes
   Sue Savage-Rumbaugh and Kelly McDonald 224

18 The ontogeny of tactical deception in humans
   Peter J. LaFrenière 238

Social or non-social origins of intelligence? 253

19 Social and non-social knowledge in vervet monkeys
   Dorothy L. Cheney and Robert M. Seyfarth 255

20 Tools and the evolution of human intelligence   *Thomas Wynn* 271

21 Foraging behaviour and the evolution of primate
   intelligence   *Katharine Milton* 285

Exploiting the expertise of others 307

22 An experimental study of social knowledge: adaptation to the
   special manipulative skills of single individuals in a *Macaca
   fascicularis* group   *Eduard Stammbach* 309

23 Invention and social transmission: new data from wild vervet
   monkeys   *Marc D. Hauser* 327

Taking stock 345

24 The experiential context of intellect   *John H. Crook* 347

25 The evolution of purpose   *Alison Jolly* 363

References 379

Index 411

Contributors of new material

Richard W. Byrne
Psychological Laboratory
University of St Andrews
ST ANDREWS
Fife KY16 9JU
Scotland

Dorothy L. Cheney
Department of Psychology
University of Pennsylvania
PHILADEPHIA
Pennsylvania 19104
USA

John H. Crook
Department of Psychology
University of Bristol
Berkeley Square
BRISTOL BS8 1HH
England

Verena Dasser
Ethologie und Wildforschung
Zoologisches Institut
Universität Zürich-Irchel
Winterthurerstraße 190
CH-8057 ZÜRICH
Switzerland

Daniel C. Dennett
Department of Philosophy
Tufts University
MEDFORD
Massachusetts 02155
USA

Alexander H. Harcourt
Department of Applied Biology
University of Cambridge
Pembroke Street
CAMBRIDGE CB2 3DX
England

Marc D. Hauser
Rockefeller University
Tyrell Road
MILLBROOK
New York 12545
USA

Alison Jolly
Department of Biology
Princeton University
PRINCETON
New Jersey 08544
USA

Peter J. LaFrenière
Ecole de psycho-education
Université de Montreal
750 Blvd Gouin East
MONTREAL
Quebec H2C 1A6
Canada

Kelly McDonald
Language Research Centre
Georgia State University
ATLANTA
Georgia 30303
USA

Katharine Milton
Department of Anthropology
University of California
BERKELEY
California 94720
USA

David Premack

Department of Psychology
University of Pennsylvania
PHILADELPHIA
Pennsylvania 19104
USA

Sue Savage-Rumbaugh

Department of Biology
Georgia State University
ATLANTA
Georgia 30303
USA

Robert M. Seyfarth

Department of Anthropology
University of Pennsylvania
PHILADELPHIA
Pennsylvania 19104
USA

Peter K. Smith

Department of Psychology
University of Sheffield
SHEFFIELD S10 2TN
England

Eduard Stammbach

Ethologie und Wildforschung
Zoologisches Institut
Universität Zürich-Irchel
Winterthurerstraße 190
CH-8057 ZÜRICH
Switzerland

Andrew Whiten

Psychological Laboratory
University of St Andrews
ST ANDREWS
Fife KY16 9JU
Scotland

Thomas Wynn
Department of Anthropology
University of Colorado at Colorado Springs
COLORADO SPRINGS
Colorado 80933-7150
USA

Contributors of previously published material

Michael W. Chance
Department of Pharmacology
The University
BIRMINGHAM
England

Nicholas K. Humphrey
Sub-Department of Animal Behaviour
University of Cambridge
Madingley
CAMBRIDGE CB3 8AA
England

Hans Kummer
Ethologie und Wildforschung
Zoologisches Institut
Universität Zürich-Irchel
Winterthurerstraße 190
CH-8057 ZÜRICH
Switzerland

Allan P. Mead
Department of Pharmacology
The University
BIRMINGHAM
England

Emil W. Menzel
Delta Regional Primate Centre
COVINGTON
Louisiana
USA

Frans de Waal
Laboratory of Comparative Physiology
University of Utrecht
Jan van Galenstraat 40
3572 LA UTRECHT
The Netherlands

1

The Machiavellian intelligence hypotheses: editorial

ANDREW WHITEN and RICHARD W. BYRNE

In this paper we clarify the different claims and hypotheses which have emerged in the three classic papers which follow in Chapters 2–4, and show why they are so exciting. As the rest of the book confirms, the idea of social intelligence is one whose time has come, but such ideas have been struggling to the surface for some time, in interestingly different forms.

Chronology of the origins of the central ideas

Nick Humphrey's (1976) essay on 'The Social Function of Intellect' is reprinted here as the last of the line of 'classic' early papers because we believe it to be the single most important seed from which much of the research reported in this book has grown. We have to step back another decade to 1966 to find in Alison Jolly's paper on 'Lemur Social Behaviour and Primate Intelligence' another clear statement on the idea of primate social intellect. Humphrey noted in a postscript that his attention was drawn to Jolly's earlier paper only after his own was written, so the two stand as independent analyses of a common theme.

More recently, Kummer (in Kummer and Goodall 1985) directed our attention back another decade to a paper by Chance and Mead (1953), which in turn is referenced by neither Jolly nor Humphrey, although in several ways it anticipates their theses! Accordingly, Kummer thence refers to the Chance–Jolly–Humphrey hypothesis. A correction to his modesty about his own role in these early developments is the inclusion, in Chapter 9, of his 1967 empirical contribution on 'Tripartite Relations in Hamadryas Baboons'—the first clear description of one primate's *use* of another as a 'social tool'.

What exactly has been claimed?

Does all this lead us to talk of the Chance–Mead–Jolly–Kummer–Humphrey hypothesis? We shall leave that line of inquiry to sociologists of science and instead simply refer to 'the Machiavellian intelligence hypothesis'. However, even to refer to *one* hypothesis obscures the fact that there are actually a number of ideas and multiple hypotheses threaded through even these few early papers, and it will help to dissect what they are saying if we distinguish three issues. We shall first just sketch all three, and then consider them in more detail one by one.

What is Machiavellian intelligence?

The most basic proposition in the classic papers is that, although most research on animal and human intellect has focused on how intelligence deals with the physical or technical world (and the very concept of intelligence has been shaped by this), in reality intelligence is applied also in dealing with other individuals. To remedy the neglect of this possibility, we need to find out *how* and in what forms intelligence works in the context of social interaction.

Is Machiavellian intelligence qualitatively different to, or more sophisticated than, technical intelligence?

The fundamental hypothesis here is that the social world contrasts with the physical world in that it is more challenging: thus intellectual capacities adapted to social life may have special and even particularly sophisticated attributes. There are claims to consider here about (a) the special nature of the social environment moulding intelligence and (b) the way in which the nature of intelligence itself may thus have been specially modified, so generating a set of Machiavellian abilities different to those of technical intelligence.

Is Machiavellian intelligence the key to understanding the evolution of primate and human intelligence?

The previous issue leads us to the hypothesis that very social species should be intellectually different to, or in some ways even superior to, less socially elaborate ones. At various points in these early papers it is suggested that it is the evolution of social intelligence which explains human brain power, and that this is a relatively recent step in a sequence of phylogenetic elaboration in social intellect which is reflected not only in human superiority over other contemporary apes, but in their superiority over monkeys, and in monkeys' superiority over prosimians.

What exactly were the 'foundation stones' papers saying about each of these three issues?

What is Machiavellian intelligence?

Humphrey uses the expression 'social intelligence', Jolly talks of 'the social use of intelligence', and we shall shortly try to tease out what they mean by these terms. However, Chance and Mead (C&M) didn't actually use the term 'intelligence'—why then did Kummer pick out the 1953 paper and why do we include it here?

What C&M claimed was that:

> the social circumstances in which the evolution of the primates must, by inference, have taken place is just such as to require the development of a differentiated control of autonomic functions.

The social circumstances they referred to are essentially those of sexual competition, where one male, attracted to an oestrous female, may simultaneously have to regulate its behaviour with respect to other, competing males. Thus, although there is no mention of intelligence, C&M were pointing out that such social circumstances present a problem—or rather, a series of problems which exhibit a continuous novelty deriving from the involvement of three autonomously moving individuals. The 'differentiated control' that has evolved to cope with such circumstances, C&M suggested, is expressed mainly in terms of neocortical *suppression* or *inhibition* of tendencies towards sexual or aggressive approach.

C&M's analysis represents a detailed and extensive contribution to the themes explored in this book. However, there is a second reason for including it. For although we must admire what was achieved with it, the data on primate behaviour—particularly under field conditions—was sparse at the time, and the C&M paper provides vivid instruction on the nature of the progress our factual knowledge and theoretical framework has made in the last three decades. Field primatology blossomed in the years between the C&M and Jolly papers, and anyone already familiar with contemporary biology and primate ethology will feel much more at home with the general 'feel' of the Jolly paper. By contrast the C&M paper has something of the strangeness of world view one experiences in reading Victorian scientific treatises.

Chance and his collegaues have, since then, repeatedly refined and extended analyses of social intellect (Chance and Jolly 1970; Chance and Larsen 1976; Chance 1980*a*, *b*), but in the earlier C&M paper an important misconception was that sexual relationships represent the essence of primate life, and the key 'problem' requiring cleverness in behavioural control. This was, indeed, the picture presented by primatology at the time, but this picture has changed. C&M talked of the need for a male to subtly and appropriately 'equilibrate' positive sexual attraction and negative motivation with respect to higher ranked competitors, but in fact primates need to balance a diverse range of competitive and co-operative options. Individuals may compete not only over mates, but (for example) over feeding resources, sleeping sites, location in the group (which may affect not only feeding, but predator avoidance), allies, grooming partners, playmates, and access to infants, and they may co-operate with others not only in mating, but in (for example) grooming and support in agonistic encounters. Moreover, these co-operative relationships are often not sexual, but based on reciprocity or kinship. All this is detailed in later chapters of this book. The implication is that if such a tangled web indeed selects for Machiavellian intellect, it should do so in both sexes, whereas the C&M analysis would predict only that male primates would be clever. Thus, the findings summarized in Kummer's subsequent paper (Chapter 9) present a nice contrast, for this early demonstration of 'social tool use' focused on the ways in which female hamadryas baboons expertly exploited the properties of adult males in threatening other females from a protected and powerful position. The later studies of Chance and his colleagues have, in fact, emphasized that both sexes may become involved in the maintenance of a prolonged state of unresolved conflict in some species of baboon and macaque.

Jolly's paper dates from about the same time as Kummer's and is much more obviously a direct forerunner to the central concerns of this book:

the social use of intelligence is of crucial importance to all social primates . . . social integration and intelligence probably evolved together, reinforcing each other in an ever-increasing spiral.

However, what Jolly means by 'intelligence' varies. Although at one point she says that by intelligence 'one usually means an ability to solve problems', she later appears to equate the term simply with 'learning'; at any rate, that is what she goes on to discuss for the remainder of the paper. Now, it has been forcefully argued that capacities for learning are widespread in the animal kingdom and, indeed, that no significant phylogenetic differences exist between non-humans in learning mechanisms (Macphail 1982, 1985). If this were accepted, what is special about intelligence-as-learning in the social sphere then comes down to two things: learning about society—about, for example, 'the rank and idiosyncrasies of all troop members' (see Chapters 6–8)—or learning from society—by observational learning (see Chapter 23).

Humphrey, by contrast, was careful to distinguish 'creative intelligence' from intelligence-as-learning. If 'an animal displays intelligence when he modifies his behaviour on the basis of valid inference from evidence' then creative intelligence may be distinguished as the capacity 'to infer that something is likely to happen because it is entailed by a *novel* conjunction of events'. This notion of solving novel problems harks back to C&M's

ideas, but is not restricted to interactions over sexual access. Although Humphrey is not even as explicit as we have been above in listing the variety of interactions and relationships within which social problems present themselves, his discussion clearly assumes this contemporary picture of primate society. We are then led to ask about what problems are set by social interaction which might mould intelligence in a special way.

Is Machiavellian intelligence different to or more sophisticated than technical intelligence?

It was a key argument in C&M's paper that this was the case. The social environment is special because it is reactive, and dealing with it thus requires a constant monitoring of its state and an appropriately timed regulation of even such behaviour as approaching, which in the case of a physical object would be straightforward. We might immediately point out that such a reactive component of the environment must include other species too—predators and prey, for example. However, C&M's argument was that in intra-group social interaction approach-avoidance conflict is likely to be particularly frequent. Whereas in many cases of approach-avoidance motivational conflict it will be beneficial for the conflict to be resolved one way or the other, C&M argued that there are circumstances where it will pay to maintain such a state of conflict for prolonged periods. The particular circumstances which they analyse at some length is what they label the 'trigonal' configuration of one male adjusting its behaviour relative to that of a female to whom it is attracted, and a male which it needs to avoid. Since the female and the second male are reactive to each other as well as to the first male, a refined ability for regulation of action is required, by contrast with behaviour directed at physical objects. We might note at this point the failure of physicists to solve the analogous 'three body problem' even with respect to the movement of physical objects; that is, at present it seems impossible to fully predict the result of the interaction between three moving bodies exerting gravitational effects on each other. However, C&M did not make explicit how important to their thesis it should be that interactions are triadic as opposed to dyadic; a trigonal interaction as they define it might involve any competitive interaction between two animals.

Jolly approaches this issue from a different starting point. It was an important contribution of her paper to point out that when people talked of monkey 'intelligence' they generally meant capacities 'measured in relation to gadgetry'. Two central messages of her paper were that, on the one hand we might make better sense of these capacities by considering the place of learning in primates' everyday life (where the main 'gadgets' are other monkeys) and, on the other hand, in social interaction some primates show learning capacities which are not matched by their competence in learning about technical matters:

there seems to be almost an excess capacity for learning about objects, possibly developed as a by-product of the all-important ability for social learning.

So Jolly doesn't claim any qualitative difference in the nature of learning in social and non-social contexts; rather, she suggested that a complex social context (see Chapters 9–11) may have been the evolutionary cradle for the otherwise puzzling intellectual feats revealed by laboratory psychologists, an issue we examine further in a moment.

However, Humphrey, in a way picking up where C&M left off, did explore the possibility that social intellect may be distinctive in its nature, with intriguing implications for human problem-solving. Independently, it seems, he came to the conclusion reached also by Jolly, that the intelligence exhibited by monkeys over laboratory gadgetry finds little obvious counterpart in their dealings with their natural physical environment: yet natural selection is not fond of truly excess capacities—so what is the natural function of intellect? What natural problems does a primate face? Answer: 'the life of social animals is highly *problematical*'. As in all learning paradigms the animal's task is to predict the future on the basis of the past, but what is special about the social context is that the data on which the predictions are to be based are highly changeable and contingent on one's own actions too. The metaphor of social interaction as a game in which the winner is the one who outwits the other is strong in Humphrey's analysis of what is special about social intellect: 'the game of social plot and counter-plot cannot be played merely on the basis of accumulated knowledge, any more than can a game of chess'—unlike most behaviour directed at the physical environment, such as foods. In particular, it may be necessary to change tactics as such a game evolves. In this way, he is pointing to primate intelligence being not just 'social', but *Machiavellian* in its origins.

Humphrey's analysis was vague in parts, but no primatologist can fail to appreciate that he was onto something which probably corresponds to something real, and eminently worthy of further scientific pursuit. One of the important tasks, which we start on in Chapter 5, is to specify more rigorously what cognitive operations might be distinctive to Machiavellian, as opposed to technical intellect. Other work has now focused on cognitive operations which can *only* be manifested in social dealings (see Chapters 12–18).

Is Machiavellian intelligence the key to understanding the evolution of primate and human intelligence?

All these papers imply that there is something special about primates in the psychological capacities discussed, and that such speciality represents a stepping stone to the unique intellectual powers of our own species. This notion stands in contrast to the assumption by generations of laboratory comparative psychologists that the most appropriate way to delineate the stepping stones was by tests with physical objects.

According to C&M, primates are special because they live in social groups in which females spend an unusually high proportion of their time in a state of sexual receptivity, thereby creating an almost continuous conflict situation for males. Since it was this situation that C&M proposed was the key to the evolution of 'differentiated control' the selection pressures to this effect would be intense for male primates—although not, as we have noted, for females, a point which C&M do not consider. However, where there is a possibility of female choice C&M's basic idea should apply to her too, and there is data which is consistent with the notion that the sexual triangle, from whichever corner it is viewed, can call forth sophisticated social skill: an example from our survey of primate deception (Chapter 16) is the female hamadryas baboon who groomed with a subordinate male only behind a rock which hid her actions from her harem leader-male, whilst allowing her to keep her eye on him.

Both Jolly and Humphrey, as we have noted, take a much wider view of the contexts of social intelligence. We have already quoted Jolly's assertions about the importance of social intelligence in the life of primates, but her evolutionary analysis goes to a finer level, and offers a nice logic. First, she points out that the lemurs she had studied exhibited a social structure and social interactions which were basically monkey-like, as was the evidence for social learning and the long period of immaturity in which this and other forms of learning could achieve so much. Yet, secondly, the performance of prosimians such as lemurs on conventional technological tests of intelligence has been inferior to that of monkeys. The conclusion, if these premises are accepted, is that 'social intelligence' can and did evolve ahead of 'object intelligence' (see Chapters 19–21).

It is then plausible that object-intelligence, such as we see in our own species today and glimpse in the everyday life of chimpanzees and the hominid fossil record, has evolved 'on the back of' an already well developed Machiavellian intelligence—what we might call the 'two-stage rocket' hypothesis of intellectual evolution. Actually, there appears to be an ambiguity in just what the two stages would be according to Jolly's analysis. According to what has been said so far, the first, prosimian-like stage would be a complex society exhibiting social intelligence (where intelligence = learning) and the second would be one with high technical intelligence also.

However, Jolly also mentions that in the lemurs, all interactions are essentially diadic, with an absence of such triadic interactions as protected threat and alliances observable in monkey society. In view of the comments already made about the role of unpredictability in social interaction generating social intelligence, and the manifest complexity added by a third social interactant, it must therefore remain a real possibility that prosimians are in some senses less socially intelligent than monkeys. If this were true then, crudely, we might say that the first rocket-stage would be just 'complex society', and the second 'Machiavellian intelligence'—indeed, at some points Jolly argues no more than that 'some social life preceded, and determined the nature of, primate intelligence'.

Whether there is a real ambiguity in Jolly's paper the reader may judge, but discussion of it does raise the possibility of a three-stage evolutionary rocket: a first stage of society without much creative intelligence (although capable of some forms of social learning); a second exhibiting creative Machiavellian (but minimal technical) intelligence; and a third showing creative intelligence in both spheres.

Such stages would, of course, never be as neat as that: and indeed, Humphrey develops an idea which complicates any simple rocket-stage hypothesis. He suggested that it might be the case that intellectual capacities which evolved to deal primarily with social transactions might often do an inappropriate job in solving physical problems. Humphrey introduces this hypothesis so engagingly that we are disinclined to summarise it more than does the quotation that 'men expect to argue *with* problems rather than being limited to arguing *about* them'. The implications of the idea are fascinating, but—like some others amongst the hypotheses we are exploring—may require some ingenuity to test!

If the evolution of general-intelligence-applicable-to-objects-if-pushed is indeed explicable as driven by selection primarily for Machiavellian intellect, we still need to explain the evolution of the latter. Humphrey suggests that the answer lies in the social complexity which is generated by long lifetimes and a group which includes several overlapping generations, and this leads in turn to the expectation of superior social intellect in humans compared to other great apes, and in the latter compared to monkeys. To date there exists no systematic demonstration that the latter differences in social intellect exist, but many field primatologists will suspect that this could yet be because a proper attempt has not been made.

Whatever the realities may turn out to be about phylogenetic differences, we should be alert to the possibility discussed by Humphrey that, without further change in extra-social selection pressures, any increase in

Machiavellian skill by one 'player in the game' will select for enhanced skill in the other, both in competitive and cooperative interaction. One can thus imagine an evolutionary spiral of Machiavellian cleverness and the question then shifts to why such a spiral should halt; perhaps there are dynamically stable states, but if so we have little understanding of them at present.

This is an appropriately speculative note on which to finish this brief introductory analysis of the papers introducing the idea of social intelligence. They are full of ideas. In the rest of the book, some of them have been refined and tested; others may be tested in future although with more difficulty, and perhaps others again will turn out to be untestable. However, the time is ripe to explore them fully.

The origins of the idea

2

The social function of intellect*
NICHOLAS K. HUMPHREY

Judging by citations, this is the paper which has fired the imaginations of most of those who have in subsequent years pursued empirical research on Machiavellian Intelligence. To quote Humphrey's own summary: 'I argue that the higher intellectual faculties of primates have evolved as an adaptation to the complexities of social living. For better or worse, styles of thinking which are primarily suited to social problem-solving colour the behaviour of man and other primates even towards the inanimate world'.

Henry Ford, it is said, commissioned a survey of the car scrap yards of America to find out if there were parts of the Model T Ford which never failed. His inspectors came back with reports of almost every kind of breakdown: axles, brakes, pistons—all were liable to go wrong. However, they drew attention to one notable exception, the kingpins of the scrapped cars invariably had years of life left in them. With ruthless logic Ford concluded that the kingpins on the Model T were too good for their job and ordered that in future they should be made to an inferior specification.

Nature is surely at least as careful an economist as Henry Ford. It is not her habit to tolerate needless extravagance in the animals on her production lines: superfluous capacity is trimmed back, new capacity added only as and when it is needed. We do not expect, therefore, to find that animals possess abilities which far exceed the calls that natural living makes on them. If someone were to argue—as I shall suggest they might argue—that some primate species (and mankind in particular) are much cleverer than they need be, we know that they are most likely to be wrong. However, it is not clear why they would be wrong. This paper explores a possible answer. It is an answer which has meant, for me, a re-thinking of the function of intellect.

A re-thinking, or merely a first-thinking? I had not previously given much thought to the biological function of intellect, and my impression is that few

* Reprinted from *Growing Points in Ethology* (1976), ed. P. P. G. Bateson and R. A. Hinde, pp. 303–17, Cambridge University Press.

others have done either. In the literature on animal intelligence there has been surprisingly little discussion of how intelligence contributes to biological fitness. Comparative psychologists have established that animals of one species perform better, for instance, on the Hebb–Williams maze than those of another, or that they are quicker to pick up learning sets or more successful on an 'insight' problem; there have been attempts to relate performance on particular kinds of tests to particular underlying cognitive skills; there has (recently) been debate on how the same skill is to be assessed with 'fairness' in animals of different species; but there has seldom been consideration given to why the animal, in its natural environment, should need such skill. What is the use of 'conditional oddity discrimination' to a monkey in the field (French 1965)? What advantage is there to an anthropoid ape in being able to recognize its own reflection in a mirror (Gallup 1970)? While it might indeed be 'odd for a biologist to make it his task to explain why horses can't learn mathematics' (Humphrey 1973a), it would not be odd for him to ask why people can.

The absence of discussion on these issues may reflect the view that there is little to discuss. It is tempting, certainly, to adopt a broad definition of intelligence which makes it self-evidently functional. Take, for instance, Heim's (1970) definition of intelligence in man, 'the ability to grasp the essentials of a situation and respond appropriately': substitute 'adaptively' for 'appropriately' and the problem of the biological function of intellect is (tautologically) solved. However, even those definitions which are not so manifestly circular tend nonetheless to embody value-laden words. When intelligence is defined as the 'ability' to do this or that, who dares question the biological advantage of being able? When reference is made to 'understanding' or 'skill at problem-solving' the terms themselves seem to quiver with adaptiveness. Every animal's world is, after all, full of things to be understood and problems to be solved. For sure, the world is full of problems—but what exactly are these problems, how do they differ from animal to animal and what particular advantage accrues to the individual who can solve them? These are not trivial questions.

Despite what has been said, we had better have a definition of intelligence, or the discussion is at risk of going adrift. The following formula provides at least some kind of anchor: 'An animal displays intelligence when he modifies his behaviour on the basis of valid inference from evidence'. The word 'valid' is meant to imply only that the inference is logically sound; it leaves open the question of how the animal benefits in consequence. This definition is admittedly wide, since it embraces everything from simple associative learning to syllogistic reasoning. Within the spectrum it seems fair to distinguish 'low-level' from 'high-level' intelligence. It requires, for instance, relatively low-level intelligence to infer that something is likely to happen merely because similar things have happened in comparable circumstances in the past; but it requires high-level intelligence to infer that something is likely to happen because it is entailed by a novel conjunction of events. The former is, I suspect, a comparatively elementary skill and widespread through the animal kingdom, but the latter is much more special, a mark of the 'creative' intellect which is characteristic especially of the higher primates. In what follows I shall be enquiring into the function chiefly of 'creative' intellect.

Now I am about to set up a straw man. However, he is a man whose reflection I have seen in my own mirror, and I am inclined to treat him with respect. The opinion he holds is that the main role of creative intellect lies in practical invention. 'Invention' here is being used broadly to mean acts of intelligent discovery by which an animal comes up with new ways of doing things. Thus it includes not only, say, the fabrication of new tools or the putting of existing objects to new use but also the discovery of new behavioural strategies, new ways of using the resources of one's own body. However, wide as its scope may be, the talk is strictly of 'practical' invention, and in this context 'practical' has a restricted meaning. For the man in question sees the need for invention as arising only in relation to the external physical environment; he has not noticed—or has not thought it important—that many animals are social beings.

You will see, no doubt, that I have deliberately built my straw man with feet of clay, but let us, nonetheless, see where he stands. His idea of the intellectually challenging environment has been perfectly described by Daniel Defoe. It is the desert island of Robinson Crusoe—before the arrival of Man Friday. The island is a lonely, hostile environment, full of technological challenge, a world in which Crusoe depends for his survival on his skill in gathering food, finding shelter, conserving energy, avoiding danger. And he must work fast, in a truly inventive way, for he has no time to spare for learning simply by induction from experience. However, was that the kind of world in which creative intellect evolved? I believe, for reasons I shall come to, that the real world was never like that, and yet that the real world of the higher primates may in fact be considerably more intellectually demanding. My view—and Defoe's, as I understand him—is that it was the arrival of Man Friday on the scene which really made things difficult for Crusoe. If Monday and Tuesday, Wednesday, and Thursday had turned up as well then Crusoe would have had every need to keep his wits about him.

However, the case for the importance of practical invention must be taken seriously. There can be no doubt that for some species in some contexts inventiveness does seem to have survival value. The 'subsistence technology' of chimpanzees (Goodall 1964; Teleki 1974) and even more that of 'natural' man (Sahlins 1974) involves many tricks of technique which appear *prima facie* to be products of creative intellect, and what is true for these anthropoids must surely be true at least in part for other species. Animals who are quick to realize new techniques (in hunting, searching, navigating, or whatever) would seem bound to gain in terms of fitness. Why, then, should one dispute that there have been selective pressures operating to bring about the evolution of intelligence in relation to practical affairs? I do not of course dispute the general principle; what I question is how much this principle alone explains. How clever does a man or monkey need to be before the returns on superior intellect become vanishingly small? If, despite appearances, the important practical problems of living actually demand only relatively low-level intelligence for their solution, then there would be grounds for supposing that high-level creative intelligence is wasted. Even Einstein could not get better than 100 per cent at O-level. Can we really explain the evolution of the higher intellectual faculties of primates on the basis of success or failure in their 'practical exams'?

My answer is no, for the following reason: even in those species which have the most advanced technologies the exams are largely tests of knowledge rather than imaginative reasoning. The evidence from field studies of chimpanzees all point to the fact that subsistence techniques are hardly, if ever, the product of premeditated invention; they are arrived at instead either by trial-and-error learning or by imitation of others. Indeed, it is hard to imagine how many of the techniques could in principle be arrived at otherwise. Teleki (1974) concluded on the basis of his own attempts at 'termiting' that there was no way of predicting *a priori* what would be the most effective kind of probe to stick into a termite hill, or how best to twiddle it or, for that matter, where to stick it. He had to learn inductively by trial-and-error or, better, by mimicking the behaviour of Leakey, an old and experienced chimpanzee. Thus, the chimpanzees' art would seem to be no more an invention than is the uncapping of milk-bottles by tits, and even where a technique could in principle be invented by deductive reasoning there are generally no grounds for supposing that it has been. Termiting by human beings is a case in point. In northern Zaire, people beat with sticks on the top of termite mounds to encourage the termites to come to the surface. The technique works because the stick-beating makes a noise like falling rain. It is just possible that someone once upon a time noticed the effect of falling rain, noticed the resemblance between the sound of rain and the beating of sticks, and put two and two together, but I doubt if that is how it happened; serendipity seems a much more likely explanation. Moreover, whatever the origin of the technique, there is certainly no reason to invoke inventiveness on the part of present-day practitioners, for these days it is culturally transmitted. My guess is that most of the practical problems that face higher primates can, as in the case of termiting, be dealt with by learned strategies without resource to creative intelligence.

Paradoxically, I would suggest that subsistence technology, rather than requiring intelligence, may actually become a substitute for it. Provided the social structure of the species is such as to allow individuals to acquire subsistence techniques by simple associative learning, then there is little need for individual creativity. Thus, the chimpanzees at Gombe, with their superior technological culture, may, in fact, have less need than the neighbouring baboons to be individually inventive. Indeed, there might seem on the face of it to be a negative correlation between the intellectual capacity of a species and the need for intellectual output. The great apes, demonstrably the most intellectually gifted of all animals, seem on the whole to lead comparatively undemanding lives, less demanding not only than those of lower primates, but also of many non-primate species. During 2 months I spent watching gorillas in the Virunga mountains I could not help being struck by the fact that of all the animals in the forest the gorillas seemed to lead much the simplest existence—food abundant and easy to harvest (provided they knew where to find it), few if any predators (provided they knew how to avoid them),—little to do, in fact (and little done), but eat, sleep and play. The same is arguably true for natural man. Studies of contemporary Bushmen suggest that the life of hunting and gathering, typical of early man, was probably a remarkably easy one. The 'affluent savage' (Sahlins 1974) seems to have established a *modus vivendi* in which, for a period of perhaps 10 million years, he could afford to be not only physically, but intellectually lazy.

We are thus faced with a conundrum. It has been repeatedly demonstrated in the artificial situations of the psychological laboratory that anthropoid apes possess impressive powers of creative reasoning, yet these feats of intelligence seem simply to not have any parallels in the behaviour of the same animals in their natural environment. I have yet to hear of any example from the field of a chimpanzee (or for that matter a Bushman) using his full capacity for inferential reasoning in the solution of a biologically relevant practical problem. Someone may retort that if an ethologist had kept watch on Einstein through a pair of field glasses he might well have come to the conclusion that Einstein too had a hum-drum mind. However, that is just the point: Einstein, like the chimpanzees, displayed his genius at rare times in 'artificial' situations—he did not use it, for he did not need to use it, in the common world of practical affairs.

Why then do the higher primates need to be as clever as they are and, in particular, that much cleverer than other species? What—if it exists—is the natural equivalent of the laboratory test of intelligence? The answer has, I believe, been ripening on the tree of the preceding discussion. I have suggested that the life of the great apes and man may not require much in the way of practical invention, but it does depend critically on the possession of wide factual knowledge of practical technique and the nature of the habitat. Such knowledge can only be acquired in the context of a social community—a community which provides both a medium for the cultural transmission of information and a protective environment in which individual learning can occur. I propose that the chief role of creative intellect is to hold society together.

In what follows I shall try to explain this proposal, to justify it, and to examine some of its surprising implications.

To me, as a Cambridge-taught psychologist, the proposal is, in fact, a rather strange one. Experimental psychologists in Britain have tended to regard social psychology as a poor country cousin of their subject—gauche, undisciplined, and slightly absurd. Let me recount how I came to a different way of thinking, since this personal history will lead directly in to what I want to say. Some years ago I made a discovery which brought home to me dramatically the fact that, even for an experimental psychologist, a cage is a bad place in which to keep a monkey. I was studying the recovery of vision in a rhesus monkey, Helen, from whom the visual cortex had been surgically removed (Humphrey 1974). In the first 4 years I'd worked with her, Helen had regained a considerable amount of visually guided behaviour, but she still showed no sign whatever of three-dimensional spatial vision. During all this time she had, however, been kept within the confines of a small laboratory cage. When, at length, 5 years after the operation, she was released from her cage and taken for walks in the open fields at Madingley her sight suddenly burgeoned and within a few weeks she had recovered almost perfect spatial vision. The limits on her recovery had been imposed directly by the limited environment in which she had been living. Since that time, in working with laboratory monkeys I have been mindful of the possible damage that may have been done to them by their impoverished living conditions. I have looked anxiously through the wire mesh of the cages at Madingley, not only at my own monkeys, but at Robert Hinde's. Now, Hinde's monkeys are rather better-off than mine. They live in social groups of eight or nine animals in relatively large cages, but these cages are almost empty of objects, there is nothing to manipulate and nothing to explore. Once a day the concrete floor is hosed down, food pellets are thrown in, and that is about it. So I looked, and seeing this barren environment, thought of the stultifying effect it must have on the monkey's intellect. Then one day I looked again and saw a half-weaned infant pestering its mother, two adolescents engaged in a mock battle, an old male grooming a female whilst another female tried to sidle up to him, and I suddenly saw the scene with new eyes: forget about the absence of objects, these monkeys had each other to manipulate and to explore. There could be no risk of their dying an intellectual death when the social environment provided such obvious opportunity for participating in a running dialectical debate. Compared to the solitary existence of my own monkeys, the set-up in Hinde's social groups came close to resembling a simian School of Athens.

Several of the other contributors to this book [i.e. 'Growing Points in Ethology', 1976—Eds.] consider the dialectics of social interaction, and do so with much more authority than I can. None of them, I think, would claim that scientific study of the subject is yet far advanced. Much of the best published literature is in fact genuinely 'literature'—Aesop and Dickens make, in their own way, as important contributions as Laing, Goffman, or Argyle. However, one generalization can, I think, be made with certainty: the life of social animals is highly problematical. In a complex society, such as those we know exist in higher primates, there are benefits to be gained for each individual member both from preserving the overall structure of the group, and at the same time from exploiting and out-manoeuvring others within it (see later). Thus, social primates are required by the very nature of the system they create and maintain to be calculating beings; they must be able to calculate the consequences of their own behaviour, to calculate the likely behaviour of others, to calculate the balance of advantage and loss—and all this in a context where the evidence on which their calculations are based is ephemeral, ambiguous, and liable to change, not least as a consequence of their own actions. In such a situation, 'social skill' goes hand in hand with intellect, and here at last the intellectual faculties required are of the highest order. The game of social plot and counter-plot cannot be played merely on the basis of accumulated knowledge, any more than can a game of chess.

Like chess, a social interaction is typically a transaction between social partners. One animal may, for instance, wish by his own behaviour to change the behaviour of another; but since the second animal is himself reactive and intelligent the interaction soon becomes a two-way argument where each 'player' must be ready to change his tactics—and maybe his goals—as the game proceeds. Thus, over and above the cognitive skills which are required merely to perceive the current state of play (and they may be considerable), the social gamesman, like the chess player, must be capable of a special sort of forward planning. Given that each move in the game may call forth several alternative responses from the other player this forward planning will take the form of a decision tree, having its root in the current situation and growing branches corresponding to the moves considered in looking ahead from there at different possibilities. It asks for a level of intelligence which is, I submit, unparalleled in any other sphere of living. There may, of course, be strong and weak players[1]—yet, as master or novice, we and most other members of complex primate societies have been in this game since we were babies.

What makes a society 'complex' in the first place? There have probably been selective pressures of two rather different kinds, one from without, the other from within society. I suggested above that one of the chief functions of society is to act as it were as a 'polytechnic school' for the teaching of subsistence technology. The social system serves the purpose in two ways: (i) by allowing a period of prolonged dependence during which young animals, spared the need to fend for themselves, are free to experiment and explore; and (ii) by bringing the young into contact with older, more experienced members of the community from whom they can learn by imitation (and perhaps, in some cases, from more formal 'lessons'). Now, to the extent that this kind of education has adaptive consequences, there will be selective pressures both to prolong the period of untrammelled infantile dependency (to increase the 'school leaving age') and to retain older animals within the community (to increase the number of experienced 'teachers'). However, the resulting mix of old and young, caretakers and dependants, sisters, cousins, aunts, and grandparents not only calls for considerable social responsibility, but also has potentially disruptive social consequences. The presence of dependants (young, injured, or infirm) clearly calls at all times for a measure of tolerance and unselfish sharing, but in so far as biologically important resources may be scarce (as subsistence materials must sometimes be, and sexual partners will be commonly) there is a limit to which tolerance can go. Squabbles are bound to occur about access to these scarce resources and different individuals will have different interests in participating, promoting or putting a stop to such squabbles. In the last resort every individual should give priority to the survival of his own genes, and following the theoretical analysis outlined by Hamilton and Trivers (see Bertram 1976; Clutton-Brock and Harvey 1976) we may predict considerable conflicts of interest among the members of any community which spans more than a single generation; the greater the number of generations present the more complex the picture becomes. Thus, the stage is set within the 'collegiate community' for considerable political strife. To do well for oneself whilst remaining

---

[1] 'Weak players grow short bushy trees, looking a short way ahead at a mass of poorly differentiated possibilities; strong players prune the tree much more efficiently and . . . construct long thin trees, looking much deeper into a few critical variations. This pruning is the heart of the problem . . . Which branches are critical, and which are redundant and can safely be cut off?'

From an article in the *New Scientist* (vol. **66**, p. 119, 1975) on the first World Computer Chess Championship. It may be that the acquisition of social skill involves the learning of standard 'gambits' and 'defences'—relatively stereotyped patterns of interaction—which allow transactions to proceed quickly and smoothly from one critical decision point to another.

within the terms of the social contract on which the fitness of the whole community ultimately depends calls for remarkable reasonableness (in both literal and colloquial senses of the word). It is no accident, therefore, that men, who of all primates show the longest period of dependence (nearly 30 years in the case of Bushmen!) the most complex kinship structures, and the widest overlap of generations within society, should be more intelligent than chimpanzees, and chimpanzees, for the same reasons, more intelligent than cercopithecids.

Once a society has reached a certain level of complexity, then new internal pressures must arise which act to increase its complexity still further. For, in a society of the kind outlined, an animal's intellectual 'adversaries' are members of his own breeding community. If intellectual prowess is correlated with social success, and if social success means high biological fitness, then any heritable trait which increases the ability of an individual to outwit his fellows will soon spread through the gene pool. In these circumstances there can be no going back; an evolutionary 'ratchet' has been set up, acting like a self-winding watch to increase the general intellectual standing of the species. In principle, the process might be expected to continue until either the physiological mainspring of intelligence is full-wound or else intelligence itself becomes a burden. The latter seems most likely to be the limiting factor; there must surely come a point where the time required to resolve a social 'argument' becomes insupportable.

The question of the time given up to unproductive social activity is an important one. The members of my model collegiate community—even if they have not evolved a run-away intellect—are bound to spend a considerable part of their lives in caretaking and social politics. It follows that they must inevitably have less time to spare for basic subsistence activities. If the social system is to be of any net biological benefit the improvement in subsistence techniques which it makes possible must more than compensate for the lost time. To put the matter baldly: if an animal spends all morning in non-productive socializing, he must be at least twice as efficient a producer in the afternoon. We might, therefore, expect that the evolution of a social system capable of supporting advanced technology should only happen under conditions where improvements in technique can substantially increase the return on labour. This may not always be the case. To take an extreme example, the open sea is probably an environment where technical knowledge can bring little benefit and thus complex societies—and high intelligence—are contraindicated (dolphins and whales provide, maybe, a remarkable and unexplained exception). Even at Gombe the net advantage of having a complex social system may in fact be marginal; the chimpanzees at Gombe share several of the local food resources with baboons, and it would be instructive to know how far the advantage that chimpanzees have over baboons in terms of technical skill is eroded by the relatively large amount of time they give up to social intercourse. It may be that what the chimpanzees gain on the swings of technical proficiency they lose on the roundabouts of extravagant socialising.[2] As it is, in a year of poor harvest the chimpanzees in fact become much less sociable (Wrangham 1975); my guess is that they simply cannot spare the time (cf. Gibb 1956; Baldwin and Baldwin 1972). The ancestors of man, however, when they moved into the savannah, discovered an environment where technical knowledge began to pay new and continuing dividends. It was in that environment that the pressures to give children an even better schooling created a social system of unprecedented complexity—and with it unprecedented challenge to intelligence.

The outcome has been the gifting of members of the human species with remarkable powers of social foresight and understanding. This social intelligence, developed initially to cope with local problems of inter-personal relationships, has in time found expression in the institutional creations of the 'savage mind'—the highly rational structures of kinship, totemism, myth and religion which characterize primitive societies (Lévi-Strauss 1962). And it is, I believe, essentially the same intelligence which has created the systems of philosophical and scientific thought which have flowered in advanced civilizations in the last 4000 years. Yet civilization has been too short lived to have had any important evolutionary consequences; the 'environment of adaptiveness' (Bowlby 1969) of human intelligence remains the social milieu.

If man's intellect is thus suited primarily to thinking about people and their institutions, how does it fare with non-social problems? To end this paper I want to raise the question of 'constraints' on human reasoning, such as might result if there is a predisposition among men to try to fit non-social material into a social mould (cf. Hinde and Stevenson-Hinde 1973).

When a man sets out to solve a social problem he may reasonably have certain expectations about what he is getting in to. First, he should know that the situation confronting him is unlikely to remain stable. Any social transaction is, by its nature, a developing process and the development is bound to have a degree of indeterminacy to it. Neither of the social agents involved in the transaction can be certain of the

---

[2] MacFarland (1976) might like to draw an isocline linking points of 'equal net productivity' in a space defined by the two axes, 'technical skill', and 'time given over to social activity'. It is, of course, intrinsic to my argument that these axes are not independent, since I am suggesting that social activity is a prerequisite of technical skill. However, the same is probably true of his own illustrative example, since a university lecturer's teaching ability is almost certainly not independent of his research ability.

future behaviour of the other; as in Alice's game of croquet with the Queen of Hearts, both balls and hoops are always on the move. Someone embarking on such a transaction must therefore be prepared for the problem itself to alter as a consequence of his attempt to solve it—in the very act of interpreting the social world he changes it. Like Alice he may well be tempted to complain 'You've no idea how confusing it is, all the things being alive'; that is not the way the game is played at Hurlingham—and that is not the way that non-social material typically behaves. However, secondly, he should know that the development will have a certain logic to it. In Alice's croquet game there was real confusion, everyone played at once without waiting for turns and there were no rules; but in a social transaction there are, if not strict rules, at least definite constraints on what is allowed and definite conventions about how a particular action by one of the transactors should be answered by the other. My earlier analogy with the chess game was perhaps a more appropriate one; in social behaviour there is a kind of turn-taking, there are limits on what actions are allowable, and at least in some circumstances there are conventional, often highly elaborated, sequences of exchange.

Even the chess analogy, however, misses a crucial feature of social interaction. For while the good chess player is essentially selfish, playing only to win, the selfishness of social animals is typically tempered by what, for want of a better term, I would call sympathy. By sympathy I mean a tendency on the part of one social partner to identify himself with the other and so to make the other's goals to some extent his own. The role of sympathy in the biology of social relationships has yet to be thought through in detail, but it is probable that sympathy and the 'morality' which stems from it (Waddington 1960) is a biologically adaptive feature of the social behaviour of both men and other animals—and consequently a major constraint on 'social thinking' wherever it is applied. Thus, our man setting out to apply his intelligence to solve a social problem may expect to be involved in a fluid, transactional exchange with a sympathetic human partner. To the extent that the thinking appropriate to such a situation represents the customary mode of human thought, men may be expected to behave inappropriately in contexts where a transaction cannot in principle take place: if they treat inanimate entities as 'people' they are sure to make mistakes.

There are many examples of fallacious reasoning which would fit such an interpretation. The most obvious cases are those where men do, in fact, openly resort to animistic thinking about natural phenomena. Thus, primitive and not so primitive peoples commonly attempt to bargain with nature, through prayer, sacrifice, or ritual persuasion. In doing so, they are explicitly adopting a social model, expecting nature to participate in a transaction, but nature will not transact with men; she goes her own way regardless—while her would-be interlocutors feel grateful or feel slighted as the case befits. Transactional thinking may not always be so openly acknowledged, but it often lies just below the surface in other cases of 'illogical' behaviour. Thus, the gambler at the roulette table, who continues to bet on the red square precisely because he has already lost on red repeatedly, is behaving as though he expects the behaviour of the roulette wheel to respond eventually to his persistent overtures; he does not—as he would be wise to do—conclude that the odds are unalterably set against him. Likewise, the man in Wason's experiments on abstract reasoning, who, when he is given the task of discovering a mathematical rule typically tries to substitute his own rule for the predetermined one (Wason and Johnson-Laird 1972), is acting as though he expects the problem itself to change in response to his trial solutions. The comment of one of Wason's subjects is revealing: 'Rules are relative. If you were the subject, and I were the experimenter, then I would be right'. In general, I would suggest a transactional approach leads men to refuse to accept the intransigence of facts—whether the facts are physical events, mathematical axioms, or scientific laws; there will always be the temptation to assume that the facts will respond like living beings to social pressures. Men expect to argue with problems rather than being limited to arguing about them.

There are times, however, when such a 'mistaken' approach to natural phenomena can be unexpectedly creative. While it may be the case that no amount of social pleading will change the weather or, for that matter, transmute base metals into gold, there are things in nature with which a kind of social intercourse is possible. It is not strictly true that nature will not transact with men. If we mean by a transaction essentially a developing relationship founded on mutual give and take, then several of the relationships which men enter into with the non-human things around them may be considered to have transactional qualities. The cultivation of plants provides a clear and interesting example: the care which a gardener gives to his plants (watering, fertilizing, hoeing, pruning, etc.) is attuned to the plants' emerging properties, which properties are in turn a function of the gardener's behaviour. True, plants will not respond to ordinary social pressures (though men do talk to them), but the way in which they give to and receive from a gardener bears, I suggest, a close structural similarity to a simple social relationship. If Trevarthen (1974) can speak of 'conversations' between a mother and a 2-month old baby, so too might we speak of a conversation between a gardener and his roses, or a farmer and his corn, and the same can be argued for men's interactions with certain wholly inanimate materials. The relationship of a potter to his clay, a smelter to his ore, or a cook to his soup are all relationships of fluid mutual exchange, again proto-social in character.

It is not just that transactional thinking is typical of man; transactions are something which people actively seek out and will force on nature wherever they are able. In the Doll Museum in Edinburgh there is a case full of bones clothed in scraps of rag—moving reminders of the desire of human children to conjure up social relationships with even the most unpromising material. Through a long history, men have, I believe, explored the transactional possibilities of countless of the things in their environment and sometimes, Pygmalion-like, the things have come alive. Thus, many of mankind's most prized technological discoveries, from agriculture to chemistry, may have had their origin not in the deliberate application of practical intelligence, but in the fortunate misapplication of social intelligence.

Once Nature had set up men's minds the way she has. certain 'unintended' consequences followed—and we are in several ways the beneficiaries (Humphrey 1973b).

The rise of classical scientific method has in large measure depended on human thinkers disciplining themselves to abjure transactional, socio-magical styles of reasoning. However, scientific method has come to the fore only in the last few hundred years of mankind's history, and in our own times there are everywhere signs of a return to more magical systems of interpretation. In dealing with the non-social world the former method is undoubtedly the more immediately appropriate; but the latter is perhaps more natural to man. Transactional thinking may indeed be irrepressible: within the most disciplined Jekyll is concealed a transactional Hyde. Charles Dodgson the mathematician shared his pen amicably enough with Lewis Carroll the inventor of Wonderland, but the split is often neither so comfortable nor so complete. Newton is revealed in his private papers as a Rosicrucian mystic, and his intellectual descendants continue to this day to apply strange double-standards to their thinking—witness the way in which certain British physicists took up the cause of Uri Geller, the man who, by wishing it, could bend a metal spoon (e.g. Taylor 1975). In the long view of science there is, I suspect, good reason to approve this kind of inconsistency. For while 'normal science' (in Kuhn's sense of the words) has little if any room for social thinking, 'revolutionary science' may more often than we realize derive its inspiration from a vision of a socially transacting universe. Particle physics has already followed Alice down the rabbit hole into a world peopled by 'families' of elementary particles endowed with 'strangeness' and 'charm'. *Vide*, for example, the following report:

The particles searched for at SPEAR were the *cousins* of the psis made from one *charm* quark and one *uncharmed* antiquark. This contrasts with the *siblings* of the psis . . . (*New Scientist*, vol. 67, p. 252, 1975, my italics).

Who knows where such 'sociophysics' may eventually lead? The ideology of classical science has had a huge, but in many ways narrowing influence on ideas about the nature of 'intelligent' behaviour. But no matter what the high priests, from Bacon to Popper, have had to say about how people ought to think, they have never come near to describing how people do think. In so far as an idealized view of scientific method has been the dominant influence on mankind's recent intellectual history, biologists should be the first to follow Henry Ford in dismissing recent history as 'bunk'. Evolutionary history, however, is a different matter. The formative years for human intellect were the years when man lived as a social savage on the plains of Africa. Even now, as Browne wrote in *Religio medici*, 'All Africa and her prodigies are within us'.

Postscript

My attention has been drawn to a paper by Jolly (1966) on 'Lemur social behaviour and primate intelligence' which anticipates at several points the argument developed here. I have not attempted to re-write my own paper in a way that would do justice to Jolly's ideas; I hope that people who are intrigued by the relation between social behaviour and intelligence will refer directly to her original and interesting discussion.

In relation to both Jolly's paper and my own the question arises how can the hypotheses be tested. My central thesis clearly demands there there should be a positive correlation across species between 'social complexity' and 'individual intelligence'. Does such a correlation hold? It is not hard to find confirmatory examples; nor is it hard to find excuses for rejecting examples which are seemingly contrary—e.g. wolves (high social complexity without the requisite intelligence?) or orang-utans (high intelligence without the requisite social complexity?). However, the trouble is that too much of the evidence is of an anecdotal kind; we simply do not have agreed definitions or agreed ways of measuring either of the relevant parameters. What, I think, is urgently needed is a laboratory test of 'social skill'—a test which ought, if I am right, to double as a test of 'high-level intelligence'. The essential feature of such a test would be that it places the subject in a transactional situation where he can achieve a desired goal only by adapting his strategy to conditions which are continually changing as a consequence partly, but not wholly of his own behaviour. The 'social partner' in the test need not be animate (though my guess is that the subject would regard it in an 'animistic' way); possibly it could be a kind of 'social robot', a mechanical device which is programmed on-line from a computer to behaved in a pseudo-social way.

3

Lemur social behaviour and primate intelligence*
ALISON JOLLY

Jolly's paper is required reading for students of Machiavellian intelligence, not only because it was the first to propose explicitly the social use of primate intellect, but because it so nicely complements the ideas later and independently developed by Humphrey. In particular, Jolly's analysis was based on a broader conception of what is 'social' in primate intelligence, and it utilized her own research and that of others to draw lessons from variations amongst primates in cognition and social behaviour.

Primates are extraordinary among mammals for their complex social relationships and their ingenuity in handling (or destroying) objects. The evolutionary trends which led to the excellence of *Homo sapiens* in these lines began long before the transition from ape to man.

All monkey species are social. Although individuals may be solitary for a time, a monkey is usually part of a group throughout his life. And when he is taken from his wild group and loosed in laboratory or house he unlatches doors, solves hardware puzzles, and carefully stuffs aquariums with brass lamps and shredded medical texts (Lorenz 1957). He can even be trained to drive tractors (*Zambia News*, 31.01.65), or show the rudiments of symbolic thought (Weinstein 1945).

Some primates are not social, however, or even particularly clever. Three great modern branches diverged from each other during the Paleocene: the Old World monkeys, or Cercopithecoidea (which gave rise to apes and men); the New World monkeys, or Ceboidea; and the Prosimii (Simpson 1945; Simons 1964; Hill 1953). Many prosimians still live solitary lives (Petter 1962), and none seem to manipulate laboratory tests like even the lowliest simian (Jolly 1964). In Madagascar, the prosimian Lemuroidea, unhampered by competition with true monkeys, have radiated to fill the ecological niches of monkeys and apes. I recently spent 11 months in

* Extracted from *Science*, 153, 501–6 (1966).

Madagascar, mainly studying two social lemurs: *Lemur catta* (L. 1708), the ringtail (400 hours of observation), and *Propithecus verreauxi* (A. Grandidier, 1867), the great white sifaka (250 hours of observation) (Jolly 1966).

To precis the findings, the lemurs, like many monkeys, form troops composed of all ages and both sexes, which is unusual among non-primate mammals. They have the cohesive bonds of contact, grooming, social play, and troop attraction to infants, although the actual grooming gesture of lemurs is different from that of monkeys. The lemurs' compressed sexual season is different from anything known in monkeys, but monkey genera also differ widely from each other in length of season. Therefore, in social behaviour, as in anatomy, it seems reasonable to say that lemurs are generally primate in structure, though with their own peculiarities.

Thus, the lemurs seem to have 'monkey-type' societies without having evolved monkey-level intelligence.

Use of intelligence

What does one mean by monkey-level intelligence? One usually means an ability to solve problems with objects, under controlled laboratory conditions. This is a limitation of history and technique. In fact, whenever a psychologist tests learning, it is of inanimate objects: symbols on the alleys of a maze, plaques covering food-wells, hardware toys or sticks, and boxes. Whether the aim is 'learning' or 'insight', whether the reward is food, sight of another monkey, or just the chance to play, intelligence is measured in relation to gadgetry.

This use of intelligence is our own forte, but not the monkeys'. Zimmerman and Torrey (1965) ruefully remark,

For example, a monkey that may require lengthy pre-training and adaptation to an apparatus as well as 20 to 100 trials to solve one two-choice object-discrimination problem will, in matter of seconds, or, at most, minutes, become thoroughly adapted to a particular dominance status when introduced for the first time to a social situation with three or four cage-mates.

Or as Washburn *et al.* (1965) conclude,

learning is not a generalized ability; animals are able to learn some things with great ease and others only with the greatest difficulty. Learning is . . . the process of acquiring skills and attitudes that are of evolutionary significance to a species when living in the environment to which it is adapted.

There are three main uses of learning or insight in the wild: toward objects, including food; toward other active species, including predators; and toward fellow members of one's own species. It is clear that the speed and subtlety of primate learning differs in these three contexts.

Social uses of learning

Monkeys, more than any other mammals except their descendants, the apes and men, learn to be social. A rhesus raised in isolation from its kind may not mate normally or rear its own young (Harlow and Harlow 1965). Primates have a long youth, compared to mammals of their size, and during this period, through association, exploration, and play, the juveniles learn the ways of the troop (Washburn and Hamburg 1965). It is even possible that primates exploit the full capacity of their brain only during youth. Man, after all, accomplishes the gigantic feat of learning to speak, and may never again face such a daunting intellectual task (Hutchinson 1965).

Social lemurs also learn much about their fellows. A lemur must learn the rank and idiosyncrasies of all troop members. They share the primate character of long youth: a 3–5-month gestation (even in the mouse-sized *Microcebus*), one young, or at most twins, each year, 1.5 years to full growth, and, in the groups I studied, possibly 2.5 years to first breeding. Hand-raised *Lemur macaco* may be strongly imprinted on humans, to the extent of never acquiring the full grooming patterns.

We have, unfortunately, few measures of the complexity of learned social relations in adult monkeys. Altmann has attempted stochastic analysis of chains of rhesus interactions (1965). However, such quantitative methods became really useful only when the completeness and accuracy of observation and the complexity of the computer program outdo the monkeys' own powers of observation and memory. Most field primatologists have not yet achieved this.

At present one is limited to making qualitative comparisons. In general, the organization of *Lemur catta* troops seems as complex as that of the troops of many monkeys. In one troop there was a linear dominance order among the five males, and long chains of interactions: approaches, spats, stink-fights, redirected aggression. The males did not, however, have a 'central hierarchy' (Hall and De Vore 1965) of friends who would support each other. Also, most single interactions could be considered as involving only two animals at a time, I did not see 'protected threat' (Kummer, Chapter 9).

After further study, it may be possible to say categorically that such subtle behaviour does exist or that it never exists among lemurs. When more primate species have been studied, it may also be clear whether such interactions are common to most monkeys, or only to the active, argumentative rhesus and baboons. Therefore, though we know that lemurs, like other primates, learn much of their social behaviour, we have no scale by which to compare their relative sophistication.

In summary, the social use of intelligence is of crucial importance to all social primates. As the young develop, they depend on the troop for protection and for instruction in their role in life. Since their dependence on the troop both demands social learning and makes it possible, social integration and intelligence probably evolved together, reinforcing each other in an ever-increasing spiral. And, although it is very likely that the learned social relations of monkeys are in fact more complex than those of lemurs, our present techniques of description emphasize the similarity between lemur and monkey social interactions.

Andrew (1962) pointed out that the mammals of both America and Afro-Asia, have, as a whole, increased in brain size since the Eocene. He suggested that the increase was due to interaction between species: as prey species grew cleverer, their predators and competitors survived only by also becoming cleverer, and vice versa. The mechanism works best with a large number of species and close competition: mammalian intelligence evolved faster and farther on the large, interconnecting continents than on Australia or Madagascar.

Most primate species ignore each other when they meet, even while feeding in the same tree. Occasionally, one group chases another (Jay 1965). *Lemur catta* actively teased the peaceful Propithecus, and investigated such varied primates as *Lepilemur* and Dr C. H. F. Rowell. One can only speculate whether such behaviour might help disperse related genera (Wynne-Edwards 1962).

Few primates eat meat, except for insects. Both baboons (Washburn, Jay and Lancaster 1965) and chimpanzees (Goodall 1965) sometimes hunt to kill, but it seems unlikely that carnivorous habits played much part in primate evolution, outside the hominid line.

Predation, however, is a major factor, evoking social defensive behaviour in primates ranging from howlers to gorillas (Jay 1965; De Vore and Hall 1965). A primate learns from his troop what to fear—not from innate recognition patterns, nor by himself narrowly escaping from every foe. Whether his enemy is a wriggling snake or the menace inside a Land Rover (Washburn and Hamburg 1965), he takes his cue from others of his own species, as much as or more than from the predator's behaviour.

Lemurs gather round to mob carnivores, *Propithecus* hiccuping 'sifak, sifak,' and *Lemur catta* yapping like terriers. However, when a hawk flies past, lemurs compulsively roar or scream. If the lemurs' response to hawks should be innate, one may see here a truly primitive mechanism. Again, though, there is no scale by which to compare the lemurs' responses with those of other primates, though Andrew's hypothesis remains the most reasonable evolutionary explanation of the difference in 'general' intelligence.

Learning and objects

Finally, there is intelligence with respect to objects. Much rigorous and skilful testing has shown that monkeys do not necessarily surpass other mammals in the ability to learn simple discriminations. Instead, rhesus monkeys excel in their ability to transfer learning from one problem to the next: they rapidly form learning sets, accept reversal tests, and so forth (Warren 1965). At least one genius rhesus succeeded in elevating this capacity to learning a 'symbol'—circle means blue (Weinstein 1945)—and *Cebus* monkeys, like chimpanzees, can use tools (Kluver 1937).

There have been two great gaps in this sort of study, though the capacities observed are rarely considered either in relation to object manipulation in the wild or to social behaviour.

In the wild, it is unusual to see the sort of intelligence toward objects one can demonstrate in captivity (Hall 1963). Not even Goodall's (1963) tool-making wild chimpanzees approach the ingenuity of Kohler's (1925) or Schiller's captives (1957), Hayes's cake-baking ape (1951), or the creative mania of Morris' chimpanzee artists with their poster-paints (1962). At a far lower level, *Lemur* in captivity actively played with new objects (Jolly 1964), whereas in the wild they apparently never manipulated inedible objects.

There are two related aspects to learning about objects: willingness to pay attention to the object in the first place, and learning capacity proper. Obviously, the capacity is there in the wild primates, or they could not show it when brought into the laboratory. On the other hand, it is the circumstances of the experiment which direct their attention to objects. There seems to be almost an excess capacity for learning about objects, possibly developed as a by-product of the all-important ability for social learning.

When primates do learn about objects, it is rarely a 'discovery', but more commonly social 'imitation' (Hall 1963; Itani 1965). (How far would most human discoverers go without first having learned from their predecessors?) In the laboratory, when a normal primate has the choice of responding to a social cue or to objects, he turns first to the social cue. During early insight tests, primates from lemur (Kluver 1937) to chimpanzee (Birch 1945) would beg from the experimenter before attempting to solve a new problem (in fact, a quite accurate assessment of the real relationships of the situation).

In the wild, preference for social cues is even clearer. The Japanese workers have repeatedly shown that wild macaques learn object relations from each other. Generally, playful juveniles 'discover' a new food or action, then the rest of the troop gradually learns by imitating those animals with whom they have close social ties. One might have predicted this for very complicated actions such as washing sweet potatoes or placer mining for wheat. However, tasting and eating a new food must be the most straight-forward case. In this situation one would expect any animal to form its own learning set, yet even this is mediated through social channels (Hall 1963; Itani 1965; Kawai 1965).

This emphasizes the whole question of the relationship of 'intellectual' skills to social learning in the individual. Harlow kept rhesus without any social contact for the first 6 or 12 months of life. Though these animals later failed to make any effective social contacts, they were able to solve an extensive battery of 'learning problems' nearly as well as the control monkeys (Harlow 1965). If the uses of learning are really so compartmentalized, this is a fascinating discovery. One hopes that either the Yerkes or the Wisconsin primate laboratories might re-examine their huge bodies of data and sum up longitudinal profiles of social and object learning in individual animals rather than just the cross-sectional results of separate experiments.

To return to the lemurs, *Propithecus* and *Lemur*, like other prosimians, fail miserably on tests of monkey 'intelligence'. Here, at least, there are standard scales of comparison, though one must take account of lemurs' preference for manipulating with their mouths, not their hands. On the whole range of tests that have been tried, from object discrimination and delayed response to 'insight' problems of opening boxes and pulling strings, the lemurs fall below other primates, even the primitive New World marmosets (Jolly 1964; Andrew 1962).

This is, in part at least, a failure to direct their attention to the relevant cues (Andrew 1962), which, in turn, is related to their willingness to manipulate objects at all. In the laboratory, lemurs may actively play. As with other primates, though, their attention is greatly modified by the social situation. A tamed *Lemur* often accepts a toy it would otherwise ignore, when the toy is offered by a human friend. On the other hand, when five caged groups of *Lemur* were given a hasp-and-pin puzzle (Harlow *et al.* 1950), four groups of two to three animals repeatedly opened the puzzle, whereas the fifth did not touch it. This fifth group consisted of two males and two females in a large cage, who had formed a sort of 'troop' with well-differentiated social roles and who indulged in much social and locomotor play. This laboratory 'troop' may have approached the situation of wild troops observed by Petter (1962) and by myself, where there was much social play, and locomotor play in springy branches, yet the lemurs were never seen to manipulate or investigate an object other than food.

The lemurs clearly lack much capacity to learn about objects, and it takes the extraordinary situation of captivity to turn their attention to objects. Yet they have evolved the basic characteristics of primate society, including relatively long youth and probably a fairly large dependence on social learning.

Primate society, thus, could develop without the object-learning capacity or manipulative ingenuity of monkeys. This manipulative, object cleverness, however, evolved only in the context of primate social life. Therefore, I would argue that some social life preceded, and determined the nature of, primate intelligence.

Summary and conclusion

Our human intellect has resulted from an enormous leap in capacity above the level of monkeys and apes. Earlier, though, Old and New World monkeys' intelligence outdistanced that of other mammals, including the prosimian primates. This first great advance in intelligence probably was selected through interspecific competition on the large continents. However, even at this early stage, primate social life provided the evolutionary context of primate intelligence.

Two arguments support this conclusion. One is ontogenetic: modern monkeys learn so much of their social behaviour, and learn their behaviour toward food and toward other species through social example. The second is phylogenetic: some prosimians, the social lemurs, have evolved the usual primate type of society and social learning without the capacity to manipulate objects as monkeys do. It thus seems likely that the rudiments of primate society preceded the growth of primate intelligence, made it possible, and determined its nature.

4

Social behaviour and primate evolution*
MICHAEL R.A. CHANCE and ALLAN P. MEAD

In this paper explicit reference was not yet made to intelligence. Fundamental ideas which the rest of this book develops are, nevertheless, to be found here. Thus with these extracts and the previous chapters, the reader has available the complete conceptual foundations of the Machiavellian intellect hypotheses.

Introduction

Zuckerman (1933) has said of the primates that their morphological characters 'are generally believed to represent a primitive mammalian condition, so that it may be truly said that the primate, except for its general tendency to cerebral development, is relatively a non-specialized mammal'. No adequate explanation has been put forward, however, to account for the development of so large a cerebrum as that found in man.

Evolutionary theory will provide an explanation of the fossil record only in terms of specific selective processes. Hence, the most pertinent question in this context is: What were the selective forces acting on man's ancestors? The evidence bearing upon this problem is of two kinds. The direct evidence is founded upon the fossil record in an attempt to identify fossil sequences. The indirect evidence comes from consideration of the features exhibited by related living primates and mammals, and entails arguments by analogy from them.

Behavioural studies in primates help us in two ways: first, by showing the behaviour patterns characteristic of primates, and secondly, by defining the circumstances of their lives, which precipitate particular stresses requiring a type of adaptive response not dependent upon the possession of specialized effector organs.

It can be shown, from existing knowledge of the natural mode of life of mammals, that the primates are subject to a large element of conflict in their

---

* Extracted from *Symposia of the Society for Experimental Biology* VII, pp. 395–439 (1953).

social relations. They differ from all other mammals, in that the particular type of social conflict is an ever-present element in the life of several species. We can infer that, in the setting of primate society, this conflict has a pronounced selective action on the breeding performance of individuals within the group and thus will have evolutionary consequences of a very high order. An attempt is made to suggest what these will be.

Theoretical framework for analysis of mammalian behaviour in free-ranging conditions

The behavioural state of the organism

We recognize three classes of stimuli—the positive or attractive, evoking approach responses in an appropriately activated animal, the negative evoking movement away, and the neutral including all stimuli not evoking approach or avoidance movements.

A positive attraction arises from either or both of two sources, and varies in strength according to the contribution from each; these are the level of drive due to internal disequilibria and the degree of attraction of the external stimulus. The sum of these produces the degree of motivation. The responses to negative stimuli are dependent on the value of the external negative stimuli, and probably also on the responsiveness of the animal to them.

Awareness of a negative sign alone may lead to flight or avoidance activity; similarly, a positive sign may lead to approach. The presence of a negative sign, together with a positive sign, can lead to the suppression of approach, i.e. to slowing of the movement towards, arrest of all movement, or retreat, all of which are regarded in this context as three degrees of the suppression of approach. The external form of the resolution between the two conflicting elements expresses the relative movement values of the positive and of the negative signs. This is expressed schematically in Fig. 4.1, which summarizes the basic elements of approach and suppression of approach situations.

In diagrams 1–6 the situation is expressed with respect to animal A; in diagram 7, one primate social situation is expressed in terms of the vectors illustrated in 1–6. In diagram 1 the positive internal motivation of the animal, and the induced motivation, by virtue of the attractiveness of the object, are added. In diagram 2 all approach movements of the animal are controlled by the predominant stimulus in that they represent an attempt to find whether the positive elements of the situation are greater than the negative, leaving excess positive elements, or vice versa (i.e. stimulus

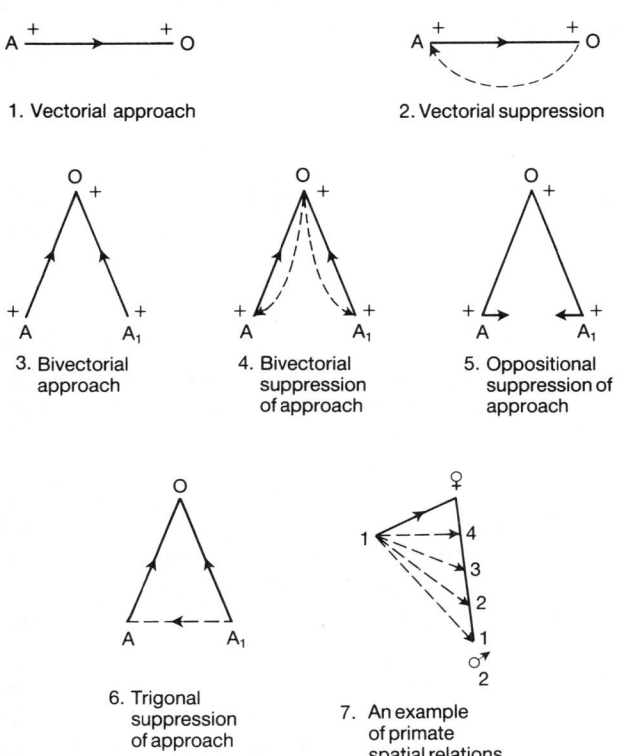

Fig. 4.1 Scheme for analysis of social relationships based on spatial organization.

ranking behaviour). This leaves a predominant sign in control of the action, except in the case of conflict where solution is not quite so simple, thus:

$A + + O + > A - + O -$ approach,
$A + + O + = A - + O -$ conflict,
$A + + O + < A - + O -$ withdrawal (Miller's avoidance; Miller 1944),

where $A + A -$ represents the internal conflict components of the animal independent of the qualities of the object in question and $O + O -$

represents the degree of induced negative and positive elements based on the degree of attractiveness or otherwise of the object.

Exactly similar is the bivectorial suppression of approach involving two animals and a common objective, except that the movement of one animal will have a bearing on the resolution of the situation by the other. Bivectorial approach is probably rare in that a situation involving two animals engaged on a common activity is unlikely to lead to complete expression of the motivation of each animal in terms of movement.

The oppositional situation is more likely to be common in cases where the object does not bear a negative sign, an oppositional vector being set up between the two animals. Stags in rut frequently express effort ranking behaviour in its simplest form when they test their strength by putting their antlers together and exerting all their strength in an attempt to push their opponent backwards. If one is much weaker than the other, he is pushed back, disengages, and retreats. If neither is strong enough to oust the other, a fight ensues (Darling 1937). The solution of such a situation is immediate, and momentary with respect to the life span of the animal. It is likely that, in nature, situations of suppression of approach involving more than one animal and a common activity, which lead to an immediate solution, will be found to be a complex involving situations 3–5 in varying degrees.

The trigonal situation is expressed in terms of the animal A, his approach being suppressed by an oppositional vector from another animal, with no immediate resolution of the situation. This demands that all movements must be made in terms of the positive common motivation and the oppositional vector.

The nature of the environment

There are two major subdivisions of the environment which influence the behaviour of animals: the reactive and the non-reactive. The reactive environment comprises all living organisms, the behaviour of which is correlated with that of the animal under consideration. It is subdivided according to whether the encounter is with a member of another species or of the same species, in which event the encounter may constitute a social event. The non-reactive environment includes all the rest of the surroundings. The characteristics of these two different aspects of the environment require the participation of different mechanisms, involved in the approach of the animal to the two different parts. The distinction is, therefore, fundamental for behaviour studies.

In approaching a non-reactive component, an essentially sequential approach is biologically appropriate, i.e. one type of act follows another, as the sensory impressions give the clues the animal receives in approaching an objective. The exact timing of the approach, in relation to the impressions derived from the object, does not have any relevance to the success or failure of the approach. In the behaviour relating an organism to the reactive environment, i.e. to the behaviour of another organism, an effective approach can only be made at a particular moment in the behaviour of the animate environment or other animal. In all these circumstances proper timing of movements involved in the behaviour during approach enhances its biological appropriateness. Suppression as well as acceleration of all movements, and a gradation of them, are thus necessary for correct timing, as well as precise spatial control of movement in an approach to the reactive environment. This involves the 'suppression of approach'.

The nature of the social environment It will be at once apparent that at times most of the animals will be involved in the same activity, and at others different animals will be occupied differently. When the animals pursue the same activity for whatever reason, the possibility of mutual interference arises, especially when the common activity is focused at a point. In these circumstances, one animal takes precedence over another, and the phenomenon of dominance arises.

The significance of a hierarchy A hierarchy is created by means of the degree of dominance between animals. This provides order by means of which the individuals' activities are related to those of the group. Wherever, therefore, the same activity exists contemporaneously in the behaviour of all the members of a group, those animals may be said to be socially integrated. This condition is usually accompanied by some degree of mutual activation. Social activities, therefore, can be inferred from the existence of dominance between animals, or where two or more animals are involved in the same activity.

The existence of a society, therefore, requires that at least during part of the behaviour pattern of the individuals their activities will coincide with those of other individuals, so that for a period the same activity sets up a system of vectors orienting the behaviour of the individuals one with another. The new element in societies is the emergence of oppositional vectors on top of already established links.

It has already been pointed out that the basis of permanent animal associations is the satisfaction of motive. These motives can be: fear of enemies, hunting for food, reproduction, and even that of gregariousness itself. Fish, equidae, ungulates, and birds live an open existence in shoals, herds, or flocks, which provide mutual protection. Wolves and many other species of canines hunt together. These associations are continuous, even though the activities they support are, in fact, periodic.

Social activation is the mutual stimulation of one individual by another, which leads to common activities, and the satisfaction of the same motive together. In this way, instead of each animal having its own cycle of motivation, overlapping in time that of another, all animals tend to be similarly motivated at the same time. Where satisfaction is achieved at a particular point in the environment, these circumstances inevitably create the conditions in which conflict arises.

Definition of conflict situations

The spatial contexts within which conflict can arise are:
(1) restriction on avoidance;
(2) vectorial, i.e. when a motivated animal meets a negative sign, as well as a positive sign, on the object to which it is oriented;
(3) trigonal, i.e. when a motivated animal receives a positive sign from one object, and a negative sign from another (this occurs when two animals simultaneously, for example, meet at a single point for the satisfaction of the same motive).

Number 1 is an essential component of frustration behaviour. Number 2 can be considered as being resolved by a simple estimate of the strengths of the two opposing stimuli, and the third requires constant equilibration in terms of spatial as well as of strength components.

Approach to the familiar and unfamiliar

We note that in general the normal environment of the rat does not provide conflict situations. That is to say, a persistent negative sign is rare, restriction on possible patterns of avoidance is rare, and the occurrence of a negative sign simultaneously with a positive sign is also rare. This is probably true of most mammals.

Mating behaviour in mammals

General

Zuckerman (1932) has recognized three types of mating behaviour in mammals. The first group is polyoestrus and includes some rodents. In them the male is always potent, and the female passes through a reproductive cycle from anoestrus to oestrus. She is receptive only during oestrus.

The second type, which is represented in the behaviour of the seals and ungulates, consists of anoestrus demarcated breeding seasons.

In the third type, characteristic of anthropoid primates, the female can be receptive throughout the whole cycle and the male is continually potent. Mating, therefore, can occur throughout the cycle, although in the primate stock there is every gradation between mating behaviour only exhibited at the discrete oestrous period and that of man where there is only slight evidence of a cyclic influence on the behaviour.

Amongst rodents which have the same type of mating relation consisting of a continuously potent male and cycling female, the rat has a cycle of 4 days, out of which oestrus occupies not more than 12 hours (Snell 1941). The cycle for the mouse occupies approximately 5 days with an oestrous period of the same length. Thus, in these two species oestrus occupies an eighth or less of the total cycle. A greater proportion of the total cycle is spent in oestrus by the primates.

'All female Old World primates experience approximately four-weekly menstrual cycles' (Zuckerman 1930). The character of the cycle can be judged in some species by the sexual swelling. In the hamadryas and yellow baboons 'the sexual skin . . . begins to swell either during or immediately after menstrual bleeding, and is suddenly absorbed soon after the middle of the cycle'. 'In the hamadryas baboon the average duration of the phase of sexual skin swelling is about seventeen days; and that of sexual skin quiescence, which is less variable in length, fifteen days.'

The significance of these figures is not immediately apparent from a comparison of their magnitude with those of a rat for example. This, however, can be revealed if a table is constructed for the proportion of time during which two males are likely to be in the presence of, and therefore, in competition for, an oestrous female, in a society in which there are approximately the same number of each sex at maturity, but in which the female's oestrous period is of different lengths.

Table 4.1 shows calculations for the smallest unit of society capable of setting up a trigonal relation between the sexes, and for twice that number, we see that, in those primate societies in which the female is in oestrus for more than one-quarter to one-third of the cycle, the males will be under mating provocation (type 1) longer than out of it, but that the reverse is true for other societies in which the oestrous interval in the female is less. In those species, therefore, where more than one-quarter of the cycle is spent in oestrus, the element of conflict is the rule rather than the exception, and when it reaches one-half, conflict is likely to be an ever-present element. Since a preponderance of females in the heterosexual groups of subhuman primates is the rule, the disparity between proportion of total time under provocation between primates and other mammals will be more pronounced than the table suggests. In these conditions the probability of continual sexual provocation and competition between males is thus very high.

Since in man the sexual propensities of the female are not dependent, as in subhuman primates, on fluctuating anatomical or behavioural features, her attractiveness for the male originates from the socio-sexual status of her behaviour. This represents the culmination of the process of the modification of sexual gestures and postures into overt forms of social communication, establishing socio-sexual status, which has originated in the primates.

Table 4.1 *Relation between proportion of reproductive cycle spent in oestrus by the females to proportion of total time under mating provocation for the males in a society of equal sex ratio*

No. of females in the group (N)	Fraction of total time spent in oestrus by any female (p)	Percentage of total time when no female is in oestrus (t)	Mating provocation time	
			Type I Percentage of time in which at least one female is in oestrus ($100 - t$)	Type II Percentage of time in which more than one female is in oestrus
2	½	25	75	25
	⅓	44	56	12
	¼	56	44	6
	⅛	77	23	1
4	½	6	94	69
	⅓	20	30	40
	¼	32	58	26
	⅛	59	41	8

Calculated from: $100 \times \left(\dfrac{N}{R}\right) \times (p)^R \times (q)^{N-R}$

percentage of time R females are together in oestrus, where the cycles occur independent of each other, when R = number of females in oestrus, p = fraction of total time spent in oestrus by any one female, $q = (1 - p)$.

The development of flexibility in the control of sexual behaviour and the consequent reduction of control by stereotyped anatomical and hormonal mechanisms as a result of the transfer of function to cortical mechanisms is shown to be a pre-eminent evolutionary process in the progress from mammals, through subhuman primates to man, in a well-documented review by Beach (1947a, b).

In conclusion, it can be stated that the majority of contemporary primates exhibit a characteristic combination of reproductive features which create the possibility of continuous mating provocation. In no other mammalian group does such continuous mating provocation occur, for all other combinations of reproductive features lead to the limitation of mating behaviour to certain physiologically defined periods, even when the animals come together in groups (*vide* red deer with seasonal anoestrus).

Therefore, we suggest that the advent of such a combination paves the way for the development of social groupings in which continuous mating behaviour, and, therefore, the continuous suppression of mating behaviour through competition, may occur. This in turn leads to the modification of sexual gestures, and even copulation itself, into a system of socially specialized forms of overt behaviour. In open terrain this will lead to the development of visual conflict and forms of socially determined behaviour in a spatial context.

Thus we believe that the following combination of factors
(1) a continuously sexually potent male.
(2) a female characterized by:
 spontaneous ovulation;
 polyoestrous cycles;
 absence of pseudopregnancy;
 no seasonal anoestrus;
 proportion of heat to cycle one-quarter or more;
 anovular cycles for at least part of the time (first year and periodically in macaque).
(3) an environment in which open terrain allows the development of a visual conflict which can be resolved in spatial terms in a social context has led to the development of the social behaviour and selective processes we are about to describe.

Sociology of the subhuman primates

Dominance

It is a valid conclusion from the evidence that sexual attraction provides the fundamental bond uniting the individuals of primate societies. The emergence of this feature to prominence in their behaviour has created primate society. This is not to deny the existence of acquired bonds between the members of a society. These in some present-day species, moreover, may persist for periods of time without primary or overt sexual activity (Carpenter 1942*b*).

The main effect of dominance is to exclude the maturing male from the group. The fact that mating behaviour is a part of a primate's repertoire of activity enabling copulation and foreplay components to be initiated at any time in the cycle, makes it likely that the relation between the sexes will be more flexible than in lower mammals. This is true of the gibbons, howler and spider monkeys (Carpenter 1934, 1935) in contrast to the baboons (Zuckerman 1932) and macaques (Carpenter 1942*a*), and a greater individual variation in reactions is evident in the individual behaviour of the chimpanzees reported by Yerkes and Elder (1936). In these circumstances initiatory, aggressive, antagonistic, and receptive behaviour may be shown by animals of either sex in their mating behaviour. On the other hand, the

Social behaviour and primate evolution 43

amount of interference between animals of the same sex for access to a member of the opposite sex is likely to be pronounced where the attractiveness of the female is high, or lasts at a moderately high level for a long time. Since the interference will be highest where attraction is most pronounced, we shall expect to find dominance exerting a marked influence when pronounced sexual activity is present. This will be more marked when more than one female is in oestrus at the same time (mating provocation type II, see Table 4.1). With lower degrees of motivation or with a rapidly varying intensity of sex drive, some degree of alternating mating may participate alongside dominance in the control of this aspect of social activity, e.g. spider and howler monkeys.

The mating provocation of primate societies constitutes what we have termed a bivectorial situation between the attractive female, and two or more males. In these circumstances we have noted that the behaviour of the two males must either be regulated by sequential mating or by some form of exclusive matings. These exclusive associations may either be the result of previous conditioning, or of dominance. The evidence from studies in the wild goes to show that dominance plays some part in many species, and a very prominent part in regulating the socio-sexual behaviour of the group in others.

The existence of relatively prolonged periods of stable relations within the group must be ascribed to the dominance relations within it, and this implies in turn that one animal is suppressing his approach with respect to another.

Spatial behaviour patterns

The factors that have been identified in primate societies provide the essential elements of conflict, as this was defined earlier. The sexual attractiveness of the female is matched and balanced by the threat of a more dominant male.

When these two social forces occur together, the region close to a dominant male where the most attractive oestrous females are situated is also the region of maximum conflict for subordinate males subject to the dominance of the overlord. This region in its simplest form is trigonal, consisting of the overlord, an oestrous female, and a subordinate male. Within such a group of individuals it should be possible to observe that in the movement of the group as a whole, the spatial relations will be maintained approximately constant in response to the attraction exerted for the males by the female, counterbalanced in the case of the subordinate animals by threat from the dominant male. Evidence of this was provided by the following observation which was made during studies of a free-ranging macaque colony. Four adults were close to the edge of an enclosure on our arrival at this colony.

The social status of these individuals was easily recognized. The situation of this group is illustrated in Fig. 4.2. The configuration shown in A was

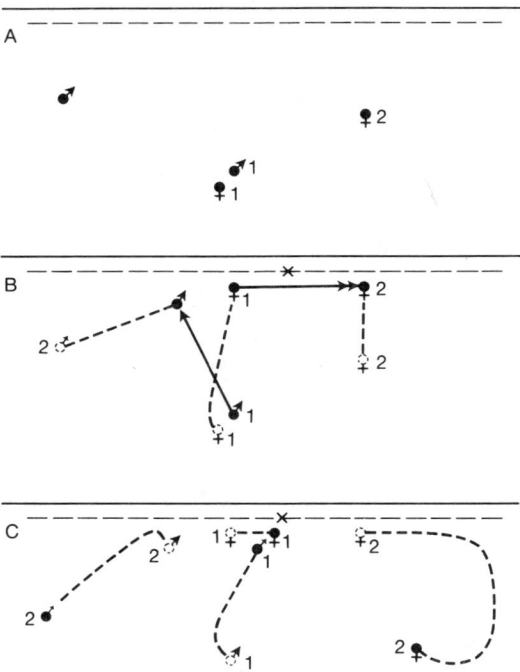

Fig. 4.2 Diagrammatic representation of observations made on a colony of *Macaca mulatta* at Dudley Zoo. Changes in position of animals within the trigonal region resulting from the introduction of food at position X. (Animals' tracks indicated by broken lines; threats indicated by double-arrowed straight lines.)

evidently one of equilibrium. The dominant male and a female showing maximal sexual skin were together copulating at intervals. A young male was in the vicinity watching this, while a pregnant female was grooming herself. A peanut was then offered as indicated in B (by a cross) through the bars. Female 1 immediately came to the bars, male 1 making no attempt to secure the peanut. Male 2 moved towards the nut and towards female 1. Female 2 was greatly interested and came to the bars. Immediately the situation became unstable. Female 2 refused to come nearer the peanut, intimidated by the more dominant female. Male 1 stood up and threatened male 2 who had moved to a position very close to female 1. The nearer female 1 moved towards the peanut the farther away from it moved female 2. The two oppositional vectors shown in B were clearly demonstrated by the behaviour of these animals, each watching the other closely throughout. Immediately equilibration occurred. Male 2 moved back towards the centre of the enclosure, stopping at about the same distance from male 1 as before. Male 1 approached female 1 who had by now obtained, and was eating, the peanut. Female 2 moved around the perimeter for several yards before moving to her original position. During the entire period of observation, the rapid movements of the eyes of the animals never ceased, and reached a peak just before the situation became once more stable, after the changes of position had ceased.

The effect of this situation on the spatial relations is clearly of a triangular character, since the farther the subordinate male is from the female, the closer it is possible for the two males to approach each other without precipitating antagonism between them, and vice versa. It is trigonal in the sense originally defined, because a negative sign provided by the dominant male inhibits the approach of a subordinate male to the attractive female. This is expressed spatially in Fig. 4.1 (diagram 7). As the subordinate male primate moves from positions 1–4, the negative sign he receives from the overlord male will increase. Clearly, there is a limiting position of proximity to the female at which the positive and negative signs represent a social equilibrium. To move beyond the limiting position increases the negative sign, and precipitates a threat or a fight as the young primate must learn. The farther away from the female he is, the more the negativity is diminished with respect to the positive, and this induces a tendency to move towards the female once more. Thus, movement of the animal does not result from stimulus ranking, which leaves a predominant stimulus; both the positive and negative aspects of the situation are in balance, and determine the locomotor behaviour of the subordinate male primate, and appear as negative and positive gradients with the properties demonstrated by Miller (1944).

We can infer, therefore,
(1) that a trigonal relation between a dominant male, a subordinate male and female develops by virtue of the sexual attraction exerted by the female for all the males;
(2) that the continual movement of the animals imposes a constant fluctuation in the spatial relations between the members of a heterosexual group which is continually assessed by the subordinate animal: (a) in order to avoid fights with the overlord, (b) in order to take any opportunities for mating;
(3) that the trigonal relation is the basis of the spatial distribution of animals in primate societies.

Carpenter considers the presence of the negative and positive elements to be general. This is true. The unique element of primate society, however, appears to be the persistent interaction between negative and positive elements in the trigonal sphere of the primate social environment, which is thereby distinguished from other mammalian environments.

In the life of baboons and macaques at least this conflict is an ever-present element requiring reassessment from moment to moment, during active foraging activities for example (Zuckerman 1932).

The evolutionary significance of the behaviour of the subordinate male

It should be noted that in primate society, a breeding premium is achieved by those animals whose attraction for the oestrous females is matched by an ability to withstand the conflict arising from the dominance of a more mature male. This ability to withstand conflict arises out of the constant equilibrational component present in their movements within the society, their stances, postures and carriage, and the absence of fighting in their behaviour. This means that these animals possess an ability to control aggressive responses under conflict. Selection of those animals which control aggressive behaviour in this region can be deduced from the disastrous effect of non-equilibrational behaviour on the fate of the maturing male primate. These are not restricted to a possible early demise, but mistakes in movements of the maturing male within the social field will affect his subsequent chances of reaching a sufficient degree of dominance to bring him within sight of the breeding position. [This is clearly a case of sexual selection as proposed by Darwin (1875) involving the intrasexual selection component distinguished by Huxley (1938).] This may be inferred from the work of Allee (1942), who has observed that success in fights makes for continued success, and failure for continued failure in fights between mice. If, as in primates, success in breeding depends upon success in a challenge, if not in a fight, then clearly those animals which do not challenge before the likelihood of success in a fight is reasonably assured will be more likely to reach a high status in the dominance hierarchy from which position most breeding is possible.

Owing to the emergence of mature mating behaviour at puberty before the maximum strength and size of the primate has been reached, failure to equilibrate for a considerable time following puberty may prevent the male reaching a social position from which he can exert his potential dominance.

Social conflict in other mammals

Unfortunately, there is insufficient information for an exhaustive review of the mating systems of mammals and their social consequences, but the information available (Echstein 1949; Asdell 1946) is sufficient for us to state that the time available to all other mammals for mating is restricted, in comparison with the primates, by virtue of a periodicity imposed on this activity by different circumstances. That such restriction exists does not rule out the possibility that social conflict may develop in all mammals, but as no other mammal is able to mate continually we should infer that the social conflict engendered in other mammals may not be precisely the same as that in primates, and when it arises it may be resolved in different ways.

The brain and social behaviour

In the present state of knowledge in neurology and related studies, the relationship between the functions of many parts of the brain and the behaviour of mammals is equivocal.

We have postulated that the enlargement of the neocortex is an anatomical adaptation to the circumstances requiring an equilibrational response, and that, therefore, the neocortex facilitates equilibration. In one of the few instances in which social behaviour has been studied following brain lesions, the subsequent behaviour of the animal is of very great interest. Immediately following either unilateral or bilateral subpial resection of the rostral cingular gyrus in the monkey, there is an obvious change in personality (Ward 1948*b*). In a large cage with other monkeys of the same size it (the monkey) showed no grooming or acts of affection towards its companions. In fact, it behaved as if they were inanimate, it would walk over them, walk on them if they happened to be in the way, and would even sit on them. It would openly take food from its companions, and appeared surprised when they retaliated, yet this never led to a fight for it was neither pugnacious nor even aggressive, seeming merely to have lost its 'social conscience' (Ward 1948*a*).

It is thus evident that, following removal of the anterior limbic area, such monkeys lose some of the social fear and anxiety which normally govern their activity, and thus lose the ability to accurately forecast the social repercussion of their own actions (Ward 1948*b*).

This example clearly illustrates that in *Macaca mulatta* the neuro-anatomical basis for the pattern of primate behaviour, the evolutionary implication of which has already been discussed, has neocortical components.

Equilibration demands of the animal an intensification of the control over its emotional responses, both facilitatory and inhibitory. It seems reasonable, therefore, to suggest that one of the major differences between a lower mammal, such as a rat, and a primate may be that in the former the control of approach and the evocation of emotion are two aspects of a single response to a negative sign, and that the capacity to differentiate these responses is limited by a physiological and possibly anatomical feature. In the primate, these two elements appear to be less rigidly associated.

Conspectus

We have identified two aspects to the uniqueness of primate behaviour. The first arises out of the fact that in those primates in which the female is receptive for longer than one-third of the cycle, mating becomes a form of behaviour possible at most times, and thus becomes of equal rank to other behavioural activities of the group. The second is the element of conflict which arises out of this situation and affects all the other activities of the group. In the evolution of the primates, these two elements combine to produce the unique type of selection. However, in studies of the behaviour of contemporary primates these two components are separable.

Consider the form of the argument. We have drawn attention to two prominent characteristic features of the primates. One is an aspect of their behaviour, and the other comprises the enlargement of a certain structure in the brain. This parallel in itself is not especially significant, since similar parallel development could be identified in the primates. What is significant is, first, that the anatomical changes in the brain are of a generalized type in a region concerned with the integration of all other brain functions; the new form of behaviour which we have described is also generalized. This means that it is possible to relate one to the other. The second point is, that these two features are also comparable in an evolutionary context, since we have adduced arguments to show that the behaviour will be selected, and we know that the anatomical enlargement has, in fact, been selected. We have, therefore, suggested a causal link between the two. Whether or not the facts we have discussed have the significance we have placed on them, requires careful consideration in any theory of primate evolution.

It must be made clear that we have not attempted to provide a complete explanation of the evolution of man. It has been our intention to provide an explanation of the ortho-selective processes in the evolution of the primates up to the stage of constitutional preadaptation on the part of some of them for man's present exploitation of the environment (see Huxley 1942). In this instance, the primary selective advantage lay in an appropriate form of behaviour; the consequent adaptation gives a greater diversity of potential behaviour and capacities.

We therefore conclude that the ascent of man has been due in part to a competition for social position, giving access to the trigonal sphere of social activity in which success was rewarded by a breeding premium, and that at some time in the past, a group of primates, by virtue of their pre-eminent adaptation to this element and consequent cortical enlargement, became pre-adapted for the full exploitation of the properties of the mammalian cortex.

5

Taking (Machiavellian) intelligence apart: editorial

ANDREW WHITEN and RICHARD W. BYRNE

The elusive, but central question of this book, 'what really is intelligence?', cannot be given a direct answer. However, we can describe some of the many facets that contribute to modern ideas on intelligence and, for each one, the consequences of intelligence being a social accomplishment, driven by selection to be Machiavellian. In doing so, this editorial chapter sets the stage for the modern empirical work that follows.

The concept of intelligence is multifaceted and that of social intelligence no less so. Research in the decade following Humphrey's paper has explored a great diversity of the components of social intelligence, to an extent that now starts to be bewildering to the new student of this topic. In this editorial chapter we therefore draw out distinctions between what appear to us to be the major parts of social intelligence, as it is studied in the research effort exemplified by the papers which follow. What are the different senses of the term 'intelligent' which allow us in an increasingly refined way to discriminate the (social) intelligence of one species from another, either quantitatively or qualitatively? And in what ways does it make sense to assess whether 'intelligence' is more sophisticated in the social domain than in the technical?

In examining such distinctions there is a danger that the meaning of 'social intelligence' will be continuously extended until it covers virtually all primate social behaviour. Let us keep our feet on the ground, then, by starting not with the primate research, but with a study of what is generally meant by 'intelligence'. Sternberg *et al.* (1981) asked people in supermarkets and railway stations to list the various behaviours and abilities which to them were 'intelligent'. A similar group of informants was then asked to rate the importance of these abilities for the concepts 'intelligence', 'academic intelligence', and 'everyday intelligence'. The same thing was done for 'experts' (researchers on human intelligence) and factor analysis was then used to reveal the major underlying components of these three types of intelligence as conceived of by the two groups of subjects. The first finding of interest to students of social intelligence is that the laypersons' responses produced a reliable factor of 'social competence' for all three types of intelligence, and this factor was second only to one of 'problem-solving skills' in the case of everyday intelligence. The notion of social intelligence generated in the preceding chapters might have seemed less revolutionary if academic students of intelligence were less blind to everyday concepts of everyday intelligence!

In fact, Sternberg *et al.* found that the 'expert' academic sample, when compared to the lay sample, made fewer distinctions between everyday intelligence, and the concepts of academic and general intelligence which it is their business to define. It is as if they saw their academic definitions extending to everyday intelligence, in a way which was, in fact, at odds with the concepts of the everyday lay sample. The ethological tradition, by contrast, is to let concepts come out of open-minded observation. Ethologists have, however, come to realize that open-minded is not blank-minded, and this is bound to be the case for such a well worn term as intelligence. The meaning and content of 'primate social intelligence' cannot come solely from watching the animals, but rather from an interaction between such observations and our already-existing concepts of intelligence. Perhaps then the best way to pursue the ethological tradition is at least to encourage the observations to interact, not with historically restricted academic definitions of intelligence, but rather with the everyday concepts of intelligence studied by Sternberg *et al.* Those concepts are diverse. They range over 'ability to apply knowledge to problems', 'awareness to world around', and 'converses well'. Such distinctions provide a loose framework for those we shall make in the next few pages and we shall refer back repeatedly to the Sternberg survey as well as to recent empirical studies. We also look to the future a little, for research on some of the components of social intelligence we are about to discuss is in its infancy.

Social knowledge

In a sense all behaviour is driven by knowledge; if an animal can do behaviour X then it must in some sense 'know' how to do X. However, there is a distinction with important implications for intelligence between this 'knowing how' and 'knowing that' (Ryle 1949). A simple example of 'knowing how' would be the rule 'on receipt of stimulus S do behaviour R'; 'knowing that' by contrast can be thought of as involving statements about the world, like 'event E1 is almost always followed by event E2'. The implication of this distinction for intelligence is that a mental system which possesses what cognitive science refers to as 'declarative representations' about the world (knowing that) has the potential for a flexible repertoire of actions, utilizing the knowledge base in a variety of ways, rather than following a limited set of 'procedural rules' (knowing how); in short, there is the possibility of 'putting two and two together' so as to generate a more intelligent response to changing circumstances. This was illustrated by Dickinson (1980), who noted that rats' learning to approach a magazine containing food-X when a tone sounded could in theory be based on either a primitive procedural representation, 'when the tone is on, approach the feeder' or on a declarative representation, something like, 'the tone causes food-X'. When, in a separate subsequent experiment, illness (actually caused by lithium chloride injection) follows ingestion of food-X with no magazine involved, contrasting representations can again be proposed to explain the rats' learning: 'when food-X is perceived, suppress eating' – or—the declarative version— 'food-X causes illness'. The problem with the procedural representations is that there seems no way they can explain the observed behaviour of the rats when put in front of the food magazine again and the tone sounded. They now do not approach the magazine (Holland and Straub 1979). Dickinson notes that such behaviour is explicable only as a result of integration of what must have been declarative representations of the two earlier sets of associated events ('the tone causes food-X' and 'food-X causes illness'), and what has come to be called cognitive learning theory is now much concerned with the role of declarative knowledge (Mackintosh 1984).

Much of the recent use of the declarative/procedural distinction in both animal and human cognitive psychology was sparked off by artificial intelligence; yet, ironically, exponents of AI have found it convenient to use an advanced procedural implementation to encode declarative knowledge (Hewitt 1968; Winograd 1970). The point of such a form of representation was, however, to enable flexible combination and modification of knowledge which the more rigid procedures Dickinson refers to could not manage.

Declarative knowlege is investigated by Cheney and Seyfarth in Chapter 19, where experimental playback of calls and other interventions are used to examine what vervet monkeys know about their environment. To take just one example, more vigilance behaviour than normal is shown when the call of a particular neighbouring vervet emanates from the wrong area; this must rest on the existence of a declarative representation along the lines of 'call X normally belongs in area Y'. As far as this component of social intelligence goes, greater intelligence amounts to greater knowledge of social facts. An important part of the Cheney and Seyfarth study is that a comparison is made between how much the monkeys know about their social world—the behaviour of other vervet monkeys, both inside and outwith the social group—and how much the monkeys know about aspects of their non-social world which can fairly be regarded as equivalent. This represents the first real attempt to test the hypothesis that there is selection for greater social than non-social intelligence in primates (see Chapters 1–4).

Possibly the first reference to 'social knowlege' in primates comes from as recently as 1982, in the title of a paper by Kummer. Contributions by his colleagues and others (Chapters 6–8 and 23) now continue the task of delineating what social knowledge primates possess.

However, is the finding that a monkey has a more or less extensive social than non-social knowledge really tackling the issue of intelligence? There is, after all, a common distinction between the individual who is truly intelligent, and another who merely knows a lot of facts. The purpose of this chapter is not to argue about what is 'really' intelligence and what is not; it is rather to tease apart capacities associated with the concept of intelligence, remaining within reasonable bounds of what is normally meant by the term. Although it may be possible to know a lot and not put this knowledge base to intelligent use, intelligence without knowledge is less likely; the point of the discussion of procedural versus declarative knowledge which opened this section was that the declarative way of representing information permits flexible or intelligent action. Note that attributes of intelligence collected in the Sternberg *et al.* survey included not only 'knowledgeable about a particular field of knowledge', but also 'able to apply knowledge to problems'. The implication of this latter concept is again that a declarative knowledge base can act as a basis for the generation of intelligent solutions to novel problems.

Discovery techniques: curiosity

If such a knowledge base underwrites Machiavellian intelligence—in particular if a *big* knowledge base permits greater cleverness—then we might expect intelligence to be marked by a tendency to *gather* declarative knowledge and, indeed, further attributes of everyday intelligence found by the Sternberg *et al.* study included 'appreciates knowledge for its own sake' and 'displays interest in the world around'. It is quite likely that one of the ways in which primates strike the observer as intelligent is simply their busy visual alertness and curiosity. Glickman and Scroges (1966) showed that primates exhibit more curiosity and exploration than other mammals tested with a range of physical objects and claimed that great apes showed the most attention. To our knowledge, this technique has yet to be modified to investigate whether more curiosity is shown about a sequence of social interactions than about a series of physical events matched for perceptual complexity, and whether any such bias is particularly marked in social primates compared to other animals.

Problem-solving, innovation, and flexibility

Extensive social knowledge only makes an individual really clever if the individual does something clever with the knowledge. The most prominent of the components of intelligence of all three types investigated in the Sternberg *et al.* study was essentially 'problem solving skill'. An extensive list of studies can be assembled to show that primates are particularly good at solving a range of problems set them on what Jolly (Chapter 3) referred to as the 'gadgetry' of the comparative psychologist's laboratory.

Sadly, the data on learning provided by comparative psychology is still such that after careful and sceptical examination, Macphail (1982, 1985) sees no reason to reject the null hypothesis, that all non-human animals are qualitatively equal in intelligence and learning ability. Passingham (1982) likewise rejects many comparisons on methodological grounds, but is, nevertheless, left with a collection of studies of 'learning set' which show both quantitative and qualitative superiority in primates compared to other mammals, and (at least in speed of learning) in apes compared to monkeys. Learning set does not entail the generation of novel behaviour so much as the use of well practised behaviour to indicate that a complex rule about a pattern of physical events has been grasped:

it is fair to conclude that both chipmanzees and rhesus monkeys can transfer strategies from a series of one type of problem to a series of related, though different, problems. They can make use of the fact that there are common principles which apply to the two series (Passingham 1982).

Compare Sternberg *et al.*'s 'identifies connections among ideas'.

Nobody has yet suggested exactly to what use this facility is put in natural conditions, let alone what role it may play, if any, as a component of social intellect in particular. It is easy to suggest that having learned the hard way about several specific relationships (with respect to, say, individuals much more dominant to oneself) it would be useful to be able to rely on some general principle learned about such interactions in dealings with new interactants; but so far as we know, such possibilities have not been investigated. For all we know, primates and other species, too, may be much better at learning such abstract or high-level rules in the social domain, than in the technical. Chapters 6 and 7 discuss the existence of abstract social categories, but we do not yet know about the relative ease with which such classifactory rules are formed.

Learning set does not entail the generation of novel behaviour so much as the grasping of a complex rule underlying a problem, an ability which can be demonstrated with quite well practised behaviour. However, even when we turn to evidence for innovations in behaviour produced to solve a problem (in line with such Sternberg attributes as 'shows creativity' and 'is a good source of ideas'), again we find primate intelligence demonstrated in problems set with objects, like the famous box, stick, and banana challenge set by Kohler (1925; see Passingham 1982, for a review). So far as we know, nobody has set equivalent social problems for primates or other species. Nor does observational study in field or captivity help us yet in tackling the question of the relative sophistication of social versus technical innovation. After 24 years of longitudinal study, Goodall (1985, in Kummer and Goodall) was able to offer a list of innovations by wild chimpanzees, including both social and non-social behaviour, which was impressive for its brevity, given the inventiveness chimpanzees have shown in captivity; moreover, Goodall did not offer strong evidence that innovations were planned solutions to a particular problem.

Such a poverty of data is not surprising given the methodological problems. On the experimental side, there is the difficulty of presenting a social problem to be solved, and maintaining it for the long period which may be necessary for an innovatory solution to emerge. In the case of naturalistic behaviour, the behavioural scientist will wish to be sceptical about whether a single occurrence of a behaviour was truly planned or just a lucky accident. Consider this example:

When Figan was part of a large group and, in consequence, had not managed to get more than a couple of bananas for himself, he suddenly got up and walked away. The others trailed after him. Ten minutes later he returned, quite by himself—and, of course, got his share of the bananas. We thought this was coincidence—indeed, it may have been on that first occasion. But after this the same thing happened again and again. (Goodall 1971)

To many behavioural scientists, the parsimonious explanation for such repetition lies simply in the reinforcement of the initial success, so even a true innovation will have difficulty in breaking through the wall of scientific scepticism; the uniqueness of a first occurence which is its central claim to invention will be written off as accident, and if repetition overcomes this objection, it is written off as simply a case of conditioning. As recently as 1985, Kummer (in Kummer and Goodall) thus despairs that

we almost completely lack an ecology of intelligence. No other dimension of behaviour has so systematically *not* been studied in the field.

There can be no doubt that this is currently a severe constraint for the study of social intellect: it is social problem solving—surely a core issue—which appears at present to be the least tractable of the intellectual components we discuss in this chapter. However, we suggest that a way forward is offered in de Waal's *Chimpanzee Politics* (1982), key segments of which are reprinted in Chapter 10. The study group of Arnhem Zoo chimpanzees was small enough and publicly observable enough that in contrast with wild populations, intensive and continuous observation inspires confidence about records of behaviour which really do represent novel Machiavellian solutions to social problems. Thus, for example,

observations in the consort context concern third individuals who attracted the attention of a dominant male. The clearest instance was Dandy, witnessing a secret contact involving an oestrus female, Spin, with whom he had a very close friendship. Loudly barking, Dandy ran to the Alpha male, who was far away, unaware of the contact, and led him to the scene where the two others were in the middle of mating. (de Waal, *ibid.*)

Even where one remains wary about each single case, scepticism wanes as records of such social problem solving multiply in de Waal's account. We may be forced to re-evaluate what shall count as that hallmark of scientific method, the replication, in the study of what by its very nature is a unique or rare behavioural event. It will often be necessary both to merge data from many different studies and to categorise behaviour at a sufficiently abstract level for a statistically robust sample to be established (see Chapter 16).

The notion of 'flexibility' represents a weaker version of creativity than innovation. In Chapter 15, Byrne and Whiten use the term 'tactical' as the most appropriate label for an aspect of deceptive competence which is suggested to be special in the behaviour of primates, in that it involves a behaviour already existing in the normal repertoire, but borrowed from the usual, honest context and used in another context where it becomes deceptive. The act is thus not innovative in the sense of being a new form of behaviour, but it is, nevertheless, an expression of flexibility in the way it is deployed, even in cases where this is only the result of trial-and-error learning (see Byrne and Whiten 1987).

The concept of innovation may thus itself benefit from further dissection in the same way as we attempt here for intelligence. There is a sense in which every situation an animal faces represents a novel conjunction of events—each sample of the same food type will, for example, require a harvesting act uniquely adjusted to its shape, size, and so on—and such expertise is possible to the extent that there is nevertheless a certain measure of predictability in the environment; each new sample is recognizable as a variant of others experienced previously. Furthermore, finding a particular food will depend on the animal's ability to remember or predict the unique circumstances of where, and perhaps when, it is available.

In Chapter 21, Milton discusses evidence that high primate intelligence has to do with prediction of particularly complex spatio-temporal patterns of food availability. In pursuing components of social intellect, what we must turn to in the next section is the ways in which novelty, and thus the nature of prediction, in social interactions compares with that in the technical domain, such as that involved in foraging.

Social expertise and social complexity

Recall that in the Sternberg *et al.* survey, an important component of intelligence concerned social competence. This makes sense because social prediction, as well as presenting the difficulties caused by each social object changing in space and time (like each plant food source), is intrinsically more complex; individuals also interact with other individuals, and this can change the behaviour which has to be expected from them. From this perspective, that individual is particularly intelligent who can handle a high degree of this complexity in social interaction (Whiten and Byrne in press).

One form complexity can take in primate society is that of triadic social interaction, mentioned first by Chance and Mead (Chapter 4) and the subject of detailed analysis in the earliest empirical paper we reprint (Chapter 9). Here Kummer describes a behaviour in which one individual is protected from an aggressor by the social configuration it expertly arranges with a dominant animal. The intellectual demands of such expertise are obvious: instead of adjusting one's orientation with respect to the single locus involved in diadic interaction, there is in triadic interaction the requirement to adjust simultaneously to the relationship between two other individuals; or to put the difference in another way, whereas diadic interaction involves one relationship between the two bodies, triadic interaction involves three relationships, one for each of the pairings in the triangle. There is an important point here, noted originally by Menzel and Johnson (1976): relationships increase in more than an additive way as the number of interactants increases; so, for example, with four interactants there are six possible relationships to be handled. Interactions clearly involving four or more individuals, in which, to generalise Kummer's definition of tripartite interaction 'n individuals simultaneously interact in n essentially different roles and each of them aims its behaviour at all of its partners', are noted in various parts of this book. A good example is in the interactions of one adult male with an oestrous female to whom he wants access, another male who is the current consort of the female, and several other males who represent manipulable potential allies (Chapter 16). What then of the earlier suggestion (Chapter 3) that prosimians do not exhibit these tripartite interactions (or show qualitatively inferior ones)? Are other groups of animals restricted in this way? Such issues are addressed in Chapter 11 (see also Barton, 1987).

We know that in tripartite interactions a primate keeps track of a number of current states of the other individuals, like their locations, orientation, and mood. In cognitive psychology, the store which keeps such a running track of events is called 'working memory' (Baddeley and Hitch 1974) and is limited in capacity. Perhaps it is working memory capacity which will be found to limit the complexity of social interactions. However, because a series of interactions comes gradually to constitute a long-term relationship (Hinde 1976, 1983), long-term memories also clearly play a role. Thus, Packer (1977) showed that pairs of male baboons could exhibit reciprocal altruism, working together so that on some occasions one harrassed another male while the other took the female the male had been consorting with, the roles being reversed on other occasions; even more importantly, preference for partners was mutual, suggesting that partnership was based on reciprocation. Seyfarth and Cheney (1984) have shown that, in a similar way, vervet monkeys remember the details of earlier help given to them and use this to bias their decisions to reciprocate at a later time. When a vervet monkey was exposed to a tape-recorder playback of a call for aid from another specific individual, the target animal showed more attention to the call if the other individual had been grooming it recently. We also know that primates take into account relationships existing between other pairs of individuals (Cheney *et al*. 1986; Chapters 6–8); and Humphrey, remember, suggested that these demands would also be greater in longer-lived species like the great apes, because of generation overlap and the sheer length of time for which relationships exist (Chapter 2). The 'memory hypothesis' of social intellect thus has two parts: the first proposes that greater Machiavellian intelligence involves a larger working memory for very recent social data; and the second part suggests that long-term memory for social events, interactions, and relationships may be more elaborate.

Perhaps we can sharpen up our conceptions of what is involved here by suggesting that, leaving aside the possibility of innovatory problem-solving discussed above, a greater degree of Machiavellian intelligence will be expressed in the capacity to act on the basis of a combination of a greater number of conditional social rules. Take the case of PA, a juvenile baboon who uttered a deceptive scream, as if a food competitor had attacked him, when in fact it had not, thus attracting the aid of the male JG so that the food competitor was chased away (Chapter 15). Here, the rule might be something like: 'If X is food competitor and if X is of lower rank to JG and if JG cannot see us, then scream'. We are thus brought back to a particular form of the notion of social knowledge, which is similar to the artificial intelligence construct of an intelligent knowledge-based system (IKBS or

Taking (Machiavellian) intelligence apart

expert system). The intelligence of an expert system derives very much from its knowing a lot of facts, and being able to provide the answer to any one of a vast array of problems set it by following a large number of if-then rules, in combination, to select relevant information from its vast data base and come up with an adaptive response.

As in the case of the number of social interactants, the number of conditions to be coped with can increase social complexity in more than additive fashion. Imagine a primate who takes into account just three conditions in social interaction, and each of these conditions has just two states; for example, whether or not the interactant is of higher or lower rank than itself; whether its own mother is present or absent, and whether the interactant has allies present or absent. The number of possible social scenarios (unique combinations of these possibilities), is s^c (where c is the number of conditions and s is the number of states each condition can take) and thus, in this example, 2^3 or 8. Imagine another primate who discriminates three states for each condition (say, whether the interactant is higher ranking, lower ranking or about the same, and so on for the other conditions) and it also takes into account two more conditions, like how altruistic the other has been, and whether the other had allies nearby. Now s^c becomes 3^5 or 243 different social scenarios. This is, of course, a very artificial example, and it is not implied that 243 different responses would need to be learned; but the exercise helps us to consider more precisely what may be involved in the oft-used, but vague term, 'social complexity'.

Discovery techniques: social play

One of the most plausible 'arguments by design' for the function of play is that it serves to gather knowledge (cf section 2 above) to allow future flexible response to relatively novel circumstances; Fagen (1976) drew the analogy of the aeroplane which is flown under the control of a computer through such exaggerated manoeuvres that it would look to be 'playing', while feedback about the effects of these behavioural extremes is used by engineers to build into the control system a model of actions and their effects which permits the consequences of new actions to be predicted.

Putting this together with what has been said in the previous section about social complexity, we would expect a species worthy of being considered Machiavellian to exhibit more social play. Smith (1982) noted that this is confirmed insofar as in chimpanzees, which he accepts as particularly socially sophisticated, triadic play has been found to constitute 25 per cent of social play (Goodall 1968), whereas in baboons it accounts for only 3–5 per cent (Owens 1975; Leresche 1976).

Levels of intentionality and mind-reading

Menzel provided the influential groundwork, both theoretical (Menzel and Johnson, 1976) and empirical (Chapter 12), but it was Humphrey (1980) who explicitly asked us to consider the nature of a 'natural psychologist'—an animal sufficiently socially intelligent to read the mind of another individual.

> If a social animal is to become—as it must—one of 'nature's psychologists' it must somehow come up with the appropriate framework for doing psychology; it must develop a fitting set of concepts and a fitting logic for dealing with a unique and uniquely elusive portion of reality. The difficulties that arise from working within an *inappropriate* framework are well enough illustrated by the history of the science of experimental psychology;

even then,

> as professional scientists, behaviourists have always had enormous advantages over an individual animal, being able to do controlled experiments, to subject their data to sophisticated statistical analysis, and above all to share the knowledge recorded in the scientific literature. By contrast, an animal in nature has only its own experience to go on, its own memory to record it, and its own brief lifetime to acquire it. 'Behaviourism' as a philosphy for the *natural science* of psychology could not, and presumably does not, fit the bill. (Humphrey 1980.)

As ethologists, Dawkins and Krebs (1978) do not see this as such a problem:

> Animals can, in principle, forecast the behaviour of other animals, because sequences of animal behaviour follow statistical rules. Ethologists discover the rules systematically by recording long sequences of behaviour and analysing them statistically, for example by transition matrices, and in the same way an animal can behave as if it is predicting another individual's future behaviour. Without committing ourselves to a view over the philosophical problems of animal mind in the subjective sense (Griffin, 1981; 1982), we may use the word 'mind-reading' as a catch-word to describe what we are doing when we use statistical laws to predict what an animal will do next.

Dawkins and Krebs' mind-reader is rather a behaviourist; should not the catchwords 'mind-reader', or 'natural psychologist' demand more than this—something closer, perhaps, to that interpretation of 'intelligence' offered to Sternberg *et al*.: 'sensitive to other people's needs and desires'? Both Premack (Chapter 13) and Dennett (Chapter 14) have provided clear statements about the intellectual demands on an animal who is more obviously deserving of the label 'psychologist'. Dennett classifies organisms into the different levels at which they might operate as intentional systems; that is to say, systems whose behaviour can be appropriately described by the use of intentional terms, which include such psychological expressions as 'believes'; 'expects', and 'wants'. The primate acting as psychologist is at least what Dennett called a second-order system, who may, for instance, *believe* that its companion *wants* those figs. For the benefit of any who doubt that higher orders of intentionality incur greater intellectual demands, we can do no better than to quote Dennett, suspecting that

you wonder whether I realise how hard it is for you to be sure that you understand whether I mean to be saying that you can recognize that I can believe you to want me to explain that most of us can keep track of only about five or six orders, under the best of circumstances.

Such dizzying orders of intentionality are no doubt beyond the intellects of primates, just as they are for us, but distinctions between the very lowest orders of intentionality have been found helpful in starting to examine observational evidence for deception and counter-deception in primates (Chapter 16); they may yet prove useful in discriminating between the social intellect of different groups of primates and other animals. The ingenious work of Premack and Woodruff (1978; see Chapter 13) involves an attempt to assess experimentally the reality of at least second-order intentionality in chimpanzees and has entailed a deep analysis of what is involved in attributing a 'theory of mind' to a non-human primate.

What would be the functional advantage of being a natural psychologist capable of second, third or fourth order intentionality? If we start with Craik's (1943) suggestion that what brains do is build a 'small-scale model' of external reality, then a more intelligent brain can be recognized by the 'accuracy and completeness of its cognitive model' (Barlow 1985; see also Jerison 1985). Now, in the case of social intellect, this modelling will have to be extended to that part of the world constituted by others. If we think of social interaction as a game between competitors, in which success depends not only on brute strength but on social skill (as section 4, above, acknowledges; see also Reynolds 1986), then as in a game of chess it is quite plausible that out-manoeuvering one's opponent will depend on how many of the possible future moves one can anticipate, and thus in turn on the adequacy of one's internal model of the operations of the other's mind. This will involve not only a capacity to represent a certain number of orders of intentionality, but to represent the rapidly branching alternatives which are raised with each anticipatory step back and forth between self and opponent, so yet again we have a potential selection pressure working on social intellect which is a geometric function, rather than an additive one, of increases in social complexity. As Humphrey noted, the evolution of more sophisticated Machiavellian intelligence in succeeding generations should create a spiralling pressure for greater and greater powers of gamesmanship. The evolutionary potential of this scenario is enormous; but then we are ultimately trying to explain a unique and staggering acceleration in brain size which took place in our ancestors. (Wynn, in Chapter 20, shows that the traditional technical explanation for this acceleration is not supported by the evidence.)

Game-playing is still not a full explanation of the utility of being a natural psychologist: why cannot expert gamesmanship derive from behaviourist, rather than psychological, analysis of others? One answer was offered by Humphrey as we have already seen. We can offer another, which is concerned with economy and efficiency. Dawkins (1976) presented a range of evidence that hierarchical organisation should be regarded as a general principle of biological control systems, and showed how the existence of a few high level units of information coding, summarizing the inputs and directing the outputs of vast branching hierarchies of lower level units, could (as in business organisations and the army) constitute an optimally efficient control system capable of rapid and appropriate response to those environmental changes it is designed to deal with. In the primate mind then, representations of the behaviour of another individual at the lower levels of a hierarchical analysis might be summarised at a higher level in what amounts to mentalistic intentional terms. Thus, for example, it might be useful for one primate's mind to be able to summarize the available information about a second individual at the level of whether or not that second individual intends to chase the first individual, or a third party. If it is this property of the second individual that is the best guide as to how the first individual should act, then it makes functional sense that the brain of the first should code the situation in this way. In Chapter 16 we argue that the existence of a mental code representing mental states of others can be inferred from the precise way in which one animal's behaviour is adjusted to that of another, and assess such evidence for primates' representation of the powers of attention of others (see also Whiten and Byrne 1988; Byrne and Whiten 1987, 1988).

Self-reflection

We are currently in the midst of a resurgence of interest in the animal mind, with both 'cognitive ethology' and 'cognitive learning theory' kicking against the traces of the essentially behaviourist traditions of their parent disciplines. The extreme of this movement is a rekindled programme to chart the dimensions of animal awareness and consciousness (Griffin 1978). Let us be blunt and say that although this is no doubt a healthy and certainly an interesting endeavour, the best attempt so far (Griffin 1981) fails to convince us that there is any prospect of our understanding being very much advanced about such issues in the near future; it is simply not obvious how we shall ever study a phenomenon which is by definition so closed to public scrutiny. This pessimism does not extend to the study of human consciousness (see Crook, Chapter 24, and Dennett, 1982).

Self-reflection is a different matter, and there is a link with mind-reading, for, as Bennet (1976) noted,

if (a language-less creature) A can think that B thinks that A thinks that P, as I have contended, then a languageless creature can manifest a belief about his *own* beliefs.

Such self-reflection, although at first sight as intangible as awareness, is in one way more manageable. Computer science is currently our most powerful source of analogies with which to understand mental processes. It is possible to see how self-reflection can make a computing machine more intelligent in a way which to our knowledge has not been done for self-awareness (what computation could not be done competently by an unaware but highly intelligent machine?). Perhaps the neatest way to think about the notion of self-reflection in a computer is that programs running on one level can act as data for higher-level programs. The computer is thus monitoring (at a 'high' level) its own computations (occurring at a 'lower' level) and the higher level can also of course implement outputs according to the outcome of its analysis of the system's own computations. Such a capacity paves the way for modification and restructuring of programs, and thus 'consciousness' becomes an answer to the question, 'if the mind runs on programs, who writes them?' (Neisser 1963; Shallice 1972; Crook 1980). However, we must wait for artificial intelligence to chart the possibilities and pitfalls of such 'unprincipled' programs which seem likely to catch themselves in a computational analogy of Russell's paradox (Byrne, in press).

It has been suggested that such a capacity for self-monitoring could have played an additional role in the evolution of social intellect, at least in the case of humans (Humphrey 1980, 1983, 1986; Crook 1980). The core hypothesis here is that self-reflection on the operation of one's own mental system allows an individual to build an explanatory model of human action—a model which can then be used to explain and predict the actions of others in a fashion superior to what would be possible if the model had to be built purely on the basis of observations of the actions of those other individuals. This is a hypothesis on a par with the Machiavellian intelligence hypothesis in its originality and potential implications for ethology and psychology (if it is true!); students of primate and human evolution should be familiar with it for this reason alone, despite the evident difficulties in testing it. Some progress has been made with the investigation of self-monitoring in animal learning, but only in a non-social context (see Beninger *et al.* 1974, and Weiskrantz 1985, for a discussion of the relationship between animal and human data). If we widen the scope of our enquiry to cover the origins and construction of socially intelligent systems in general, there are grounds for more optimism about testing the hypothesis. In artificial intelligence, we shall in future be able to ask whether self-monitoring does, indeed, help in the construction of machines which are socially intelligent. In human developmental psychology we can ask about which comes first in development; understanding of one's own, or other people's psychological processes?

Imitative learning

Another of the Sternberg attributes of intelligence was 'interest in learning and culture.' Recall that Jolly, too, treated cultural learning as an important part of primate social intelligence (Chapter 3). So it must be if we interpret social intellect more broadly than we have so far; if we treat it as 'adaptive capacity', then observational learning from others makes important contributions to this capacity, and Hauser treats this topic in detail in Chapter 23.

Observational learning in some form or other is widespread in the animal kingdom and, for our present purposes, it is important to make one major distinction amongst the several that can be made between different forms of observational learning (Passingham 1982; Galef, in press). Stimulus enhancement is the process whereby one animal's attention is drawn towards a particular object, or locus in the environment, by the behaviour of another, but the form of any behaviour done there does not owe anything to the observational process. In the case of tits opening the tops of milk bottles, for exampale, it seems that the explanation is that tits learn from others to approach the bottles, but then learn to open them by trial and error (Hinde and Fisher 1951; Krebs *et al.* 1972). By contrast in true imitation the form of the behaviour is copied. Primatologists often tend to use the terms interchangeably. Yet although imitation is clear in apes (at least in chimpanzees; e.g. Hayes and Hayes 1952) there is actually minimal experimental evidence for imitation in other animals, including monkeys (Passingham 1982). There is now even concern that the best known case of observational learning in primates, the sweet potato washing of Japanese macaques (Kawamura 1963) did not involve imitative copying, but rather a mixture of stimulus enhancement and reinforcement (Green 1975; see also McGrew and Tutin 1978).

If imitative copying is indeed intellectually demanding and restricted only to animals with a high index of encephalisation [see Passingham (1982) for an account of imitation in dolphins], just why should this be? Bruner (1972) reviewed evidence that imitation was restricted to apes and suggested that it involved two sophisticated abilities. First, it involves the imitator in a particularly difficult mental transformation. Its perception of the behaviour seen in others must be translated into motor acts which, even if it can see them, will from this new point of view look quite different to the model behaviour originally observed in others. Bruner likens this to the problem facing the child in mastering deixis—

as in learning that when I say *I*, it is not the same as when you say *I*, or that *in front* of me is not the same as *in front* of you.

The second ability Bruner describes as 'construction of an action pattern by the appropriate sequencing of a set of constituent sub-routines to match the model'. He is referring to a facility to decompose and recombine actions; we are back again to flexibility, self-monitoring, and other related concepts we have already discussed in this gallop through what we regard as the principal contemporary issues for examination in the study of social intellect.

It will be apparent that the concept of social intelligence generates an enormous range of fascinating research questions. As these are pursued in the succeeding chapters, more and more flesh is added to the skeletal account of the scope of social intellect which we have given here. It will remain for Alison Jolly in the final chapter to assess just how far we have progressed since the idea of social intellect emerged explicitly in her seminal paper of 1966.

What primates know about social relationships

6

Do monkeys understand their relations?
ROBERT M. SEYFARTH and
DOROTHY L. CHENEY

Do the abstractions which human observers use in analysing the social interactions of primates—kinship, dominance, and reciprocity, for example—exist also in the minds of monkeys and apes? In this chapter Seyfarth and Cheney present new experimental and observational data and review other recent evidence which directly tackles this question.

Introduction

A scientist studying social behaviour typically uses observational data in an attempt both to describe and explain patterns of interaction from an observer's point of view. Obviously, this is an essential place to start, but studying animals exclusively from the human perspective is likely to raise more questions than it answers. Descriptions of foraging, mating, and other social interactions, for example, often suggest that animals employ 'strategies' and make sophisticated judgements when interacting with their environment. Indeed, many ethologists find it difficult even to describe what they have observed without attributing motives to their subjects, or assuming that the animals are making use of concepts like kinship, a dominance hierarchy, or reciprocity (Cheney *et al.* 1986). Does this mean that non-human species are actually capable of such conceptual thinking? Or have we, by virtue of our own language and cognitive abilities, exaggerated the complexity of social systems which are in fact relatively simple when seen from a perspective other than our own?

In this paper we examine whether non-human primates actually make use, in their social interactions, of some of the complex mental operations that are implied by descriptions of their behaviour in the wild and suggested by the results of some recent laboratory tests.

As an example of data from the field that prompt us to consider the mechanisms underlying behaviour, consider the pattern of 'redirected aggression' observed by Judge (1982) among pigtail macaques and by Smuts (1985) among baboons. In both these species, when an individual receives aggression from a higher-ranking opponent the victim frequently redirects aggression toward a previously uninvolved bystander. More often than expected by chance, the targets of such redirected aggression are individuals that have a strong behavioural association with the original aggressor. This suggests that macaques and baboons—like the ethologists who study them—recognize the associations that exist among others (see also Chapters 7 and 19). The data also raise the possibility that animals can make judgements about the similarity or difference between different social relationships.

Complementing these field observations are some intriguing tests that suggest complex cognitive abilities in captive apes. For example, Premack (1976, 1983) trained the chimpanzee Sarah to make same/different judgements between pairs of stimuli, and to solve problems in analogical reasoning. For example, the relationship 'can opener: can' was judged by Sarah to be the same as the relationship 'key:lock'; the relationship 'paintbrush: painted surface' was judged to be the same as the relationship 'pencil: marked paper'; while in tests of proportions a cylinder ¼ full of water was judged to be the same as ¼ apple, but different from ¾ apple. In each case Sarah was first trained using a small number of exemplars and then performed at levels significantly above chance using novel objects (Gillan *et al.* 1981; Woodruff and Premack 1981; see also Savage-Rumbaugh *et al.* 1980). Note that two abilities underlie the solution of such analogies. First, an individual must respond to stimuli not on the basis of their physical features but according to some more abstract property, for example, the concept of 'opening' when the exemplars are a can opener and a can. Secondly, an individual must compare two abstract relations and make a same/different judgement between them, again independent of the particular objects that may be involved (Premack 1983).

These results lead one to ask whether there are any cases in which free-ranging animals respond to stimuli according to some property other than their physical features. If so, can we find any evidence that the animals go further, and behave in ways that imply an ability to recognize the relation between relations? Such an ability would suggest that free-ranging primates classify objects not just according to their physical properties, but also according to some abstract feature. Recent research suggests that monkeys do have abstract representations, at least of their social world (see also Chapters 1 and 5). In the following sections we examine some evidence of such abilities. We first describe some experiments that suggest that free-ranging adult vervet monkeys classify different vocalizations on the basis of their meaning rather than their acoustic properties. We then describe some observations indicating that vervets recognize the similarity between certain kinds of social relationships, regardless of the individuals involved.

Study area and subjects

Subjects for the research described below were vervet monkeys (*Cercopithecus aethiops*) living in Amboseli National Park, Kenya. The park consists of open plains, arid *Acacia* woodlands, and permanent swamps surrounded by thick, bushy vegetation. Throughout this area groups of vervet monkeys live in territories that average approximately 23 hectares in size and are aggressively defended against incursions by the members of neighbouring groups.

Study animals live in three social groups that have been under continuous observation by ourselves and colleagues since 1977 (Cheney *et al.* in press). During this period groups have numbered between one and seven adult males, between one and eight adult females, between one and 11 juveniles (animals aged 1–5 years), and between one and 11 infants (animals less than 1 year).

Within each group, individuals can be assigned dominance ranks that accurately predict the direction of agonistic interactions in a variety of different contexts (e.g. Seyfarth 1980). Immature animals acquire ranks immediately below those of their mothers (Cheney 1983; Lee 1983), so that adult females and immatures who are close genetic relatives occupy adjacent ranks. Females become sexually mature at approximately four years of age (Cheney *et al.* in press) and generally remain in their natal groups throughout their lives (but see Hauser *et al.* 1986). Males become sexually mature at around 5 years of age, at which time they usually transfer from their natal group to another. Male transfer may occur only once or a number of times during a male's lifetime (Cheney and Seyfarth 1983).

Judgements based on abstraction

Although the exact form of abstractions used by humans to compare word meanings are not understood (e.g. Anderson 1978), the fact that they do so is not in dispute. In tests of analogical reasoning by chimpanzees it also seems clear that some form of abstract representation must be involved (Premack 1983). Premack further contends that the ability to form such abstract representations is enhanced by, and may require, language training. His claim is not that chimpanzees naturally lack the ability to reason abstractly. Instead, he believes that all primates possess the potential for such skills, but only chimpanzees subject to language training are able to realize this potential.

A test to determine whether free-ranging monkeys compare vocalizations on the basis of their meaning thus represents a test of two separate hypotheses: first, whether any non-human species can make semantic judgements similar to those that occur in language; and secondly, whether non-human primates require language training in order to make such judgements. Below we describe an experiment that addresses these two questions.

East African vervet monkeys give three different vocalizations when interacting with the members of another group: a 'wrr', a 'grunt', and a 'chutter', (Struhsaker 1967 a, b; Cheney and Seyfarth 1982a, b). Although the three calls are accoustically different (Fig. 6.1), all appear to have a broadly similar meaning, and are given only in the presence of another group. To test whether monkeys attend to these calls on the basis of their meaning, we designed an experiment that addressed the question: if a monkey repeatedly hears animal A's 'wrr' to another group when there is no other group present and hence ceases to respond to that call, will it also cease responding to A's 'chutter'? We reasoned that, if two calls have similar meanings, and if monkeys use meaning to judge the relationship between calls, habituation to one call should produce habituation to the other. Alternatively, if monkeys use some other features (such as acoustic similarity) to judge the relation between calls, these features, and not the calls' referents, should determine whether or not habituation is transferred to another call.

Each experiment was conducted over a two day period (for a more complete description of the experimental protocol, see Cheney and Seyfarth 1988). On day 1, we played animal A's inter-group chutter (or, in one trial, an inter-group grunt) to a subject in order to establish the duration of that subject's response in the absence of other inter-group calls. The following day, we played animal A's 'wrr' to the same subject at least eight times. Each 'wrr' was separated by an average of 30 minutes. Then, approximately 30 minutes after presentation of the eighth 'wrr', we played animal A's 'chutter' to determine whether habituation across call types had occurred.

Our second series of trials examined whether subjects would transfer information about the reliability of an inter-group call from one individual to another. Thus, on day 1, we played as a control the inter-group 'chutter' of individual B. This was followed on day 2 by at least eight presentations of individual A's 'wrr'. After which individual B's 'chutter' was played again.

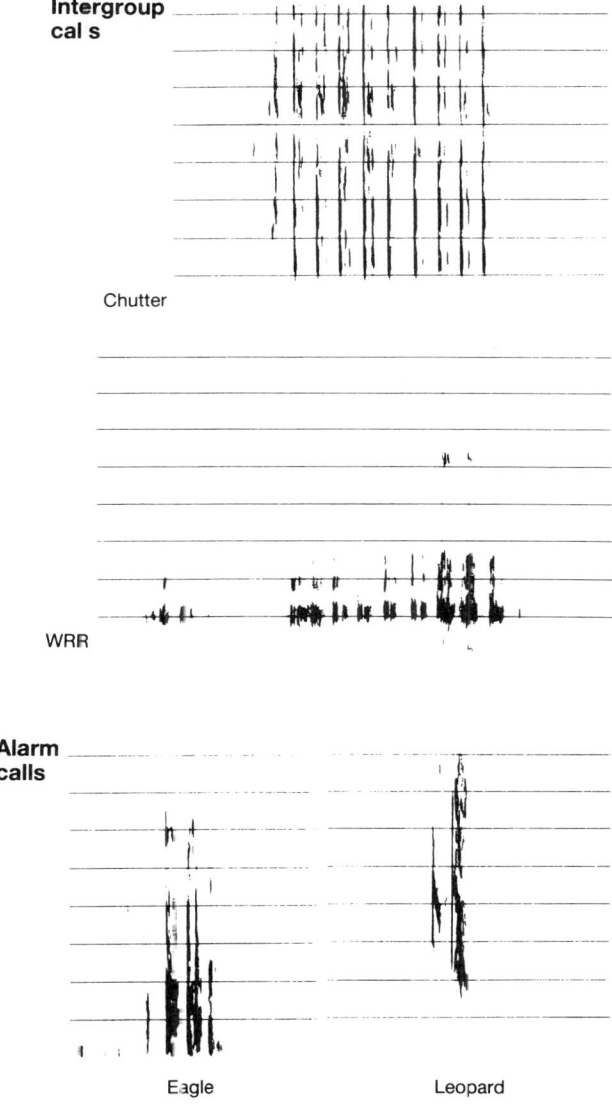

Fig. 6.1 'Chutter' and 'wrr's' given by vervets during inter-group encounters (top) and alarm calls given by vervets to leopards and eagles (bottom). Y-axis shows frequency, in units of 1 kHz; X-axis shows time: duration of the 'chutter' is 500 msec. The two inter-group calls were given by adult female SN and the two alarm calls were given by adult female AM.

The third series of experiments examined whether habituation to a given individual would also occur if the signalling individual remained the same but the call's referent changed. The calls chosen as stimuli were alarm calls given to leopards and eagles. In these experiments we first played, for example, individual A's eagle alarm call. Then on day 2 we played the same individual's leopard alarm call eight times, followed again by its eagle call. Leopard and eagle alarm calls were used alternatively as habituating stimuli and test stimuli.

In the fourth series, both the referent and the signaller were changed. For example, individual B's eagle alarm was played on day 1, followed on day 2 by eight presentations of individual A's leopard alarm, after which B's eagle call was played again.

Results provided clear evidence that vervet monkeys use meaning to make judgements about the relation between two vocalizations. When compared with control trials, all 10 subjects showed a decrease in the duration of their response to the 'chutter' of a given individual following repeated exposure to that individual's inter-group 'wrr' (two-tailed Wilcoxon matched pairs test, $t=0$, $P<0.01$; Fig. 6.2). This decrement was less marked when a different individual's grunt or chutter was played. In this case six of the 10 subjects decreased their response while three increased it. Overall, subjects showed a significantly greater decrement in response when played a call with the same referent from the same individual than when played a call with the same referent from a different individual (Wilcoxon test, $n=10$, $t=2$, $P<0.01$; Fig. 6.2). Thus, both meaning and caller identity appear to affect the transfer of habituation from one vocalization to the next.

In contrast, when the meaning of the two calls was different there was no transfer of habituation, regardless of whether the signaller was the same or different. Subjects who had habituated to presentation of another individual's eagle alarm showed no change in their response to a leopard alarm from either the same or a different individual (Fig. 6.2).

In summary, results suggest that under natural conditions, in the absence of any special training, vervet monkeys make judgements about the relationship between two vocalizations on the basis of their referents, or meaning. Such judgements require that an animal both recognize the relationship between a call and its referent, and compare two referents. Results thus argue against the view that non-human species require language training in order to compare two stimuli on the basis of some abstract relationship (Premack 1983). They further suggest that in earlier analyses of vervet monkey communication we may have underestimated the extent to which animals make use of abstract representations.

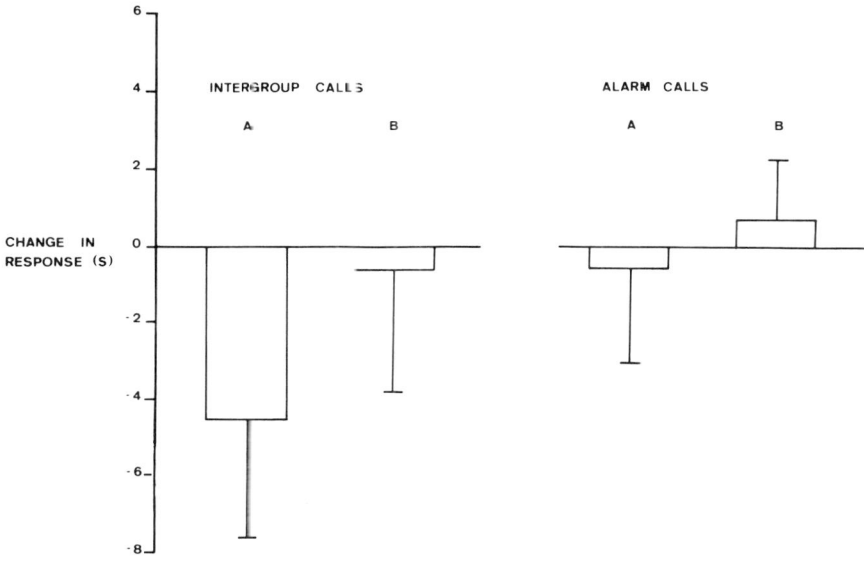

Fig. 6.2 Changes in the responses of subjects to a novel vocalization after habituation to an acoustically different vocalization with: the same referent given by the same individual (intergroup call A); the same referent given by a different individual (intergroup call B); a different referent given by the same individual (alarm call A); or a different referent given by a different individual (alarm call B). Histograms show means and standard deviations of changes in response durations. $N = 10$ subjects in conditions A and B and nine subjects in conditions C and D.

The recognition and comparison of social relationships

Recognizing relationships

Two experiments demonstrate that vervet monkeys recognize the relationships that exist among others. In one test, three adult females (all of whom had offspring in the group at the time) were played the scream of a 2-year-old juvenile. One of the females was the juvenile's mother. Mothers looked toward the speaker or approached the speaker for longer than control females, indicating that they recognized the voice of their offspring.

In contrast, control females responded by looking toward the mother, often before she herself had made any movement. Control females appeard to recognize the relationship between a particular scream, a particular juvenile, and a particular adult female (Cheney and Seyfarth 1980, 1982b).

In a second experiment, subjects in one group of vervet monkeys (group B) were played a vocalization from a member of another group. The call was played either from an appropriate location (for example, a call by a member of group A played from a speaker located in group A's territory), or from an inappropriate location (a call by a member of group A played from a speaker located in group C's territory). When the call came from the inappropriate location the responses of subjects were significantly stronger, indicating that they recognized the relation between a particular call, a particular individual, and a particular social group (Cheney and Seyfarth 1982b, and Chapter 19).

Recognizing the relation between relationships

The experimental data presented above provide evidence that vervets recognize each other as individuals, and that they associate individuals with other animals or social groups. Do they, however, recognize that particular sorts of social relationships are characterized by similar properties? Data on the pattern of redirected aggression suggest that they do, and that they are capable of making judgements about the relationship between different social bonds.

As in many other primate species, when one vervet monkey receives aggression from another it often 'redirects' aggression toward a third, previously uninvolved individual. Observations of redirected aggression over two 7-month periods have suggested, however, that such aggression does not occur at random. Males and females of all ages were significantly more likely to threaten a given individual if they had recently been involved in an aggressive dispute with that individual's relative (Fig. 6.3; Cheney and Seyfarth 1986). Like the experiments described above, this suggests that vervets can recognize that certain individuals associate regularly with each other.

More important, males and females over 3 years of age (but not younger animals) were significantly more likely to threaten a given individual when that individual's close kin and their own close kin had previously been involved in a dispute (Fig. 6.4). Thus, for example, a fight between A and B significantly increased the probability that a relative of A and a relative of B would later be involved in a fight. This result held only for animals over 3 years of age, and not for younger juveniles (Cheney and Seyfarth 1986). Adult vervets, therefore, seemed able to recognize some similarity between their own close associates and the close associates of others.

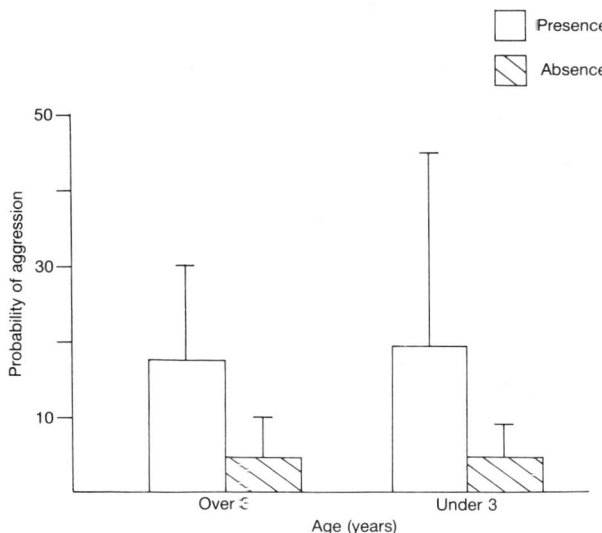

Fig. 6.3 The probability that vervets would threaten other individuals in the presence and absence of a fight with those individuals' close kin within the previous two hours. Histograms show means and standard deviations for individuals over and under the age of 3 years in two social groups. For further details, see Cheney and Seyfarth 1986.

Recognition of the close bonds that exist between other individuals could easily result from simply associative learning, and can be explained without invoking any sort of abstract representation. To recognize that certain social relationships share similar properties, however, an animal must either memorize all the relationships to which it has been exposed and evaluate them according to some criteria, or classify different types of social relationships in such a way that they can be compared independently of the particular individuals involved. In either case—but particularly the latter—some sort of abstract representation is required (see Chapter 7).

Discussion

Monkeys make good ethologists

An ethologist studying a group of primates begins with description. The aim, however, is not simply to describe every grooming bout, vocalization, or copulation that was observed, but instead to assemble data on

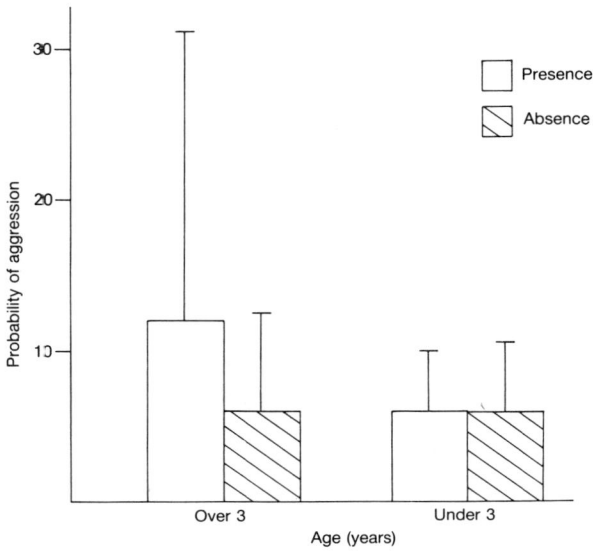

Fig. 6.4 The probability that vervets would threaten other individuals in the presence and absence of a fight between those individuals' close kin and their own close kin within the previous two hours. Histograms show means and standard deviations for individuals over and under the age of 3 years in two social groups.
For further details, see Cheney and Seyfarth 1986.

different interactions among different individuals, and using these data to produce generalizations that allow descriptions of the relationships observed (Hinde 1976). Constructed in this way, a relationship is an abstraction because it is defined relative to other relationships and is independent of the individuals or actions it represents. The abstraction exists in the ethologist's mind.

The thesis of this paper is that monkeys make good ethologists, because they appear to go beyond simple observation to make judgements about their fellow group members based on abstract features such as social relationships and the meaning of vocal signals. Consider, for example, the processing of vocal communication. When an ethologist studies animal communication, the goal is not only to describe a signal's physical properties, but also to deduce its meaning—the information it conveys to others. Vocalizations, for example, can be grouped according to acoustic properties, which are concrete and can be measured, but they may also be classified according to meaning, which is more abstract. In some cases there is a correlation between meaning and acoustic structure (Morton 1977), while in many other cases—such as the alarm calls, grunts, and chutters of vervet monkeys—establishing a call's meaning is not aided by knowledge of the call's physical features (Seyfarth *et al.* 1980; Cheney and Seyfarth 1980; Cheney and Seyfarth 1988).

Our experiments suggest that vervet monkeys can judge the similarity between two vocalizations according to their meaning, and that such judgements in some cases override judgements based on caller identity or acoustic structure. A monkey who has determined that individual X is unreliable when signalling about a particular referent will transfer its skepticism to all other calls by the same individual that concern the same referent, regardless of any acoustic difference between calls. Like humans drawing up a list of synonyms, the monkeys act as if they can divide vocalizations into abstract categories based on meaning—categories that cannot be reduced to or explained in terms of the physical features of the objects within them.

The monkeys' ability to respond to stimuli on the basis of an abstraction also appears in their judgements about social relationships. Specifically, data on redirected aggression suggest that individuals observe the behaviour of others and that when they have seen an aggressive interaction between a member of their own family and a member of family X, preferentially choose another member of family X as the target of their own aggression. Like good ethologists, the monkeys appear to recognize some similarity between the relationships that exist within different families, independent of the particular individuals involved. To do so they must compare animals not according to their physical features or a specific type of interaction, but according to an underlying relationship that has been abstracted from a series of interactions over time.

Ethological skills require experience

Ontogenetic data on the pattern of redirected aggression suggest that monkeys require some experience before they can recognize the similarity between different social relationships. Vervet monkeys of all ages, for example, appear to recognize that two members of the same family are close associates. Only animals over 3 years of age, however, appear to recognize the similarity between relationships in other families and relationships in their own (see Fig. 6.4; see also Cheney and Seyfarth 1986). These data suggest the hypothesis that knowledge about the social environment is acquired gradually, through levels of increasing complexity. Initially, animals recognize others as individuals; at a later age they form associations among others; and at a still later age they form associations among associations, recognizing that despite the different individuals involved some relationships share similar properties.

The attribution of emotions

Thus far, we have been concerned primarily with the ways in which monkeys can be said to classify objects according to some abstract feature such as meaning or a relation rather than simply the objects' physical properties. It is also interesting to speculate about the monkeys' ability to attribute motives, emotions and intentions to others (see Chapters 12–14). We think of attribution as a complex cognitive (perhaps uniquely human) ability, because it demands that individuals have some conscious awareness of other individuals' behaviour and possibly also of their own. In non-human primates deception provides some evidence for attribution, since it suggests that individuals have an understanding of the consequences of their own behaviour on the responses and beliefs of others (see below; see also Chapters 15, 16, and 19; Woodruff and Premack 1979; de Waal 1982; Cheney *et al.* 1986).

Do the redirected aggression data also imply an ability to attribute intentions or emotions to others? If A fights with B, and then later threatens B's relative, we cannot rule out the possibility that A attributes certain emotions felt by B toward her relatives, and recognizes that one way to retaliate further against B is the threaten B's relative. Similarly, if A's relative (A1) later threatens B's relative, she may do so at least partially because she knows that this is an effective means of annoying B. Even if their knowledge of the relationship between B and B's relatives has been acquired through associative learning alone, why should either A or A1 threaten B's kin except to retaliate against B and assert family A's dominance over family B?

One interpretation of redirected aggression, then, rests on the assumption that B experiences certain emotions toward its relatives and that other animals recognize these emotions in B. This brings us full circle, to a situation we first described in our introduction. In order to explain the pattern of redirected aggression in vervet monkeys we may be forced to assume that the animals can attribute motives and intentions to one another—and yet we have no direct evidence that the monkeys can actually do this. Moreover, it is possible to argue that the attribution of motives in others demands some awareness of motives in oneself, although, again, the data are by no means sufficient to prove this. Clearly, more observations and experiments are needed before we can firmly establish whether attribution exists in the minds of monkeys, or only in the minds of their beholders (see also Chapter 14).

Some caveats

The habituation/dishabituation procedure described in this paper is widely used in psychological tests of preverbal infants (e.g. Eimas *et al.* 1971). The test determines whether subjects can perceive a particular distinction and it assumes that if the distinction can be perceived the subjects will make use of it in their daily lives. There is some evidence, however, that infant chimpanzees who can perceive a relational distinction when tested using the habituation procedure may nonetheless be unable to make use of the same distinction in a match-to-sample test (Oden *et al.* in press). This suggests that habituation data alone cannot be used to argue that vervets recognize or are aware of their ability to make abstract judgements when assessing social relationships. Instead, the habituation data suggest that they possess the requisite perceptual skills, while the data on redirected aggression are needed to show that these skills are actually put to some use. Even the data on redirected aggression, however, do not provide clear evidence that the animals recognize an analogy, or that they understand that they can make judgements about the relation between relationships. For example, it is entirely possible that the vervets' ability to recognize the similarity between two sets of relationships is restricted to social contexts, and cannot be generalized to other domains.

Thus far, we have drawn attention to the parallels that exist between the solution of analogical reasoning by laboratory chimpanzees and the recognition of social alliances by vervet monkeys. Although we believe the parallels are significant, at this stage in our work we know little about the cognitive mechanisms used by the monkeys, and it does not necessarily follow from our results that the animals are using an analogical format to represent their knowledge of social relationships. It is worth remembering, for example, that when laboratory chimpanzees were trained to recognize an abstract relationship between two (or more) sets of stimuli, they could subsequently make similar judgements using entirely novel objects (Savage-Rumbaugh *et al.* 1980; Premack 1983). The ability to transfer performance under such conditions provides stronger evidence for the use of abstract representations than that presented in this paper. These apparent distinctions between the field and laboratory data suggest three alternative hypotheses.

First, vervet monkeys in the wild, like laboratory chimpanzees, are capable of recognizing relations between relationships even when dealing with entirely new stimuli. Our inability to demonstrate this is entirely methodological, since it is difficult to collect precise data on the behaviour of free-ranging monkeys when they encounter completely new situations.

Secondly, vervet monkeys can recognize relations between relationships, but only when using a restricted set of stimuli, namely their social companions. If this were true, it would suggest that monkeys do not recognize the abstract relationships among sets of objects, but instead memorize a large number of associations between particular stimuli. The monkeys would be well-equipped for life in their particular social group, but would require time to adjust to major social upheavals.

Thirdly, as Premack (1983) has argued, primates like vervets are potentially capable of making judgements based on abstract representations, but do not make use of this ability in the wild, and require a form of language training before it will appear. Furthermore, even if free-ranging monkeys can be shown to behave as if they were capable of recognizing the relation between relationships, they may not be able to transfer this ability to other situations. Such as ability demands that individuals understand an abstract rule whose application is not restricted to specific contexts.

Whatever the outcome of tests among these hypotheses, they remind us that, while research on captive animals can clarify a species' abilities after training and under test conditions, field research can tell us how such skills emerge in the absence of training, and (more important) what they are good for.

The demands of social life

Life in a social group presents monkeys with a variety of problems. Not only must they be able to predict the behaviour of others, but they must also be prepared for complex events such as the formation of alliances or attempted deception. (See Chapters 5, 9–11, 15–18). Some alliances, like those between mothers and offspring, are highly predictable, while the likelihood of others can change from hour to hour, day to day, or gradually over longer periods of time depending on the formation of a close bond (e.g. Smuts 1985), prior grooming (e.g. Seyfarth and Cheney 1984), or some other factor.

In order to compete, survive, and reproduce an individual must therefore make judgements about the *relationships* that exist among others (Hinde 1979). This could be done by memorizing all group members as well as the pattern of interactions among each pair. Observation and memorization would suffice in small groups whose membership rarely changes, but it would place sharp limits on an individual's ability to anticipate behaviour as either group size or changes in group composition increased.

Alternatively, a group-living monkey could classify the relationships it observes into categories. This method has two advantages. First, it allows individuals to identify types of relationships quickly and predict the behaviour of others based on partial information. Thus, a monkey who joins a new group, or whose group receives an influx of outsiders, can make valid predictions about behaviour without having to observe interactions between each pair of individuals. Secondly, as group size increases, forming categories (and making judgements based on these categories) provides an increasingly efficient method for memorizing the characteristics of relationships and predicting what individuals are likely to do next.

In a similar manner, group life requires that individuals quickly and accurately assess the meaning of each others' signals. Given evidence that, in some cases, animal signals provide recipients with false information (reviewed in Krebs and Dawkins 1984), considerable research in animal communication now attempts to establish how signallers might attempt to deceive others and when they should try to do so. In group-living species, for example, deceitful signals may be less easy to detect if they are given only rarely, or if an individual is reliable in some contexts but not in others. Concurrently, there should be selection for recipients to transfer information about the reliability of a particular signaller from one context to another.

The detection of unreliable signals will, therefore, be influenced by the ways in which animals assess and compare the meaning of signals. Deception will be constrained if recipients can recognize a false relation between a call and its apparent referent, and it will be further constrained if recipients can transfer information about the reliability of a signaller's calls from one context to another. In the case of vervet monkeys, and probably other primates, selection may have favoured the ability to recognize 'spheres' of meaning and to transfer information gained in one sphere to other, related ones. Individuals who have come to recognize that one type of call by a given signaller is unreliable appear to transfer their skepticism to other calls of broadly similar meaning, but not to calls whose referents are different. Thus deception, if it occurs, seems less likely to be detected when contexts are changed between successive deceptive acts than when contexts remain the same.

Acknowledgements

We thank the Office of the President, Republic of Kenya for permission to conduct research in Amboseli National Park. We also thank Bernard Musyoka Nzuma for field assistance. Helpful comments on an earlier draft were provided by C. Allen, V. Dasser, C. G. Gallistel, R. A.

Hinde, M. Kelly, P. Marler, D. Premack, J. Sabini, and C. Snowdon. Research was supported by NSF (grants BNS 82–15039 and 85–21147) and NIH (grant 19826).

7

Mapping social concepts in monkeys
VERENA DASSER

The work reviewed in the previous chapter is invaluable because it analyses the scope of social cognition in the natural environment. Under such conditions, however, it is often difficult to impose experimental control to discover whether certain concepts are abstract in the full sense of applying to novel examplars. This is exactly what Dasser's work does: the two chapters thus represent a complementary pair.

Introduction

The social structure of primate social groups appears to be based to some degree on the ability of the group members to assess interindividual relationships. Existing evidence suggests that monkeys may achieve this assessment by forming concepts of social relationships (Bachmann and Kummer 1980; Cheney and Seyfarth 1985). However, while non-social concepts have been identified in primates and birds by controlled laboratory experiments, such testing of concepts of the social environment has not yet been done. From observations on free-ranging primates and from field experiments, it is usually difficult to infer the precise cognitive mechanism underlying the behaviour (Dasser 1987 a; Chapter 6). Laboratory subjects, on the other hand, usually do not live in a large social group. It cannot, therefore, be tested whether they recognize some aspects of the social structure so characteristic of a monkey group for the human observer. What we need is a socially experienced monkey that is a permanent member of a naturally structured group as well as a subject performing in laboratory tests.

In this paper, I describe some experiments carried out with two long-tailed macaques that met the above given criteria for a suitable subject. The experiments examined whether these monkeys recognize affiliative relationships among their group members independently of the particular individuals involved. I adopted a method devised by animal psychologists for studying conceptual performance in animals in the laboratory. A concept is defined here as a set of rules which enable an individual to recognize objects or some other, more abstract entities, as belonging to a particular category. To have a concept, thus, means to be capable of categorizing entities and of drawing inferences about novel instances; they will have all or most of the properties of their category.

I will first describe two experiments indicating that the subjects differentiated mother-offspring pairs from other pairs of group members regardless of which mother and which offspring was involved, and regardless of the age of the offspring (Dasser 1987 *a*). I then report some additional data that suggest that the ability to categorize social affiliations may go beyond the discrimination of mother-offspring pairs *v*. all others. The general experimental design was to train each subject on a few examples of the particular, experimenter-defined category of social affiliation, using colour slides of group members, and then to test its response to other group members, also instances of the same category. Operationally then, a concept denotes the ability to generalize from the stimulus monkeys used in training to other group members.

Subjects and method

The subjects were two 5-year-old *Macaca fascicularis* females, Rini and Riche. They were members of a captive group of 40 animals of both sexes and all ages. All members have lived in this group since birth. They were permanently housed together in a large indoor/outdoor compound. The subjects were only isolated for one daily session up to 90 minutes, 5–7 days a week. Their training history included 1-year of adaptation to the experimental situation and exposure to a variety of slides of familiar objects, as well as experiments testing their ability to recognize group members on colour slides (Dasser 1987 *b*).

The subjects were individually trained and tested in one part (about $1m^3$) of the indoor compound. The main part of this chamber was the experimental panel (108 × 82 cm) which was used with both subjects. It consisted of clear plexiglass onto which three daylight projection screens were arranged in a triangle standing on its base. A response button was attached at the lower edge of each screen and a feeder tube was located below the centre screen through which juice could be delivered automatically. A more detailed description of the panel is given in Dasser (1987 *a*, *b*). Colour slides were projected from behind and were 30 × 20 cm in size. All equipment was controlled by a computer which also recorded the experimental events and the responses of the subjects. The subjects could hear other group members during their sessions and partly see them through small openings in one side wall, but they could not interact with each other. This presence of group members proved critical to a high motivation of the subjects who were in no way forced to enter the chamber and to work on their task.

The slides showed the group members in the outdoor compound with various backgrounds and in different postures and distances, just as the subjects were used to seeing them. The slides used as alternative choices in a trial in all experiments were always matched with respect to part of body shown, and size of the stimulus animal(s) in the slide, and generally for body posture, background, and illumination. The faces were always visible and shown in life size if no other body parts were on the slide; whole bodies were up to about three times smaller.

Procedures

I used two different procedures: Rini was trained in a simultaneous discrimination procedure (the SD-experiment) and Riche worked on a simultaneous matching-to-sample task (the MTS-experiment). Basically, each experiment consisted of two phases: during the training phase, the subjects were trained to respond to a few instances of the mother-offspring affiliation represented by slides showing pairs of group members (SD-experiment) or single individuals (MTS-experiment). In the subsequent transfer phase, the subjects were tested in unique transfer trials on other instances of mother-offspring pairs of the group. Since the number of transfer trials was limited by the number of available mother-offspring pairs and suitable slides, and since the motivation of the subjects varied widely, it was felt that the critical transfer trials should only be presented on days when the subjects were 'at their best', i.e. highly motivated and attentive to the stimuli, and that only very few should be given at a time. Such good days were determined by the subjects' performance on the training examples before any novel mother-offspring example was presented.

Left/right position of the positive alternative varied according to a pseudo-random balanced sequence (Gellermann 1933; Fellows 1967). The subjects indicated their choice of one of the alternatives presented at the lower screens by once pressing the response button underneath it. The slides remained illuminated until the subjects responded. The choice of the positive alternative slide was rewarded with honey water in all trials, including transfer trials. A wrong response resulted in a time-out interval after which the next trial began. No human observer was present when transfer trials were given.

Mother-offspring pairs

On each trial, Rini was simultaneously presented with two slides projected onto the lower screens. The rewarded one showed a mother and her offspring, of any age, and the negative alternative showed another pair of group members. One instance of a mother-offspring pair was given as training example: a mother and her juvenile daughter, represented by different slides. Five different negative pairs were used. On the transfer trials, novel mother-offspring pairs and novel negative non-mother-offspring pairs were administered. A total of 14 slides each showing a different mother-offspring pair was available and, thus, 14 transfer trials were conducted. The negative pairs showed any other two group members. They were usually not matched to the mother-offspring pair with respect to the sex/age combination. However, care was taken to avoid any consistent differences between the positive and the negative pairs other than the mother-offspring tie.

In the MTS-format, the mother was represented as the sample on the centre screen, and one of her offspring and another stimulus animal of the same sex and age as the offspring were given as the positive and negative alternative, respectively. During training, Riche saw two mother-offspring pairs and a single slide of each of these four stimulus animals. Thus, she was trained on the following combinations: mother 1 with offspring 1 and offspring 2; mother 2 with offspring 2 and offspring 1. Both offspring were juvenile females. Here, 22 other instances of mother-offspring pairs were presented in unique transfer trials.

The high number of correct choices by both subjects (Fig. 7.1) suggests that they solved the transfer trials by categorizing pairs of group members. Mother-offspring pairs were differentiated from any other pair, regardless of the age of the offspring and the particular individuals involved. Cues other than the relation between individuals do not plausibly account for the result (Dasser 1987 *a*). The subjects apparently used information gained during their daily group life in the present experimental situation. The most likely explanation is that the subjects chose the 'stronger' of the two affiliative relationships on the basis of the behaviour of the real mother and her offspring. If this interpretation is correct, picking out the mother-offspring pair may have been an easy task because two-thirds of all negative pairs summed over both subjects were composed of group members of different matrilines whose affiliations were known to the experimenter from proximity data, agonistic interactions, and grooming frequences to be weak in comparison with that of the mother-offspring pairs. Behavioural associations of individuals are generally stronger within matrilines than between matrilines. One might, therefore, expect that the subjects would find it more difficult to distinguish mother-offspring pairs from other pairs composed of group members of the same matriline. However, only one error occurred on 12 such transfer trials. It seems, then, that mother-offspring pairs can be distinguished from any other pair of group members.

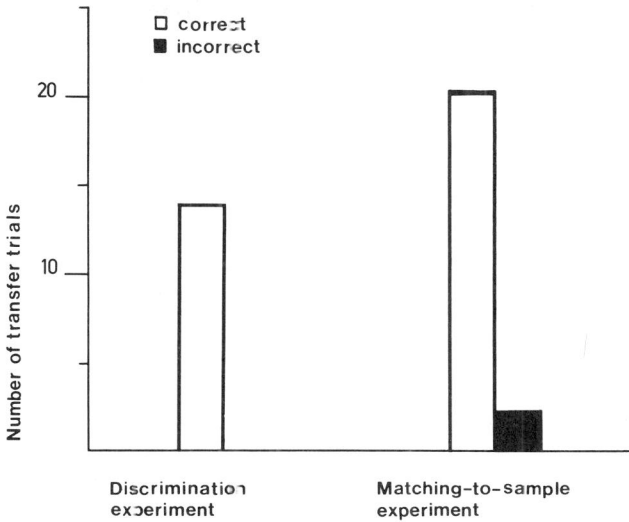

Fig. 7.1 The number of correct and incorrect responses on transfer trials in the simultaneous discrimination experiment and the matching-to-sample experiment.

Other categories of affiliative relationships

What of other categories of social affiliation? Does a monkey's ability for categorization go beyond the distinction mother-offspring v. non-mother-offspring? An additional experiment was carried out with Rini to determine whether she is capable of differentiating pairs of maternal siblings from other pairs of group members, in particular from pairs of individuals belonging to one matriline.

The pairs were again presented to the subject using colour slides, and the same simultaneous discrimination procedure was used as in the previous experiment. Each pair of siblings was paired with an exemplar of one of the following types of negative pairs resulting in three types of discriminations: a sibling pair v. either a mother-offspring pair (type 1 discrimination), or a pair of matriline members otherwise related (type 2 discrimination), or a pair of unrelated group members from different matrilines (type 3 discrimination).

On each training trial, Rini was confronted with either a type 1 or type 3 discrimination. Only one training example of the positive category was presented, namely one pair of adult sisters represented by two different slides. Two different slides of a mother and her juvenile son, and two slides showing one of the sisters once with an unrelated female and once with an unrelated male served as negative alternatives. On each transfer trial, Rini was tested on one of the three types of discrimination on non-training pairs. A total of nine transfer trials of each type was presented. Using the same subject in successive experiments carried with it the disadvantage that all or some individuals and pairs and, at a higher level of abstraction, social categories may have a history as positive and/or negative stimulus animals or stimulus categories which may influence the responses of the subject. Care was taken, therefore, to reduce this possibility as much as possible. In particular, for the choices on the most interesting type of discrimination (type 2), any such influence was excluded by the selection of stimulus pairs not previously used. A possible influence of previous exposure to the stimulus monkeys on the results of the other types of discrimination will be discussed below. Individual slides were all novel though.

The overall result, 19 correct out of 27 transfer trials ($P=0.05$, binomial test, two-tailed) suggests that Rini again used a rule of categorization, that is, a concept, which enabled her to pick out the sibling pair. She was hardly ever wrong unless the negative pair was a mother and her offspring (Fig. 7.2). It seems that in spite of the training on the negative mother-offspring pair in the present experiment, Rini still stuck to her first training on the positive mother-offspring pairs. The perfect performance (100 per cent correct) on type 3 discriminations was not surprising in view of the results of the mother-offspring experiment where all but one of 24 pairs of unrelated group members summed over both subjects were correctly rejected. It seems that the subjects easily discriminated between two categories of affiliation if one of them involved group members from different matrilines. The most interesting discrimination of this experiment was that of type 2, the discrimination between the two categories of within-matriline pairs, sibling pairs v. matriline membership. With only one error in nine transfer trials, Rini passed what was probably the most demanding of her tasks. She was able to pick out the sibling pair regardless of the composition of the negative pair: four pairs showed aunts (uncles) and nieces (nephews), two pairs exemplified cousins, one showed a grandmother and her granddaughter, another one a great-grandmother and her great-granddaughter. The one pair preferred to a sibling pair was a great-aunt and her great-nephew. This is quite impressive, but it should be mentioned that this significance hinges on the correctness of a single

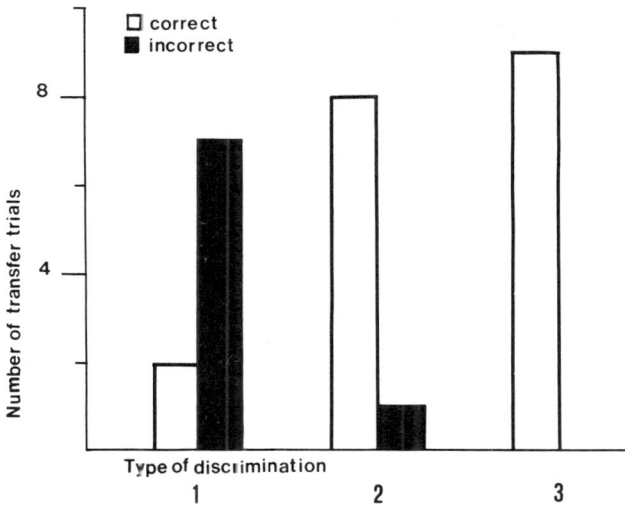

Fig. 7.2 The number of correct and incorrect responses on transfer trials of each type of discrimination. Type 1 discrimination: mother-offspring pair *v.* sibling pair; type 2 discrimination: pair of otherwise related group members *v.* sibling pair; type 3 discrimination: pair of unrelated group members *v.* sibling pair.

choice, for although eight correct out of nine is significant, seven out of nine is not. In conclusion, it would appear that here as in the mother-offspring experiment, Rini differentiated between different categories of affiliation.

Discussion

Kin recognition in the sense that related individuals are distinguished from unrelated ones is a widespread phenomenon in animals. It usually manifests itself in preferential associations of relatives. There is another type of kin-recognition which was described as non-egocentric by Cheney and Seyfarth (e.g. 1985) and which thus far was only suggested by a few primate studies, namely the discrimination of kin-associations between others. For example, Cheney and Seyfarth (1980) showed that vervet females associated screams of particular juveniles with particular adult females, their mothers.

It is this latter type of discrimination with which the present experiments were concerned. However, it is important to note that although the experiments involved genetically defined categories of relationships, they did not aim at showing whether the subjects recognized the genetic relationships between group members, but rather they attempted to show whether monkeys have concepts of social affiliation enabling them to categorize relations between known individuals under controlled laboratory conditions. The results suggest that monkeys have such concepts. I do not wish to imply that the subjects formed these concepts during the experiment. This seems extremely unlikely given the small number of training examples and the high degree of generalization asked for in the transfer trials. In fact, there seems to be no way of teaching a monkey the concept of affiliation with the method used, regardless of the number of training instances given; and perhaps there is no way at all. Rather, the data provide evidence of concepts possessed by the subjects prior to any experimental experience. The results of field experiments on vervet monkeys reported by Seyfarth and Cheney, in Chapter 6, strongly confirm this conclusion.

Pigeons in a study by Cerella (1979) generalized from a single example of a white oak leaf to novel oak leaves. How does this performance compare with the result of the present experiment? While this generalization from one instance of a concept to the whole equals the high degree of generalization found in this study, the kind of concept tested does not appear to be equivalent, nor does the mechanism underlying the generalization. One oak leaf necessarily shares common features with other oak leaves and at the same time looks different from the non-oak leaves presented as negatives. The picture of one mother-offspring pair, however, is not likely to share more common features with the pictures of other mother-offspring pairs than with the pictures of other pairs of group members. A possible exception here may be the physical resemblance between close kin. It was, however, excluded as a major source of the matching performance of Riche for she was only successful in the transfer trials with recent slides. In control trials using older slides, taken 2–3 years before, Riche failed to choose the correct offspring significantly above chance level (Dasser 1987 *a*). While individual appearance in immature monkeys changes over the years, the relative resemblance between the pictures of a mother and her offspring, and the picture of a non-offspring does not depend on the age of the picture.

Generalization from one example to the concept as a whole seems therefore to involve more than the perception of features common to the positive and/or negative stimulus slides. Mothers and offspring, or sibling pairs, share common properties on the level of their behavioural association.

The experiments do not show, of course, that humans and monkeys share the same social concepts, but that certain human-defined concepts may have an analogue in the monkey's view of its social world.

Acknowledgements

This research was supported by the Swiss National Science Foundation (Grant No. 3.359.082 and No. 3.242.85 to Dr H. Kummer). I am grateful to H. Kummer for valuable discussions throughout this project and to my colleagues for their technical assistance.

8

The cognitive demands of children's social interaction with peers
PETER K. SMITH

Following a common primate pattern, the human child graduates from a primary attachment phase typically focussed on parents to a phase when attention is directed more towards other young children. This chapter examines the principal relationships apparent in children's peer groups—friendship and dominance—and reviews what we know about the cognitive sophistication involved, and how it is achieved. There is evidence that, in several respects, cognition is more advanced in social compared to non-social contexts.

For some decades, developmental psychologists have looked at the links between cognitive development in children, and their social development and experience. The growth of the research area of 'social cognition' bears witness to this. By and large, however, cognitive development has been taken as primary, facilitating or allowing certain forms of social behaviour. Such an emphasis is certainly inherent in the work of Piaget and the neo-Piagetians. Although Piaget wrote that 'cognitive and affective or social development are inseparable and parallel' (Piaget and Inhelder 1969, p. 117), and although, in his early work on moral judgement, Piaget emphasized the importance of peer interaction in bringing about a shift from heteronomous to autonomous morality, nevertheless the predominant view has been that a level of cognitive development has been a necessary condition for a level of social cognition and of moral judgement. As Kohlberg (1976, p. 32) expressed it

first, a person attains a logical stage, say, partial formal operations . . . next he attains a level of social perception or role taking . . . finally he attains Stage 4 of moral judgment.

Furthermore, the kinds of tasks used by developmental researchers in assessing the competence of a child, have most generally involved non-social objects. Even one of the most dominant strands in research on social cognition, that of visual perspective taking, has involved the perception of non-social objects (as in the '3 mountains' experiment).

Yet, this 'object-centred' view of the child's development has been undergoing a gradual change in recent years. This is not just in introducing more 'social' tasks into the 'cognitive' framework (though this has often happened), but to some extent, and more radically, re-interpreting the child's development in terms of the demands of and response to social experience. Some developmental theorists are now arguing the primacy of the social world, especially in infancy and early childhood. The ethological tradition of work in child development has also produced some interesting correspondence to findings in other primates, referred to in this volume.

This chapter traces some of the main developments in social experience with others through late infancy and early childhood, emphasizing particularly the knowledge or skills which underlie social interaction. Where possible, comparisons of similar abilities between social and non-social domains are made. Finally, there is a discussion of how social knowledge is obtained.

Main trends in social development through infancy and early childhood

Through the first 6–9 months of life, the human infant is very interested in social stimuli. Humans or non-social stimuli resembling humans in various ways (such as cardboard face-like stimuli or mobiles which move in response to the infant's movement) evoke the most attention and the most positive affect response (Watson and Ramey 1972; Brooks-Gunn and Lewis 1978). Also, actions by humans (such as mouth opening or tongue protrusion) can evoke imitative responses by infants from a few days of age onwards (Meltzoff and Moore 1983; Kaye 1982).

Although at first all human or human-like stimuli are responded to positively, by 7–9 months of age the infant is characteristically forming attachments to one or a small number of primary caregivers (e.g. mother, father). As conceived in the attachment-theory framework of Bowlby (1969), an attachment relationship gives an infant the security to explore the environment, both social and non-social, with the attachment figure serving as a 'secure base' to return to.

Following a common primate pattern, young children tend to venture further from attachment figures as they grow older (Rheingold and Eckerman 1969) and to tolerate longer absences more securely. There is a rapid growth in the frequency and complexity of peer interactions in the period from 1 to 3 years of age. In the context of modern urban societies, most interactions of children which are with non-adults, are with peers in toddler groups, nurseries and schools. Social competence with peers seems to be facilitated by a secure primary attachment (Bretherton 1985). The actual modes of peer interaction seem to be partly carried over from earlier and concurrent interactions with adults, and partly learnt intrinsically in the experience of interacting with age-mates (Vandell 1980). However, the impact of older siblings in the home has been under-researched. Recent evidence from Dunn and Kendrick (1982) and others suggests that the challenge of coping with older (or younger) siblings may indeed be a strong impetus for growth in social understanding and the ability to manipulate others, coupling as it does strong motivation with a high degree of familiarity and mutual experience. Konner (1975) has also suggested that in traditional societies, in which mixed age groups are more common, older children may play an important transitional role in the move from adult-infant to peer-peer interactions.

Through the years of middle childhood, the peer group is certainly one of the most important social contexts for children. Much peer interaction is friendly and playful, but the elements of dominance and social manipulation, always obvious to observers, have received increasing research attention in recent years.

Social experience and cognitive development in later infancy, and the transition to peer relationships

An early landmark in the development of social behaviour is the understanding that there are particular other social beings. Referred to as 'person permanence', this has been assessed in terms of infants' recognition of particular others and search for them when a person disappears from view. Person permanence implies an internal representation of a social being, corresponding to its continuity in time and space.

Closely related to the concept of 'person permanence' is that of 'object permanence', applied to non-social objects. Both are achieved during the sensorimotor period (up to about 18 months of age), with major progress in the degree of permanence achieved (i.e. the success of search strategies) towards the end of the first year. A few studies have compared the onset of person permanence and object permanence, and indeed these provide perhaps the best direct comparison between cognitive abilities in social and non-social domains, in childhood (cf. Chapter 19).

Bell (1970) tested 33 infants, aged 8½–11½ months, for levels of person and object performance. In person permanence tests they searched for the mother behind furniture or curtains, in object permanence tests they searched for an object under a cloth. Bell found that 23 infants showed more advanced person permanence than object permanence with only 7 infants showing the opposite tendency. Similarly, Paradise and Curcio (1974) found that out of 30 infants aged 9 months, 23 were ahead on person permanence and only five on object permanence.

A problem with these and similar studies is that the person/object permanence distinction is confounded, first by whether the search has to be *behind* a screen or *under* a cloth; and second by the degree of familiarity of the person (mother) or object (unfamiliar toy). In a careful series of studies, Jackson *et al.* (1978) controlled these aspects. When task demands were properly equated, the 'decalage' or disparity between person and object permanence was reduced. Nevertheless, the authors reported from their longitudinal data that, although no decalage was most common, 'on those occasions when decalage occurred, it occurred significantly more often in favour of persons rather than objects' (*op.cit.*, p. 6). This finding occurred in spite of equating the social and non-social stimuli for practice and familiarity effects. In actual development, however, interest and familiarity with social stimuli might be one of the mechanisms favouring a primacy of impact of social over non-social events in cognitive development.

What other aspects of social experience might be related to early person permanence? Bell (1970) found that a secure, unambivalent attachment relationship to the mother was characteristic of infants showing advanced person permanence scores. However, the dependent variable (security of attachment) was obviously confounded with the independent variable (person permanence level, assessed by search for mother); poorly attached infants might be less motivated to search (see Smith 1979, for a review).

Two other factors have been suggested as crucial in the development of early social-cognitive skills. These are the developing sense of self and the experience of contingent responsiveness.

The importance of the development of a sense of self, as a reference point for understanding others, has been hypothesised by Lewis and Brooks-Gunn (1979), although as they point out, the idea goes back to Baldwin (1897) and even to some of Piaget's writings. They advanced three principles regarding early social awareness:
(1) any knowledge gained about the other also must be gained about the self;
(2) what can be demonstrated to be known about the self can be said to be known about the other and what is known about the other can be said to be known about the self;
(3) social dimensions are those attributes of others and self which can be used to describe people.

Lewis and Brooks-Gunn (1979) found that infants aged as early as 9 to 12 months were capable of some differentiation between pictorial representation of themselves, and others (e.g. by differential smiling). There is also some empirical evidence that the self is used as a reference point, when through the second year of life infants seem to be more interested in (and less fearful of) other children than adults, and possibly same-sex rather than opposite-sex persons. Lewis and Brooks-Gunn suggest that a knowledge of and (positive) valuation of oneself leads to differentiation of and preference for certain social objects.

However, another very powerful influence in the infant's social world is contingency of response. This refers to the infant getting a response to its own behaviour, at an appropriate time and of at least a roughly appropriate nature. An adult picking up a crying infant would be an example of contingent responsiveness; so would a person talking to an infant when it vocalizes. Many studies have now demonstrated that contingent responsiveness is generally reinforcing and enjoyable to infants (e.g. Watson and Ramey 1972; Goldberg 1977). It is also thought to lead to generalized expectancies about the behaviour of other social beings and to enable characteristic interaction patterns with particular others to develop.

Contingent response can be provided by non-social objects (e.g. Watson and Ramey 1972; Gunnar 1980), but it is normally and primarily provided by other social beings. Indeed, an influential current conception of the infant's social development is that of an apprentice learning social skills and acquiring social knowledge through the carefully paced and appropriately chosen responses of parents and other older persons (Kaye 1982; see Trevarthen 1982, for a view putting even stronger emphasis on the social competence and social motives of infants). Contingency of response is also an important component of the more subjective concept of 'sensitivity', which has often been related to the development of secure attachment relationships in the first year of life (Ainsworth *et al.* 1978).

The shift from parent attachment to peer interaction

The two factors we have just considered may shift in balance through the first few years of life, possibly underlying the shift in social orientation from one predominantly to adults (in the infancy period) to one in which interaction with peers plays a much larger role (by the preschool and school years). If Lewis and Brooks-Gunn are right [and this is a view which clearly links with Humphrey's (1986) about the role of conscious awareness in understanding others' behaviour] then it will be easiest for infants and young children to understand peers. Indeed, considerable interest in peers is noticeable by the second year of life, but this is still counterbalanced by the greater ability of adults and other persons to provide contingent responses and, effectively, to allow their own behaviour to be controlled predictably by the infant. In infancy, interaction with peers is thought to be difficult to sustain, since peers cannot at that age provide appropriate and contingent feedback to social initiations. As Bronson (1974) put it, 'feedback given by peers is too variable and often too delayed to allow the baby to develop firm predictive expectancies of links between his behaviour and the actions of a peer'. (*op. cit.* p. 255).

Rather, social interactions in early infancy are sustained by the ability of adults (or other persons) to 'scaffold' the interaction, providing the appropriate choice and timing of response to suit the infants needs and intentions in a way which another infant cannot yet do. This 'scaffolding' is important, but decreasingly so, well into the second year of life (Kaye 1982). Although adult support and understanding obviously has a part to play throughout life, the asymmetries of adult-infant interaction have clearly lessened by the preschool years, and children are able to interact with each other in competent and appropriate ways. By the third year of life, young children have sufficient social skills to maintain sustained peer interactions, and peers may then be *easiest* to interact with on the basis of understanding others in relation to one's knowledge of oneself.

Social dimensions used in infancy and early childhood

We have seen how infants achieve permanence through the sensorimotor period, often ahead of object permanence. As concepts of particular persons become more stable, persons can begin to be categorized along social dimensions. Brooks-Gunn and Lewis (1978) argued that the three earliest social dimensions learned are familiarity, age, and gender. They argue that these are concurrently developed in relation to oneself and to other persons.

The salience of familiarity is apparent from the body of research which shows that infants behave differentially to familiar and to strange adults by around 7–9 months of age (when attachment relationships have usually formed), if not earlier (Scarr and Salapatek 1970; Sroufe 1977). Differential reactions to familiarity with peers seems to develop slightly later. Jacobson (1980) found that awareness of an unfamiliar peer (compared to a familiar peer) developed between 10 and 12 months. Greater previous experience with the familiar peer predicted an earlier peak for wariness.

Wariness has usually been thought to require the cognitive capacity to relate present events to schemata stored in memory. In the case of strange objects, this seems to develop at around 8–10 months (Schaffer 1974). While direct comparisons are difficult to make (since task situations have been different), the development of wariness to unfamiliarity would seem to come first with adults, then with non-social objects, followed by peers. The experience of contingent interactions with adults may well have provided the experiential basis for laying down schemata of familiar interaction patterns. For reasons already explained, such schemata would be more difficult to form with peers at this early age, and perhaps more so than for non-social objects which would be less interesting, but more predictable, in their response.

Familiarity can be applied as a dimension to both social and non-social objects. The dimensions of age and gender are much more limited to the social domain; thus, direct comparisons to the non-social domain are not possible. It is clear, however, that age and gender are used very early on in a categorical way.

Infants aged 6–9 months of age discriminate between the live approach of a child and of an adult, and infants aged 9–12 months can differentiate between photographs of baby and adult faces. Brooks-Gunn and Lewis (1978) suggest that height, movement, voice cues, and extreme differences in facial and hair cues (in baby and adult photographs) serve as indices of age. Verbal age labels (e.g. baby, mummy, daddy) begin to be used correctly by 18–24 months of age.

By the preschool years, age can be used as a criterion for classification, and for seriation or ranking. Edwards and Lewis (1979) found that 3½-year-olds could successfully sort head and shoulders photographs of persons into four categories: little children, big children, parents and grandparents. Furthermore, two studies (Taylor *et al.* 1982; Jones and Smith 1984) have found that 4-year-olds can rank order photographs by age. Taylor *et al.* (1982) used a paired comparisons method, while Jones and Smith (1984) asked children to successively pick out the oldest person from an array of photographic stimuli. In the latter study, 30 out of 34 subjects (mean age 4 years 2 months) achieved significantly accurate results.

Classification and seriation have been thought of as cognitive abilities characteristic of the concrete operational period (6 years onwards) in the framework of Piaget's theory. Subsequent research in non-social domains has found that Piaget underestimated the ability of the pre-operational child, and that if task demands are made more easy, interesting and familiar, then tasks such as seriation can be achieved earlier. For example, transitive inferences (necessary for seriation) can be achieved in the non-social domain at 4 years of age (Bryant and Trabasso 1971), but only if the materials are very familiar so that memory difficulties can be overcome. In ordinary life, social stimuli may well be particularly familiar and interesting, so that seriation becomes possible first in social domains. This argument is reinforced by work on the perception of dominance hierarchies in the peer group, to be discussed shortly.

Differentiation of social objects by gender also occurs early. Nine- to 12-month-olds respond differentially both to photographs of female and male strangers (Brooks-Gunn and Lewis 1981) and to direct approach by male and female strangers (Smith and Sloboda 1986). Verbal age labels (e.g. mummy, daddy, boy, girl) begin to be used correctly by 18 months of age, and gender identity is fully achieved by the third year (Thompson 1975). Gender stability (realizing gender is a stable attribute over time) is achieved by 4–5 years, but gender constancy (realizing gender is invariant, despite changes in clothes, appearance or activity) only by 7–9 years. Marcus and Overton (1978) found that gender constancy in the face of transformation of pictorial representations of boys and girls generally was achieved after conservation of quantity using non-social objects (e.g. plasticine, soy beans).

Familiarity, age, and gender continue to be used as social dimensions throughout life. However, two other social dimensions become especially important in the peer groups of preschool and school age children: friendship and dominance.

Cognitive demands underlying social interaction in the peer group

The two main dimensions of children's peer groups which have been studied by researchers, are friendship (or affiliation) and dominance. We know something about how children themselves think about friendship, from interviews in which they are asked to describe what a friend is, or what their expectations of friends are. We also know what children know about the social structure of their group. So far as friendship is concerned, this has been studied primarily by sociometric methods—asking children to nominate friends or rank classmates for liking; or observing which children play with or associate with certain others. Dominance has been studied by asking children to nominate, or rank classmates for, toughness or strength; or observing which children win against certain others in threat or conflict situations. In both areas of study, interview data from children usually agrees reasonably well with data from direct observation.

Interviews with children about the concept of friendship (e.g. Selman 1981) suggest that the earliest concept, prevalent in the 3–7-year age group, is simply of someone who plays with you or who lives near you. The notion of a friend as helping you is apparent in the 4–9-year age range, but ideas of reciprocity only appear in the 6–12-year period. These findings would not appear to suggest an important role for friendship knowledge in cognitive development; since, for example, the notions of reciprocity in friendship appears to await Piaget's stage of concrete operations, at 6 years. However, direct observations do suggest that, at a behavioural level, reciprocity may be operating well before this age. Strayer *et al.* (1979), observing a group of 26 children aged 3–5 years, found that children were selective in their distribution of altruistic acts, and that there was a significant correlation between initiation and receipt of altruistic behaviours; this suggests that reciprocity was occurring, although as Strayer *et al.* (1979) recognize, their data were not sufficiently detailed to establish the existence of such reciprocity at the level of dyadic relationships.

What is certainly startling (at least to those who have forgotten what childhood is like) is the extent of the knowledge that young children have of the social structure of their group. The concordance between rating nominations of friendship and observations of associative play show that children know who their own friends are, which is not surprising. However, they can also remember who their friends were earlier in the term, and can say who other children's friends are, to a considerable degree of accuracy, by 3 and 4 years of age (cf. Chapter 6).

Children's memory for previous friendship was reported by Delfosse and Smith (1979) on a sample of 15 children of mean age 4 years 2 months. They had had an average of 10 months nursery experience together. Each child was asked 'who did you play with a lot today', 'who did you play with a lot yesterday', 'who did you play with a lot last week', and 'who did you play with a lot at the beginning of nursery' (i.e. at the beginning of term, after the long summer vacation). A picture sociometric technique was used. The results were compared with direct observations of group play for the time period in question.

It was found that the children scored well above chance level for all the questions. Over 50 per cent of 'yesterday' responses, over 40 per cent of 'last week' responses, and 25 per cent of 'beginning of nursery' responses were completely accurate (i.e. the child's first nomination matched the child most often played with, from the observational data). Some level of success could have been obtained if children just recalled their present companions (since friendships are moderately stable at this age), but subsidiary analysis showed that this was not the case; it seemed that true recall strategies were being used.

This study points to the salience of friendships and of the child's social world for memory process. Memory of peers and their attributes, behaviour, and dispositions could be very important for a preschool child in a nursery or playgroup. In a related study in the same class of children, Smith and Delfosse (1980) found that children could also nominate who the best friend was of other children in the class. Over 60 per cent of these responses were completely accurate (i.e. the child's nomination of X's best friend matched the child whom X had played with most over the last 4 weeks). All 15 children could do this task at above chance levels of accuracy.

These results show some impressive knowledge and recall ability about friendships. Unfortunately, it is difficult to compare these abilities with those in non-social domains. For example, comparable studies on long-term recall of naturally occurring non-social events, have not been reported.

Some more precise comparisons can be made in the case of young children's knowledge of dominance relationships within a preschool class. Children can be asked to rank other children in their class for 'toughness' or 'strength'. There is generally a high consensus on this, and this consensus ranking agrees well with observations of the outcome of dominance encounters. Thus, children are accurate in perceiving the dominance relationships of other children; they are less accurate in assessing their own position, however, systematically over-rating their own place in the dominance hierarchy (Omark and Edelman 1975; Sluckin and Smith 1977; Pickert and Wall 1981).

In the case of dominance (unlike friendship) it is clear that there is an objectively perceivable 'rank order' in the class which children can accurately reproduce. Sluckin and Smith (1977) observed two groups of preschool children, and found that in each group there was an observable dominance hierarchy. They interviewed 20 children, aged 3:2 to 4:10 years of age. Eight children (age 3:2 to 4:10 years) could given statistically reliable rank orders (as assessed by retesting), and the rank order of these children agreed highly and statistically significantly with each other.

As pointed out earlier, this ability to rank order accurately is equivalent to the logical ability of seriation. Piaget (1952) distinguished three stages in the development of this ability. In the first stage, the child always fails to make a complete series or order. During the second stage, the child produces a complete series, but cannot systematically correct mistakes. In the third and final stage transitive inferences are used to produce a correct or near-correct rank order; that is, if A>B and B>C, then A>C. Transitive inferences are exactly the sort of cognitive process involved in the construction of a dominance hierarchy. This is especially so as not all the dyadic dominance relationships might be observed directly; to obtain an accurate linear hierarchy, it is likely that a child would need to 'interpolate' certain positions using inferential procedures.

In the Sluckin and Smith study, the degree of concordance for dominance ranking was 0.61 (Kendall's W) for four children in one class (mean age 4:4 years), and 0.67 for four children in the other class (mean age 4:7 years). The high degree of concordance certainly suggests, though it does not prove, that children were at the third of Piaget's levels and that inferential procedures were being used.

Although Piaget did not believe that children could carry out transitive inferences until 6 years of age, Bryant has found that younger children can make transitive inferences in non-social tasks if enough practice is given. Bryant and Trabasso (1971) carried out a study using coloured rods of different lengths with 20 children of mean age 4:5 years (and older children), and subsequently with another 25 4-year-olds. They found that transitive inferences could be made, at about an 80 per cent level of success, in this age group. Bryant and Kopytynska (1976) found that 5- and 6-year olds could actively and spontaneously use transitive inference procedures in non-social tasks.

Edelman and Omark (1973) and Omark and Edelman (1975) suggested that the developing dominance hierarchy in children's groups might form part of the necessary social experience for transitive inference procedures to develop, and the work of Sluckin and Smith (1977) showed that children's social experience in preschool could provide a suitable experiential matrix for this to occur. Bryant's work suggests that similar abilities can be manifest at about the same time (4-year-olds) in non-social situations. However, it should be borne in mind that the Bryant and Trabasso (1971) task was a very artifical, experimental one, involving specially constructed apparatus, clearly distinct stimuli (five different coloured rods of lengths 3, 4, 5, 6, and 7 inches), and many repeated practice sessions of unambiguous dyadic stimulus comparisons. In other words, the experimenter had to go to a lot of trouble to demonstrate transitive inference in the non-social domain (and even then, we do not have evidence for its active, spontaneous use until 5 years of age).

Thus, it remains plausible that a child's social experience of dominance relationships provides the earliest normally encountered matrix of experience suitable for developing inferential procedures. Hierarchies seem to be present in 3- and 4-year-olds, and can be verbally reproduced by many 4-year-olds.

Social interaction skills with peers

Children use their knowledge of social structure, in particular their knowledge of affiliation and dominance within their social group, to assist in coping with problematic social situations (see Chapter 5). A taxonomy of critical problematic social situations in younger school-age children has been developed by Dodge et al. (1985). Two of the most problematic social situations identified in that study were children's peer group entry attempts, and being confronted with provocation by a peer. These kinds of situations, and children's strategies in coping with them, have been studied by a number of investigators (e.g. Corsaro 1981; Sluckin 1979; Dodge et al. 1986).

Corsaro (1981) carried out an ethnographic study of a nursery school, with children aged 2:11 to 4:10 years. He put particular emphasis on understanding the role of friendship in social situations. One common concern was seen to be protection of interactive space. Friendship could be used as a basis for allowing some children to enter this interactive space, but to exclude others (e.g. Child 1: 'You can't play!' Child 2: 'Yes I can. I have some animals too.' Child 1: 'No you can't. We don't like you today.' Child 3: 'You're not our friend.') Other justifications for exclusion included reference to arbitrary rules (e.g. 'You can't play with bare feet!' or 'this is only for girls') or specific claims of ownership (e.g. 'I had that first'). However, such exclusion was often negotiable (e.g. 'You were my friend a minute ago'). In about half the cases, the new child eventually joined the group.

Corsaro saw his findings as consistent with Selman's (mentioned earlier). In general, friendship at this age was seldom based on the children's recognition of enduring personal characteristics of playmates, but often served specific functions such as gaining access to play groups and building solidarity with several playmates as a means of maximizing the probability of successful entry (see Chapters 9–11). Sluckin (1979) carried out a similar sort of study on 5–10-year-olds in a primary school, with particular emphasis on how children dealt with social problems confronting them. Social problems might include exclusion from a game, teasing, bullying, ownership disputes, and methods of choosing roles in games. Sluckin distinguished about a dozen different types of solutions to such problem situations. These included verbal ritual (e.g. 'bagsee mine'), standard rules involving values of the playground community (including friendship or rules such as segregation by sex), restating the situation (e.g. 'we're playing tag not bulldog'), situational discounting (e.g. 'it's not real, but pretend'), arguing about implications of definition for legitimate action, denying agency (e.g. 'it wasn't me'), normalization (e.g. 'everybody does it'), conventionalization, castigation, actor discounting (e.g. Child 1: 'Hey, you're out'; Child 2: 'It doesn't matter, we're only little'), and denying it happened (e.g. 'you didn't tig me').

Dodge et al. (1986) have used a laboratory paradigm to examine in more detail both peer entry group strategies, and strategies used when confronted with provocation by a peer, in 7–9-year olds. They conceptualized these social situations in terms of a 'social exchange model', in which a child processes the informational aspects of a social situation and responds, this response is interpreted by the other child, and so on. Each bout of the interaction is conceived of in terms of five stages of information processing by the child whose turn it is to act: (1) encoding the social cues presented, (2) interpreting them accurately and meaningfully, (3) generating potential behavioural responses, (4) evaluating these responses and selecting the optimal one, and (5) enacting the chosen response in behaviour.

This model of processing social information has been criticized by Gottman (1986), who commented that it should not be the 'dry cognitive event' suggested by the model. As Dodge *et al.* (1986) themselves point out, the similarities betwen social and non-social information processing should not be pressed too far. A person (unlike most objects) is a highly changeable and unpredictable stimulus. Thus, 'the task in social cognition is to interpret another person's 'intent", a concept that applies only to interpersonal exchange' (*op. cit.* p. 63, see also Chapter 14).

The ways in which social knowledge is gained

How is knowledge of the social structure of the peer group obtained, and how are judgements of 'intent' made? Undoubtedly, some of the processes considered earlier are involved, such as the comparison of the other with oneself, the categorization of others in terms of familiarity, age, and gender, and the prediction of behaviour based on contingencies occurring in prior interactions with particular others. Three other processes will be discussed briefly here: Social attention, social referencing, and social play.

The concept of attention structure has been used by ethologists studying children's groups, in a way similar to its original application in primate societies (Chance and Jolly 1970). For example, Vaughn and Waters (1980) ranked the children in a preschool class of 4- and 5-year-olds, in terms of the absolute number of looks and glances received by each child. They also obtained a dominance hierarchy from observational measures, and a popularity ranking from a picture sociometric assessment. They found a high correlation between the attention structure ranking and the popularity ranking, and lower, but positive correlations with measures of dominance. In similar studies by Abramovitch (1976, 1980), Anderson and Willis (1976), and LaFrenière and Charlesworth (1983), more substantial correlations of attention structure and dominance were obtained. In general, both dominance and popularity seem to be correlates of the amount of visual attention received, and this would be consistent with the hypothesis that much information about dominance, and about friendships, in the social group can be gained by a child watching other children. Further work by Abramovitch (1980) suggested that any sort of social interaction was attention attracting, and relatively more so than was movement or object play. She concluded that attention structures were primarily social in function. It is also the case that some kinds of social interaction are more attention getting than others. Hold (1976); Hold-Cavell (1985), in studies of German kindergartens, documented various attention-getting activities, including: initiating behaviour (e.g. starting a game) and organizing behaviour (e.g. distributing roles), and specifically 'showing-off' behaviour, such as calling loudly, climbing on furniture, wearing unusual clothes. Interestingly, Sluckin and Smith (1977) found that the observed initiation of aggression correlated more highly with children's perception of dominance, than actually winning a conflict; perhaps the initiation of aggression is a particularly attention-getting behaviour, as Hold's results also suggest.

If a child's observation of two persons interacting can be shown to influence his or her later interaction with one of them, then this is evidence that visual attention does indeed affect social behaviour. Such 'indirect effects' have been shown experimentally in the case of infants' reactions to strangers. Feiring et al. (1984) found that 15-month-old infants would react more positively to a strange adult after seeing their mother interact positively with the stranger, than if the mother ignored the stranger. A child may also deliberately seek out a third person's reaction to a social event. For example, an infant may turn to look at the mother if a stranger enters the room, apparently to gauge her reaction. These kinds of behaviours are referred to as 'social referencing', this being a 'process characterized by the use of one's perception of other persons' interpretation of the situation to form one's own understanding of the situation' (Feinman 1982, p. 445). Social referencing seems to be selective to important or familiar figures, and to be more likely in ambiguous situations.

Children may also obtain social information about others by directly interacting with them, and indeed information about one's own status in the group (vis-à-vis affiliation or dominance, for example) can only be gained this way. Knowledge of one's dominance position can be gained most obviously by fighting (in most children's groups, parents and older siblings are not present, so that mediation in fights is infrequent and relatively unbiased). However, it has been suggested that social play has an important function for gaining social knowledge. For example rough-and-tumble (play-fights) might function either as a means of making affiliative bonds, and/or as a means of learning one's dominance position or one's status in a social group in a relatively safe way (see Smith 1982, for a general discussion of the functions of social play).

Recent research on rough-and-tumble play in childhood suggests that in the preschool and early school years its function is more likely to be affiliative than related to dominance. At this age, rough-and-tumble is usually clearly distinct from serious fighting. Observations of partner choice for rough-and-tumble bouts indicate that from preschool up to around 7-9 years of age, children choose friends (as assessed by independent sociometric measurement), rather than other children who are matched with them (or consistently stronger or weaker than them) in dominance ranking (Smith and Lewis 1985; Humphreys and Smith 1987). However, by 11 years, children do seem to choose as rough-and-tumble partners other children who, besides often being friends, tend to be similar in dominance ranking but slightly lower in ranking than the child who initiated the bout (Humphreys and Smith 1987).

The interpretation of this latter finding may be that children already know the dominance ranking in their group from other sources of information, and are using at least some bouts of rough-and-tumble to enforce it or to publicly display their dominant position in a non-hostile (and therefore less potentially injurious) context. Neill (1976) has further suggested that in 12–13-year-old boys, some rough-and-tumble bouts apparently involve one child deliberately hurting another one within a supposedly 'playful' context. These bouts did not usually become serious fights, but they did seem to be examples of play being used in a consciously manipulative way by at least one of the participants. Such manipulative use of playground conventions in other domains was also noted by Sluckin (1981; see also Chapter 18).

Conclusions

Recent work on the development of the infant and young child has reinstated the social world of the child as probably having primary importance. It seems likely that the understanding of permanence, the ability to selectively attend and to recall past events, and to categorize, compare, and seriate, are particularly practised in the social domain, due to the familiarity and salience of social objects, and their importance to the child. In so far as direct comparisons are possible, cognitive abilities seem to be available as early or earlier in the social domain, compared to the non-social domain, in the period of infancy and early childhood.

Such evidence is not so far available for middle childhood and beyond. Very sophisticated social skills are already utilized in peer relationships in this period (e.g. Sluckin 1981), but in so far as comparisons are available, the social domain does not emerge as the leading influence in cognitive development. For example, Marchand (1974) found that 5–10-year-olds, given tasks of conservation of discontinuous quantity, class inclusion, multiplication of classes, and multiplication of relationships, performed better with non-social objects than with social stimuli (pictures of social interactions). We have already noted how Marcus and Overton (1978) found that gender constancy lagged behind conservation of quantity, in 7–9-year-olds. In a wider study of children's knowledge of sex and gender, Goldman and Goldman (1982) constructed scales which were scored in a Piagetian manner. They found that a stage corresponding to concrete operations was not reached until around 11 years in this domain, and suggest that 'a reason for this operational time lag on the Piagetian scale may be a reflection of social immaturity' (*op cit.*, p. 375).

These findings for older children may reflect the influence of the educational process, which in Western societies emphasizes thinking and skills in non-social domains much more than in social ones. They may also in part be influenced by the nature of the tasks used; in the studies just cited, the 'social' tasks are in fact pictorial stimuli or verbal questions, modelled on 'non-social' tasks, which may not give a true or optimal picture of children's knowledge or skills in the social domain.

An under-researched area, for example, is children's knowledge of and practice of kinship terms and kin-related behaviour (Heider 1976). Furthermore, it is only recently that it has been generally accepted that 'social intelligence' is an important component of what we mean by intelligence and must be incorporated fully into theories of cognition (e.g. Sternberg 1984). Further research may well bring social experience and knowledge into a leading role in cognitive development through the life span, as it already has done in infancy and early childhood.

Social complexity: the effect of a third party

9

Tripartite relations in hamadryas baboons*
HANS KUMMER

Where intelligence is not just 'social', but truly Machiavellian, its cornerstone is the use of a third party's behaviour for personal gain. Kummer's closely observed and pioneering hamadryas studies show how these baboons learn to handle complex juggling of interactions in 'protected threat', whether the third party has an interest in co-operating or not.

Analysing the interrelation of three rapidly unfolding individual sequences of behaviour is a task which, if properly carried out, requires recording methods of higher sensitivity for timing minute events than those available so far in field studies. While tripartite behaviour in other baboons and macaques has been repeatedly described in qualitative terms (especially by De Vore 1962; Altmann 1962), its interpretation on the basis of mere description by others seems too daring to attempt. This chapter, therefore, will be limited to the species which I could observe in the Zurich Zoo from 1955 to 1958, and near Erer-Gota and at other locations in eastern Ethiopia in 1960 and 1961. Fortunately, hamadryas baboons display all the major tripartite patterns observed in other species. This was especially true in a captive hamadryas colony in which more complex tripartite patterns had developed than were found in free-living troops (Kummer 1965).

In this chapter, no emphasis will be placed on minute descriptions of the behavioural sequences, since most of them have been described earlier, together with definitions of the component behavioural units (Kummer 1957). The aim here is to link the main facts together in ontogenetical order and to give a working hypothesis of their genesis—namely, that the first and basic triangular relationship in a primate's life is the protection of the infant by its mother against another group member's curiosity or aggression, and that many of the tripartite relationships among adults stem

* Reprinted from *Social Communication among Primates* (ed. S. A. Altmann), University of Chicago Press, Chicago (1967).

from it. For the sake of brevity, terms like 'threat' and 'aggression' will be used as labels for sets of behavioural units commonly considered to have those functions.

A tripartite relation as it is understood in the following is not merely a behavioural sequence in which three monkeys participate, since hardly any interaction between two animals remains unaffected by the presence of others. Rather, it is composed of sequences in which three individuals *simultaneously* interact in three *essentially different roles* and *each of them aims its behaviour at both* of its partners. Thus, many of the more common scenes in primate social life are excluded. If a subadult male is attacked by an adult male and afterward redirects his counteraggression onto a low-ranking female, this is not a tripartite relation, since the events occur at different times. The co-operation of two animals in threatening a third one or in searching for a lost infant also must be excluded, since the co-operators do not appear in essentially different roles. All tripartite behaviour described in this chapter is displayed in agonistic contexts. It is certain that other tripartite and more complex relations exist, for example, among the males directing the troop's movement, but they will not concern us here.

The relationship of an infant, an aggressor, and a protective mother, though transitory, fulfils the above conditions. It is known from many primate species that a mother will slap at a female trying to touch her infant, or that she will threaten an aggressive playmate when her infant screams. Mason (1964) has noted that the filial attachment of the infant may later be transferred from the mother to another adult (for instance, a dominant male). The evidence of this shift is especially convincing in hamadryas groups. The question arises whether the triangular relationship of an infant, its protector, and an antagonist is equally preserved into the juvenile and adult life. According to the evidence gathered on the hamadryas groups, the answer is yes, although the originally simple behaviour takes on a number of forms as the infant grows and sexes differentiate.

A hamadryas infant frightened by another will run to its mother and grasp her fur, whereupon the mother may threaten the aggressor. However, a baboon, at 1 year of age, very often plays with other juveniles in a group out of sight of its mother. Such a play group of hamadryas juveniles usually forms around a subadult or young adult male. He does not take part in the play, but now and then screaming, frightened players run into his arms, whereupon he threatens the aggressor, acting as a substitute for the several mothers in succession. Instead of just clinging to the male's fur, the infant sometimes turns around and looks at the aggressor, still screaming, and now it seems to learn that sitting close to a male and screaming at another animal is likely to induce submission in the latter and provoke attack by the male protector. Thus, the infant or young juvenile is able to *establish* the triangle with himself in the role of the protégé and with the male in the original role of the protective mother. However, the male is not that juvenile's protector exclusively, as its mother was, for he will also attack it in certain situations in favour of another protégé. Therefore, the large male becomes an ambivalent figure, mainly protective toward the infant sitting close to him, but on the other hand aggressive if another infant manages to sit closer. In contrast to the certainty of the roles in the infant-mother-aggressor situation, it is no longer certain which will be the protected, and which will be the attacked. The consequences of this situation are easily observed in behaviour. First, the juvenile wavers between fleeing toward the adult male and fleeing away from him. Secondly, the place in the arms of or close to the protective male is now an object of competition among the participants in an agonistic encounter, and this situation is new for the infant who, until now, was the only protégé of its mother. In spite of these complications, juvenile and subadult hamadryas baboons continue to run toward an adult male in most serious encounters, and the females even do so throughout their lives.

The conflict about the two flight pathways results in several displaced and redirected forms of grooming the male's coat (Kummer 1957), which need not concern us here, except for the fact that even adults, not only infants (Harlow and Harlow 1963), may under stress be strongly attracted by fur. The other difficulty, namely, the competition of two potential protégés and the need for protection even against the protector, is the cause of the social techniques which we are to consider now. In the strongly cohesive one-male groups of hamadryas baboons, several females are conditioned to follow one male at all times. Females wandering away are brought back by the group leader's neck bite. For each female, thus, there exists but one protector, and seeking out another male would lead to severe fights in which the female would be physically torn back and forth between the two males. This protector, however, is also her main potential aggressor, and thus the situation is basically the same as it was in the play group around the subadult male.

In some one-male groups the competition of the different females for the male's protection is an almost constant matrix of interactions. The females in this respect are especially resourceful. The most simple form of becoming the protégé is to arrive first at the male's side. This, however, is far from a safe means; therefore, in addition to screaming, even a 2-year-old female will present to the male. It is almost certain that presenting protects against attacks by a dominant male.

At this stage, the competitive atmosphere of the situation is quite obvious: the female closer to the male now tries to threaten her opponent away from the male, staying as much as possible between the two, while presenting to the male. Both females may sometimes run in circles around the male, the closer one always holding her place on the radius of the more distant one while presenting to the male. This whole pattern of so-called protective threat in front of the group leader is the fundamental form of aggressive encounter between adult hamadryas females.

In the next most complex step, the experienced adult female encounters an opponent of roughly equal or lower dominance status than herself. She now proceeds to perform her own attacks away from the male, aiming her bites typically at the tail of the other female. The latter will not flee, but will stay as close to the male as the attacking female will let her. Between the attacking, even a very dominant female will each time withdraw and present before the male.

This whole technique probably must be learned, as the following shows. A juvenile female sometimes runs such attacks away from the male against an adult female much above her in the dominance order. In these instances, when the juvenile aggressor is safely away from the protector, the dominant female may counterattack and displace the juvenile from her central position, thus reversing the original pattern.

What is the reaction of the male to all this? Usually none. But if he attacks, it is usually the more distant of the opponents which is bitten. The following sequence observed in Ethiopia may exemplify the effectiveness of staying close to the male—even if he does not interfere at once. A very aggressive old female ran attacks against a subadult female who was clinging to the back of the group leader. The aggressor passed the two very closely and tried to slap at the young female, but did not reach her. After 8 seconds the male rose to his feet and stared at the aggressor, the young female still clinging to his flank. Now the aggressor ceased to attack and withdrew.

During phases of changing group structure in the zoo colony, females sometimes would initiate protective threat without apparent reason against an opponent which was not even aware of the aggressor. Now, it was never observed that a male attacked an unaware opponent. The technique developed by several females of the colony to overcome this problem was pulling the opponent's tail before threatening.

Adult female IV quietly sat cradling her infant. Subadult female 3a lingered around without being noticed. Suddenly 3a started screaming, rushed past IV pulling her tail violently, and ran on to their group leader. Female IV jumped to her feet, screamed, and ran toward the male, where 3a was already in the protected threat position. The male aimed a slight brow-lifting threat at IV.

The last step of development, in which the female protégé practically took over the male's role, was only once observed in captivity. The old, experienced female Vecchia had recently entered a young and inexperienced male's group. In situations which usually provoke a group leader's attack and neck bite, this male consistently failed to react. The old female usually was sitting behind the male. In most instances where the male failed to attack, she ran forward and carried out 'his' attack with a typical neck bite, which normally is the group leader's privilege. Immediately, she would withdraw behind the male again. Vecchia also mounted females when they presented to the male and bit them when they mated with the male. The following sequence is an illustrative example:

> Without apparent reason, a female, Sora, of the inexperienced male's group attacked the subadult female of the senile group leader nearby. Two seconds later, both females were attacked, but while the subadult female was chased by her senile group leader, Sora drew a neck bite from Vecchia. The subadult female then followed the senile male and groomed him; Sora ran to her inexperienced group leader, but there Vecchia constantly got between the two and finally took Sora's grooming.

Let us examine briefly the probable mechanisms underlying this entire development.

First, the growing female infant seems to transfer the mother's role onto a male. Throughout adult life, a female under extreme stress will cling to the male's back or be embraced by him in the way of the infant. Another piece of evidence supporting the transfer hypothesis is that the hamadryas female enters the consort relation with her group leader not as a fully mature consort, but as an infant of about 1 year of age. These initial groups, which make up almost one-fifth of all one-male groups in our Ethiopian population, consist of one immature female and a young adult male. They exhibit typical maternal behaviour, whereby the group leader takes on the mother's role toward his young consort. The initial groups seem to develop into normal one-male groups as the female grows up.

Future studies may support the speculation that the hamadryas one-male group evolved on the basis of a transferred mother-infant relation. This transfer, however, is not sufficient to keep the adult female closer and more constantly within the male's neighbourhood than even a 1-year-old infant stays with its mother. Additional means of attraction are needed. First, the 'following' response of the female is reinforced again and again by the male's attacks and neck bites. In the field these attacks occur several times per hour in the first days of the male's relation with the female infant, while later they occur once every few days. Besides this, the hamadryas male's coat may have evolved as a supernormal stimulus (see Tinbergen 1951) aimed at a persisting infantile attraction to fur. In captivity, the tendency of all subadults and adults to inspect or groom the male's coat during stress was so pronounced that this behaviour even was redirected to the ground when inhibited by strong flight motivation, and again when the group leader became senile and his hair became short and brown.

A second mechanism underlying the phenomena of protected threat is social facilitation. The probability that a primate threatens or attacks an object is higher if another group member has already done so. Hamadryas baboons sometimes just stare and slap in the direction of their neighbour's threat, even if they cannot see the object of this threat because of an obstacle. In protected threat, the central female provokes the male's threat partly by means of social facilitation. Altmann suggests that threats away from the dominant animal appease the latter.

Thirdly, the aggression away from the dominant animal may partly be redirected aggression originally aroused *against* the dominant male.

A fourth mechanism is also probably responsible for the behaviour of the experienced female acting in place of her group leader. Imagine an animal A clearly aiming communicative gestures at B who, however, does not react. Now, a third animal, C, with a very low threshold for exactly this reaction, stays near B. It is likely that under certain circumstances C will react instead of B. Two examples may illustrate this. A mother did not react to the screams of her infant which had slipped into a crevice in front of her. Another female having no infant of her own and sitting close by, after a while grasped the infant and gently pushed it to the mother's breast. Another time, an adult female ran to groom the male when a juvenile female belonging to the same group looked away from the group leader who threatened her for staying too far away. The adult female thus did what the younger one, which was probably her daughter, should have done. Similarly, the female applying neck bites in place of the inexperienced group leader probably may have had a lower threshold for this male reaction than even the male himself. That she performed as a male is not unusual, since there are no gestures strictly limited to one sex in hamadryas baboons.

There is a further aspect of the protected threat situation. By placing itself exactly in front of the dominant male, the protégé makes it almost impossible for the opponent to threaten or attack back, since any such gesture would automatically be aimed at the male also. It looks as if the protégé would graft its own gestures upon the protector, thus exploiting the protector's physical features of dominance. Even if the dominant figure is most unlikely to act at all, as in the example of the neck-biting female, its features still can be used as an impressive backdrop.

Finally, there is a possibility that one monkey, for instance the inexperienced group leader, can by slight intention movements induce another monkey, for example the old experienced female, to carry out an attack against a third. There is no evidence at all for this in our material, but experimental investigation should not fail to consider the possibility, especially since it is of importance in human tripartite relations (see Russell and Russell 1961).

So far, we have considered the development of tripartite relations in hamadryas females. Although the males follow the same course up to an age of about 2 years, they never develop the full pattern of protected threat. As subadults, the males often show another kind of tripartite behaviour when under threat from an adult male. It is again the triangle of mother, infant, and aggressor which is simulated, but typically, the subadult male adopts not the infant role, as the females do, but the role of the mother, while the dominant male appears as the aggressor.

In a typical sequence the frightened subadult male grasps an infant and embraces it, turning sharply away from the adult male. He also may invite an infant or juvenile to jump on his back by lowering his hindquarters, and then carry it in front of the adult male or away from him. A 5-year-old subadult may thus carry a large 4-year-old. Often, the infant which is embraced or carried is not at all frightened or even aware of the threat, but is definitely pulled onto the stage by the subadult male, as in the following sequence from the zoo colony:

Subadult male 4a saw that adult male α looked at him. Male 4a started lip-smacking and approached, finally grimacing and screeching intermittently with knees bent. Male α merely stared at him. Two feet from them a 1-year-old female explored the wire net, turning her back to the males. Suddenly, 4a looked at her, pulled her down, and embraced her. Gradually, his screeching diminished; the female jumped off and he ran away.

Three-year-old females often mated with males of the same age—always hidden by some rock from their group leader's view (see Chapter 16). At intervals, both would run to a place where they could see the group leader and there the male sometimes embraced the female with repeated grasps on her flank, grimacing at the group leader; or he carried her on his back to the group leader's place.

The evidence is not sufficient to decide whether such maternal behaviour reduces the probability of the dominant male's attack against the subadult male, in spite of the fact that adult males attack two animals simulating a mother-infant pair only half as often as they attack single animals. Itani (1963) gives some evidence which supports the idea of such a protective function, describing a subleader in a troop of Japanese macaques who 'by hugging an infant succeeded in being tolerated by the females and leaders' in the central part of the troop. While the original function of maternal behaviour is to protect the infant which is embraced or carried, a protective effect here is sought in the maternal role. This is especially evident when a subadult hamadryas male flees *toward* a dominant male. He will not dare to groom the male or to cling to him as a female would, nor will he present. He is trapped in front of the male, looking at his face and screaming more intensively every second until finally he may draw a bite from the adult male. If, however, he manages to invite a nearby infant to jump onto his back in this situation, he will at once run off with it. Carrying an infant triggers the flight, and thus the carried infant breaks the magic circle of attraction toward the dominant male.

Subadult males never used the protected threat, and females under threat never picked up an infant. One reason for the strictness of this rule is that subadult males are not members of the one-male groups. They are never attacked for leaving a group leader, and their staying close and presenting to adult males has little, if any, appeasing effect. Their tendency to flee *away* from an adult male is much stronger than the tendency to flee *to* him—so much so that they do not even touch him. If they choose the flight pathway toward the male, they groom the ground near him or merely look at his fur and chew. According to Mason (1964), clinging and perhaps grooming have a 'stress-reducing' effect. Both are impossible for the subadult males, at least in the infant role. Their resort, that is, mothering an infant, may have a similar effect in reducing stress, especially since embracing allows both animals to cling to something. It is interesting to note that group leaders when under stress themselves during fights within the troop, would embrace one of their females.

Since hamadryas males begin their career as group leaders in a maternal role toward their infant females, it would be expected that the males of this species are more strongly motivated toward maternal behaviour than the males of other species. This is in fact observed: from sexual maturity onward, subadult hamadryas males frequently catch very young infants, cuddling and carrying them up to half an hour. Also, motherless infants are invariably picked up and 'adopted' by young adult males having no females yet. These patterns are never observed in females, and in males they abruptly vanish once they have their own females. It is not surprising, therefore, that subadult males under threat reduce their stress by maternal behaviour, for which they have a strong motivation anyway.

The most obvious class of tripartite relationships in hamadryas baboons, then, can be described by the roles of a protected, a protector, and an antagonist. First in ontogeny, and probably in evolution, these roles are taken on by the infant, its mother, and an aggressor. As the infant grows, these roles are preserved, but they shift to other actors. The situation is at once complicated by the fact that several individuals transfer the role of a protective mother to the same dominant male and that thereby they are forced to compete for his protection. This situation is first encountered by the juvenile when he plays with others around a single subadult male. It is again experienced by the females sharing the protection of one group leader. The genesis of the one-male groups suggests that a transferred mother-infant relationship was an important root of their evolution. Consistent with this view is the strong maternal motivation of hamadryas males, which is revealed throughout the subadult and early adult age until they meet the needs of their first consorts.

This hypothesis barely covers more than one aspect of the complex phenomena. Two points, however, emerge from the phenomena themselves: it is, first, astonishing how often the behaviour patterns of the mother-infant relationship appear in tripartite relations of subadult and adult hamadryas baboons. Secondly, the tripartite behaviour in some instances comes close to exploitation: a primate may learn to use for his own protection, or for increasing the effect of his aggression, another one who primarily is not involved in the events.

10

Chimpanzee politics*
FRANS DE WAAL

The sophistication of Machiavellian intelligence shown by chimpanzees is a quantum leap ahead of that of even Kummer's baboons; nowhere is this more clearly illustrated than in studies on the Arnhem colony. In these extracts, de Waal gives examples which argue for the use of an enriched description of the animals' behaviour and describes a clear case of the practical application by chimps of principles described by Machiavelli in The Prince.

Daring interpretations

Diametrically opposed to the concept of instinctive and impulsive animal behaviour is the concept of conscious, premeditated action. There are, of course, many animals which are probably totally unaware of the consequences of their social behaviour. Does a male cricket, for example, know that his chirping attracts females? And yet that is the function of his signal. Higher animals, however, do seem to know the effects of their signals. Great apes, in particular, behave so flexibly that we get the impression that they know exactly how others will react, and what they can achieve as a result. Their communication looks very much like intelligent social manipulation, as if they have learnt to use their signals as instruments to influence others.

Example 1 On a hot day two mothers, Jimmie and Tepel, are sitting in the shadow of an oak tree while their two children play in the sand at their feet (playfaces, wrestling, throwing sand). Between the two mothers the oldest female, Mama, lies asleep. Suddenly the children start screaming, hitting, and pulling each other's hair. Jimmie admonishes them with a soft, threatening grunt and Tepel anxiously shifts her position. The children go on quarrelling and eventually Tepel wakes Mama by poking her in the ribs several times. As Mama gets up Tepel points to the two quarrelling

* Extracted from *Chimpanzee Politics*, Jonathan Cape, London (1982).

children. As soon as Mama takes one threatening step forward, waves her arm in the air and barks loudly the children stop quarrelling. Mama then lies down again and continues her siesta.

Interpretation In order to understand this interpretation fully, it is important to know two things: first, that Mama is the highest-ranking female and is greatly respected; and secondly, that conflicts between children regularly engender such tension between their mothers that they too come to blows. This tension is probably caused by the fact that each mother wishes to prevent the other from interfering in the children's quarrel. In the case of the example above, when the children's game turned to fighting, both mothers found themselves in a painful situation. Tepel solved the problem by activating a dominant third party, Mama, and pointing out the problem to her. Mama obviously realized at a glance that she was expected to act as arbitrator.

Example 2 Yeroen hurts his hand during a fight with Nikkie. Although it is not a deep wound, we originally think that it is troubling him quite a bit, because he is limping. The next day a student, Dirk Fokkema, reports that in his opinion Yeroen limps only when Nikkie is in the vicinity. I know that Dirk is a keen observer, but this time I find it hard to believe that he is correct. We go to watch and it turns out that he is indeed right: Yeroen walks past the sitting Nikkie from a point in front of him to a point behind him and the whole time Yeroen is in Nikkie's field of vision he hobbles pitifully, but once he has passed Nikkie his behaviour changes and he walks normally again. For nearly a week Yeroen's movement is affected in this way whenever he knows Nikkie can see him.

Interpretation Yeroen was playacting. He wanted to make Nikkie believe that he had been badly hurt in their fight. The fact that Yeroen acted in an exaggeratedly pitiful way only when he was in Nikkie's field of vision suggests that he knew that his signals would only have an effect if they were seen; Yeroen kept an eye on Nikkie to see whether he was being watched. He may have learnt from incidents in the past in which he had been seriously wounded that his rival was less hard on him during periods when he was (of necessity) limping.

Example 3 Wouter, a young male chimpanzee of almost three, gets into a quarrel with Amber and screams at the top of his voice. At the same time he advances aggressively towards Amber. His mother, Tepel, goes over to him and quickly places her hand over her son's mouth, smothering his screams. Wouter calms down and the quarrel is over.

Interpretation Noisy conflicts attract attention. If they last too long, one of the adult males will come over and usually put an end to them. When a bluffing male approaches, Wouter will automatically seek refuge near his mother. This means that she runs the risk of receiving the punishment which was meant for her son. Tepel wanted to avoid running the risk by shutting Wouter up before things went too far.

This is not the only known instance of enforced silence. I have also seen a mother place a finger over the small mouth of her baby when the latter started barking aggressively at a dominant group member from the safety of her lap. Once again this was probably due to the mother's reluctance to get drawn into difficulties because of a social *faux pas* committed by her child.

Example 4 Dandy is the youngest and lowest ranking of the four grown males. The other three, and in particular the alpha male, do not tolerate any sexual intercourse between Dandy and the adult females. Nevertheless every now and again he does succeed in mating with them, after having made a 'date'. When this happens the female and Dandy pretend to be walking in the same direction by chance, and if all goes well they meet behind a few tree trunks. These 'dates' take place after the exchange of a few glances and in some cases brief physical contact.

This kind of furtive mating is frequently associated with signal suppression and concealment. I can remember the first time I noticed it very vividly indeed, because it was such a comical sight. Dandy and a female were courting each other surreptitiously. Dandy began to make advances to the female, whilst at the same time restlessly looking around to see if any of the other males were watching. Male chimpanzees start their advances by sitting with their legs wide apart revealing their erection. Precisely at the point when Dandy was exhibiting his sexual urge in this way, Luit, one of the older males, unexpectedly came round the corner. Dandy immediately dropped his hands over his penis concealing it from view.

On another occasion Luit was making advances to a female while Nikkie, the alpha male, was lying in the grass about 50 m away. When Nikkie looked up and got to his feet, Luit slowly shifted a few paces away from the female and sat down, once again with his back to Nikkie. Nikkie slowly moved towards Luit, picking up a heavy stone on his way. His hair was standing slightly on end. Now and then Luit looked round to watch Nikkie's progress and then he looked back at his own penis, which was gradually losing its erection. Only when his penis was no longer visible did Luit turn around and walk towards Nikkie. He briefly sniffed at the stone Nikkie was holding, then he wandered off leaving Nikkie with the female.

Females sometimes give away their clandestine mating sessions by emitting a special, high scream at the point of climax. As soon as the alpha male hears this he runs towards the hidden couple to interrupt them. An adolescent female, Oor, used to scream particularly loudly at the end of her matings. However, by the time she was almost adult she still screamed at the end of mating sessions with the alpha male, but hardly ever during her 'dates'. During a 'date' she adopted the facial expressions which go with screaming (bared teeth, open mouth) and uttered a kind of noiseless scream (blowing from the back of the throat).

Interpretation In all these examples sexual signals are either concealed or suppressed. Oor's noiseless scream gives the impression of violent emotions which are only controlled with the greatest of effort. The males are faced with the problem that the evidence of their sexual arousal cannot disappear to order, but they too have their solutions.

The cheek of Luit actually sniffing at the weapon Nikkie held in his hand only goes to show how sure he was that the alpha male would find no cause to proceed against him. This behaviour is in marked contrast to an incident I once witnessed between two male macaques. The alpha male met another male several minutes after the latter had secretly mated. Alpha could not possibly have known anything about this, but the other male acted unnecessarily timidly and submissively. His behaviour was so exaggerated that, if the alpha male had had a chimpanzee's social awareness, he would certainly have realized what the matter was.

Luit's behaviour after his abortive adventure was very different. There was no trace of a 'guilty conscience'. Chimpanzees are masters of pretence and will seldom put an idea into the head of the unsuspecting.

Luit's new policy

In 1976, Luit, one of the adult males in the colony, successfully challenged the dominant male, Yeroen. During a period of several months Luit provoked and intimidated Yeroen, who invariably turned to the females to recruit their support. Yet, Luit had support as well, namely from the third adult male, Nikkie. The result was a battle between Luit and Nikkie, on the one hand, and Yeroen and the females, on the other. Since the females' willingness to get involved decreased over time, due, no doubt, to the frequent punishments they received from Yeroen's rivals, the battle was eventually decided in Luit's favour. He quickly grew into the alpha position. Like Yeroen before him, Luit constantly had his hair slightly on end, so that he appeared big and powerful. He looked magnificent.

However, it was not only Luit's outward appearance and the way he bluffed which had changed, he had also adopted a brand-new *policy*. (The word 'policy' is used here to denote a consistent social behaviour with a view to achieving a certain aim, quite apart from whether this behaviour was determined by innate tendencies—intuitive policy—or by experience and insight—rational policy—or both. For example, a mother chimpanzee who defends her infant each time it is threatened or attacked also conducts a policy of a kind: a policy of protecting her offspring.) To begin with, Luit, assisted by Nikkie, followed a policy which led to Yeroen's dethronement. As soon as this particular power take-over was behind him, Luit's social attitude altered totally. His new policy seemed to be aimed at a completely different objective, namely to stabilize his newly acquired position. He changed his attitude towards the adult females, towards Yeroen and towards Nikkie.

Luit's new attitude towards the females was obvious from his behaviour whenever serious quarrels broke out. For example, on one occasion a quarrel between Mama and Spin got out of hand and ended in biting and fighting. Numerous apes rushed up to the two warring females and joined in the fray. A huge knot of fighting, screaming apes rolled around on the sand, until Luit leapt in and literally beat them apart. He did not choose sides in the conflict, like the others; instead, anyone who continued to fight received a blow from him. I had never seen him act so impressively before. This particular incident took place in September 1976, only a few weeks after he had become leader. On other occasions he put a stop to serious conflicts less heavy-handedly. When Mama and Puist were locked in a fight he put his hands between them and simply forced the two large females apart. He then stood between them until they had stopped screaming.

Besides such impartial interventions Luit also intervened on behalf of one or other party. Once again, however, his policy changed. Instead of a *'winner-supporter'*, he became a *'loser-supporter'*. The term 'loser-supporter' is used to describe a third individual who intervenes in a conflict on the side of the party who would otherwise have lost; for example, if Nikkie attacked Amber, Luit would intervene to help Amber chase Nikkie away. Without Luit's assistance, Amber would never have beaten Nikkie. If Luit's interventions were purely arbitrary, he would be found to support the losers about 50 per cent of the time and the winners the other 50 per cent of the time. In fact, after his rise to power Luit began to show solidarity with the weaker party. Before, he had supported losers 35 per cent of the time, but after his elevation the figure increased to 69 per cent. The contrast between these two figures reflects the dramatic change in Luit's attitude. A year later Luit's support for the losers had increased still further to 87 per cent.

It was not surprising that Luit, as the alpha male, should set himself up as the champion of peace and security, and try to prevent conflicts escalating by supporting the losers. This form of behaviour is referred to as the *control role* of the alpha male, and is found in many species of primates. Less is known about the possible importance of this role to the male himself. There are indications among macaques that the group leader's protective role and his strong ties with the females exclude other males from the central position in the group. This naturally furthers the stability of his leadership. Rivals who are not prepared to be frightened away meet with a great deal of resistance. One such example was described by Irwin Bernstein, who observed a change of leadership in a group of macaques. He concluded 'that young males of superior fighting ability cannot usurp power without the support of a sizeable portion of a group'.

It is conceivable that there is a connection between the protection offered by a dominant group member in his control role and the support he receives in return when his position is threatened. In other words, an alpha male who fails to protect the females and children cannot expect help in repulsing potential rivals. This would suggest that the control role of the alpha male is not so much a favour as a duty: his position depends on it. If looked at in these terms Yeroen's fall could be explained by his inability to protect the others effectively against the aggresion of Luit and Nikkie. Luit's behaviour can be interpreted in the same light. To begin with he demonstrated just how little the females could count on Yeroen's support by attacking them or being heavy-handed with them in Yeroen's presence. Later his attitude changed completely and he himself adopted the role of protector.

About 4 months after the struggle for dominance between Yeroen and Luit had ended, the females started to support Luit. During the winter of 1976 the number of female interventions in conflicts between Yeroen and Luit were nine out of ten times in favour of Luit. The same was true of their interventions in conflicts between Luit and Nikkie. Luit's alpha position had taken on a broader basis. Only Gorilla refused to desert Yeroen. In the period when the others were changing camps a great deal of tension arose between Gorilla and Puist. The reason for the almost daily fights between the two females was probably the emergence of Puist as by far the most important of Luit's supporters. In the autumn we sometimes saw Puist rush to Luit's aid in a conflict, but Mama would realize what Puist was going to do and nip any such initiatives in the bud by chasing Puist away. In the winter Mama was less prepared to interfere (she was pregnant), so that the way was clear for Puist and the others to give vent to their 'true feelings'. However, developments did not stop there, because in time Mama also switched her allegiance to Luit. Now the new leader was able to call on this powerful and influential female when the other males placed him in difficult positions.

If Luit had been the leader of a group of macaques, the support of the females might have been sufficient in itself, but among chimpanzees the tendency towards coalition among males is so strong that a leader must always allow for the possibility of other males ganging up on him, if his group contains two or more males other than himself. While Yeroen was at the pinnacle of his power this problem did not present itself. Yeroen had built up his position on the support of the females; Luit, the other adult male, was forced to live slightly apart. He was often seen on his own, far away from the rest of the group. At that time the group's structure strongly resembled that of a small macaque colony, with one absolute leader as the magnificent focal point and potential rivals forced into a peripheral position. Yeroen's leadership only ended when Nikkie had grown up sufficiently and had developed into a possible coalition partner for Luit. The subsequent dominance process not only affected and altered the individual rank of some of the leading group members, but it also affected the prerequisites of stable leadership. The new alpha male, Luit had to contend with not just one but two rivals. There was no point in Luit's trying to ban both Yeroen and Nikkie to the periphery of group social life. That would have been tantamount to political suicide, because the two ostracized males would have joined forces against him. The only course left for Luit was to try to convert one of the two males to his cause; he chose Yeroen.

Luit's choice proves just how much friendships among chimpanzees are situation-linked. After all Luit had banded together with Nikkie against Yeroen and the females. Now he turned everything upside down, and sided with Yeroen and the females against Nikkie. Whereas previously Nikkie's attacks on the females had done much to further Luit's own cause and Luit had at times even supported Nikkie against the females, now Luit intervened between Nikkie and the females sometimes even before an incident had developed. When Nikkie raised his hackles and approached a female, swaying slightly, ready to attack her, Luit would act by placing himself beside or in front of the aggressor, so that Nikkie did not dare go any further. On other occasions Luit would actually attack Nikkie, and slap and kick him until he ran away screaming. Luit's attitude to Nikkie had hardened and the number of conflicts between them increased.

The greatest source of conflict between them, however, was not the females, but contact with Yeroen. Each of them sought the ex-leader's friendship and would not allow the other one to sit by him. Whenever the other two were relaxing anywhere near each other Nikkie would start hooting and displaying some distance away, and he would continue to do so for some minutes until Yeroen got up and walked demonstratively away from Luit. Yeroen offered little resistance, but he was often doubtful as to exactly where he stood. First of all, he would walk away from Luit, which made Nikkie start displaying all over again. When this happened Luit would place himself between Yeroen and Nikkie or he would chase Nikkie away, thus defending his contact with Yeroen.

Later still Luit began to take a more active interest in contacts between Yeroen and Nikkie. Because of his position he was able to act resolutely and effectively, usually by attacking Nikkie. This in turn meant that Luit usually won the competition and formed the stronger tie with Yeroen. By the end of the winter Luit had succeeded in establishing a situation whereby he himself had far more contact with Yeroen than Nikkie. To this extent Luit's strategy had been completely successful, but there was one factor he could not control and it was upon this factor that the stability of his position as the alpha male depended, namely Yeroen's own attitude.

Collective leadership

The following year, in a second take-over, Nikkie managed to secure a strong position for himself. He was 'greeted' by Yeroen, Luit, and the other group members and was thus the formal leader of the colony. However, there was something lacking in his leadership. He met with great resistance from the females and they found it difficult to defer to him. He was unpopular and his authority was not readily accepted. He was 'greeted', groomed, and obeyed, but not in the same matter-of-course way as the previous two leaders. He was feared rather than respected.

When Luit became leader he also became a loser supporter. He received the support of the females and rose so high in their estimation that they 'greeted' him more often than Yeroen. Earlier on I explained these developments in the following way: a leader receives support and respect from the group in exchange for keeping order. The same phenomena occurred after the second power take-over, but the great difference this time was that these qualities were not seen in the leader. It was not Nikkie, but his coalition partner, Yeroen, who defended the peace and consequently won general respect. This development surprised me more than anything else. Up to that point I had thought that this role as well as formal dominance had to be assumed by a single individual. Instead, whereas Yeroen and Luit had been sole leaders, Nikkie *shared* leadership with another individual.

The policing was done by Yeroen. Not counting the many times he and Nikkie intervened in each other's conflicts, Yeroen was a loser supporter 82 per cent of the time and Nikkie only 22 per cent of the time (measured in 1978–9). Nikkie was still, despite his position as the alpha male, a winner supporter. Initially, following Nikkie's rise to the top, Yeroen worked against him so effectively that Nikkie could not really be said to be in control. For example, when the young leader, his hair on end, prepared to intervene in a conflict between two females or when he actually did intervene, Yeroen would immediately turn on him and chase him away, sometimes aided by the two females. It may well have been Yeroen's resistance to Nikkie's policing efforts which in 1979 still prevented Nikkie from gaining complete control.

In Nikkie we had a leader who was regularly set on by alliances of females. What was even more surprising was that Yeroen encouraged the females in their resistance to his own coalition partner, although his encouragement became much less than it had been, with the result that the length and fierceness of these incidents decreased. The other side of the coin was that Yeroen was the male with the best political credentials as far as the females were concerned. The females turned against him for the duration of Luit's leadership, but after Luit's dethronement they returned to Yeroen's camp; they supported him more fervently than they did Luit, and in no sense did they support Nikkie.

The group's respect was accorded *en masse* to Yeroen. The females and children 'greeted' him almost three times as often as Nikkie and five times as often as Luit. When Yeroen and Nikkie had ended one of their many joint displays and side by side they approached a group of sitting apes, it was quite normal for these lower-ranking members to stand up, and hasten to 'greet' and kiss Yeroen while seeming to ignore Nikkie's presence. However, respect for Nikkie eventually began to rise. In the course of 1980, Nikkie started to receive the same number of 'greetings' as Yeroen, two full years after he rose to dominance.

Sometimes it seemed that Nikkie was being used as a figurehead, and that Yeroen—experienced as he was and extremely cunning—had him in the palm of his hand. The broad basis for leadership rested not under Nikkie, but under Yeroen. The older male had a coalition with the females to pressurize Nikkie and a coalition with Nikkie to keep Luit in check. Seen in these terms the situation appeared to represent a comeback for Yeroen. Luit had deprived him of the support and respect he had hitherto enjoyed, but by pushing a youngster forward Yeroen seemed to have succeeded in reacquiring both.

This picture was not altogether correct. Yeroen had had to sacrifice much for his 'comeback'. It was true that Nikkie did not dominate him at all times, but he was strong enough for Yeroen to 'greet' him. If Yeroen refused to recognize Nikkie's position—and Yeroen certainly did refuse in the first few months of Nikkie's leadership—this gave rise to severe conflicts between them and the coalition was in serious danger of collapsing. Nikkie was dependent on Yeroen, but the converse was equally true. Furthermore, Nikkie enjoyed the sexual privileges due to his rank. Nikkie occupied the top position, while Yeroen fulfilled the control role and had the authority which goes with it.

Nikkie's position was not an easy one. Compared to him, Yeroen and Luit were almost all-powerful, thanks to the collaboration of the females. The important difference between Nikkie's leadership and the old order was that Nikkie stood on the shoulders of someone who was himself very ambitious. The ensuing problems are familiar enough in the human world.

Machiavelli wrote about the relative powerlessness of this kind of leader. If in the following quotation from *The Prince* we translate 'nobility' by 'males of high rank' and 'common people' by 'females and children', then we see that Nikkie's 'principality' is indeed very different from the 'principality' of his two predecessors:

He who attains the principality with the aid of the nobility maintains it with more difficulty than he who becomes prince with the assistance of the common people, for he finds himself a prince amidst many who feel themselves to be his equals, and because of this he can neither govern nor manage them as he might wish.

11

Alliances in contests and social intelligence
ALEXANDER H. HARCOURT

The core of the skill involved in Machiavellian intelligence is the ability to make and keep alliances with the right individuals. Harcourt reviews evidence that complex patterns of alliance make primate contests more intellectually demanding than those of other animals and, in arguing that this has led to primates' greater relative brain size, challenges students of non-primates to produce better comparative data.

Summary

Primates have larger brains for their body size than do other mammals and, in laboratory tests, are apparently more intelligent than other mammals. I suggest that these differences might be linked to the fact that, on current evidence, primates support one another in contests, i.e. form alliances, more often than do non-primates. Alliances are far more complex social interactions than are two-animal contests, requiring more information and more complex information for their successful conclusion.

Even in the simplest form of alliance, protection of vulnerable kin against another group member, numerous decision rules have to be followed, which are contingent and can conflict with one another. For example, in a fight between a dominant non-relative and a subordinate relative, the advantages of protecting the relative have to be offset against the possibility of valuable later reciprocation from the dominant animal in return for current support in the contest. In addition, rapid tactical decision making is required. All these statements apply equally to dyadic contests. However, an alliance automatically involves at least three animals, instead of just two. Therefore, the information processing abilities required for success are, I argue, far greater; complexity is geometrically, not arithmetically, increased with the addition of further participants in an interaction (see Chapter 5).

Furthermore, and in marked contrast to two-animal contests, alliances require comparison among others, not simply with self. In a straightforward contest, a simple comparison of own with adversary's competitive ability is required; alliances necessitate comparison of others' competitive ability, as when animals choose whether to support one of two group members, both of whom are subordinate to themselves. Also, and again in contrast to dyadic contests, successful alliances require the ability to incorporate information about animals not initially involved in the interaction, for not only does the competitive ability and readiness to fight of the adversary have to be judged, so does the presence, readiness and competitive ability of its potential supporters.

Currently, analysis of the level and nature of the complexity involved in intelligent decisions during alliance formation is largely *post hoc*. If we are to go much further in our understanding, we need to direct at studies of social intelligence the extent and ingenuity of experimental analysis that has so far been directed almost solely at non-social intelligence. At another level of investigation, we will not have a good theory of the evolution of social intelligence until we understand, in addition to the mental constraints, the environmental and social constraints on alliance formation, especially in non-primates.

Introduction

If brain size of adult mammals is plotted against their body size on a log-log scale, all but two orders fall on a single straight line (Jerison 1973; Martin 1981, 1983). The exceptions are the toothed whales and the primates, which have unusually large brains for their body size. The contrast is even more striking at birth, when primates have brains that are twice the weight expected of a normal mammal of similar body weight (Sacher 1982; Martin 1983). At some time between the radiation of the earliest insectivore-like primates in the late Cretaceous and the second radiation of the strepsirhine-like 'primates of modern aspect' in the mid- to late Paleocene, an extraordinary evolutionary event took place (Sacher 1982). An enlargement of relative brain size also occurred within the primates, for living prosimians have larger brains for their body size than do fossil ones and living simians have larger brains than do the living prosimians (Gurche 1982; Jerison 1983; Martin 1983).

What, then, are primates doing with their brains that non-primates are not? The answer that 'Trivial Pursuits' gives to the question, 'What is the most intelligent animal after man?' is 'A dolphin'. If relative brain size, and hence cortical volume (Jerison 1973; Passingham 1981; Macphail 1982), correlates with intelligence, or the capacity to process information (Jerison

1973, 1983), the answer should probably have been 'A primate'. Certainly, despite the inherent difficulties of measuring intelligence, most people are forced to conclude that primates appear to perform better than do non-primates in laboratory tests of intelligence (Passingham 1981; Macphail 1982, p. 288). The question then becomes, 'What are primates doing with their superior intelligence that non-primates are not?'.

Primates do not obviously live in a more complex physical environment than do non-primates: baboons mingle with impala on the African savannah and leaf monkeys forage in the same trees as do squirrels in Asian forests. Nor do primates seem to use the environment in a more complex way or process more information in their use of the environment. They do not, for example, have larger home ranges and, while they might have a more diverse diet, the correlation between diet and brain size within primates (Clutton-Brock and Harvey 1980) might be determined more by dietary influenced differences in metabolic rate (Martin 1983; Harvey *et al.* 1987) than by requirements for information processing (Clutton-Brock and Harvey 1980; Milton, Chapter 21).

Superficially, a primate's social environment does not appear more complex than a non-primate's: the range in size and composition of primate groups (Clutton-Brock and Harvey 1977 *a*, *b*) matches that seen in antelopes (Jarman 1974), for example. However, several authors have now suggested that primates, including humans, use the social environment in more complex ways than do non-primates (see Chapters 1–4, and Holloway 1975), or in more complex ways than they do the physical environment (Cheney and Seyfarth, Chapter 19). Neither Jolly (Chapter 3) nor Humphrey (Chapter 2) were very specific about what problems in a primate's social life were more complex than a non-primate's and, therefore, benefitted more from intelligence in their solutions. Chance and Mead (Chapter 4) were. Conflict and, therefore, social behaviour between individuals is more frequent in primate than non-primate groups, they argued, and in primate groups, behaviour between any two individuals is influenced by the presence and behaviour of other group members, so making social interaction a complex process. Kummer (Chapter 9) took the argument a stage further and showed, not simply that two interacting individuals reacted to the presence and behaviour of a third, but that they reacted in complex ways, and indeed apparently deliberately incorporated the third individual in their interactions.

Several of Kummer's examples were of alliances in contests, that is, of one animal supporting another, the recipient, in a fight with a third, the opponent (Fig. 11.1). (In the literature on support in contests, no agreement has been reached on terms. Alliance, coalition, aid, support, and so on, are used in different and sometimes contradictory ways.) I want here to raise the possibility that the efficient use of such alliances might be the

Alliances in contests and social intelligence   135

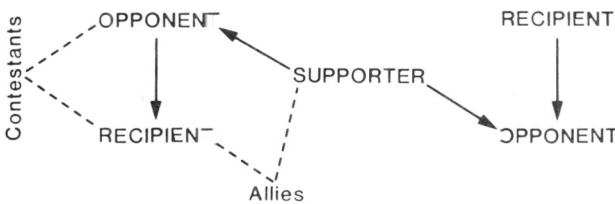

Fig. 11.1 Participants in an alliance. Solid arrows = direction of aggression. By threatening one Contestant (the Opponent), the Supporter is assumed to have automatically supported the other (the Recipient of support), independently of the direction of aggression between the contestants.

social phenomenon that separates primates from non-primates, and hence possibly the selection pressure responsible for the evolution of intelligence. I argue this for four reasons. The first two are related to the argument that primates are different from non-primates with respect to the use of alliances in contests. The second two concern the complexity of contests with alliances compared to contests without them.

1. Reports of alliances in contests are far more common in the primate than the non-primate literature. In Old World primate groups, up to 30 per cent of contests involve the intervention of a third animal (Harcourt 1987). As a result, information on alliances is easy to find, and papers have been devoted solely to the subject. In the non-primate mammalian literature by contrast, accounts of alliances are rare. One exception concerns males simultaneously acquiring groups of females and defending them against subsequent challenges, lions (*Panthera leo*) being perhaps the best studied example (Bygott *et al.* 1979; Packer and Pusey 1982). Another, that in flocks of swans and geese on their winter feeding grounds, parents frequently support offspring in contests with others (e.g. Boyd 1953; Scott 1980), as do adult female hyaenas (*Crocuta crocuta*) (Frank 1986).

2. Where reports do exist of non-primates supporting one another in contests, they are almost exclusively of individuals supporting vulnerable close kin, usually parents supporting offspring. By contrast, primates of many different ages and degrees of relatedness support one another in a wide variety of situations, indicating that the process of alliance formation, and the competitive decisions involved, are far more complex in primate than non-primate groups.

3. It seems logical that contests involving three animals should be more complex and require more intelligence for their successful execution, than two-animal (dyadic) contests, in the same way that mathematical formulae with three variables are usually more complex than ones with just two (see Chapters 1 and 5). Thus, analysis of two-person prisoners' dilemma types of mathematical games of co-operation is far advanced (Axelrod and

Hamilton 1981; Parker 1984), but mathematicians have only just begun to unravel the complexity of three-person prisoners' dilemma games.

4. Finally, contests that involve alliances do not remain as three-person games, for if one contestant can be supported, so can the other. Decisions about whether or not to initiate contests or alliances, with whom, and whether to escalate them (Maynard Smith and Parker 1976; Parker 1984) need then to take into account not just the opponent's competitive ability and its readiness to defend the resource in question, but also the competitive ability of its supporters and of the subject's own supporters; in addition, the availability of the two sets of supporters, and their readiness to intervene are further variables to be incorporated (Kaplan 1977).

I first briefly show that partners in both primate and non-primate societies can benefit from forming alliances. Then in the rest of the chapter I argue that in a situation where the outcome of a contest depends on the successful establishment of alliances, the decisions that an animal has to make are more complex than those required in direct contests; and that primates do, indeed, seem to make these complex decisions. As well as describing the potential complexity of the interactions, I outline their possible beneficial consequences, because the potential complexity can often best be appreciated if the goal of the alliance is also understood.

Efficacy of alliances

Contests often occur over resources. If there are, indeed, beneficial consequences to alliances on which natural selection has acted, we should be able to detect them, at their simplest level, in improved access to resources. Such is the case, both for primates and non-primates (Fig. 11.2).

Alliances in primate groups are complex social interactions

Numerous, contingent rules determine the establishment of alliances

Participants in dyadic (two-animal) contests need to process information about the potential adversary's consanguinity (to avoid endangering close kin), and its competitive ability (to avoid endangering themselves; Maynard Smith and Parker 1976). In effect, two relatively simple social rules have to be followed. Even at the simplest level of alliance, merely protecting vulnerable kin, I suspect that additional rules govern decisions, and that the sorts of information that have to be processed and efficiently acted on are more complex.

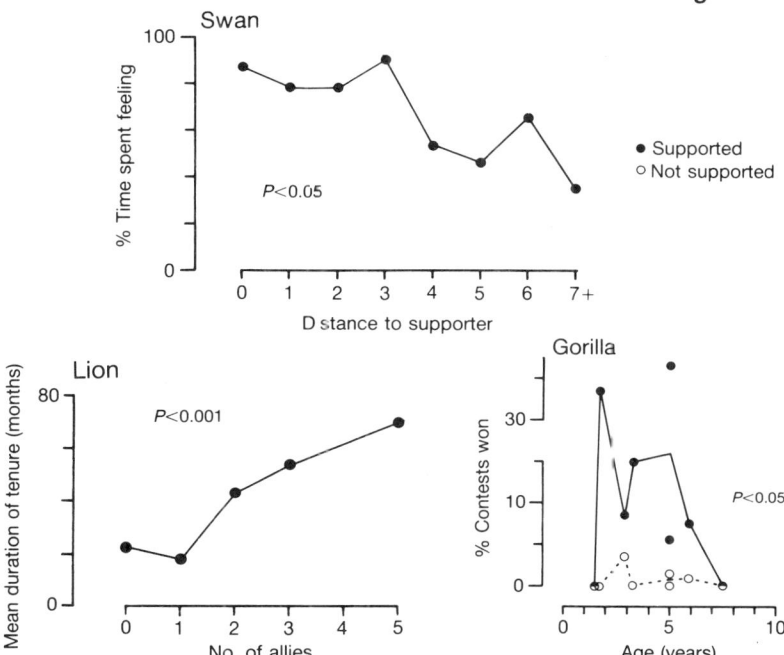

Fig. 11.2 Alliances are advantageous. (a) Immature swans spend more time feeding (percentage of total observation records in which feeding) when near supporters (parents) than when far from them (measured in 'swan-lengths'). (Data from Scott 1980, Fig. 2; 17–236 records per distance, except 7+, where $N=3$; $r_s = -0.76$; $P < 0.05$.) (b) Male lions hold prides for longer when with several allies. (Data from Bygott et al. 1979, Fig. 1a; 2–10 records per number of allies. (Data from Bygott et al. 1979, Fig. 1a; 2–10 records per number of allies; mean durations of tenure are minimum values, because completed tenure was usually not known; $r_s = 0.71$, $P < 0.001$.) (c) Immature gorillas gain access to food in contests with dominant opponents more often when the contests are intervened in (Supported) than when they are not. Circles = individuals' values ($n = 9$). (Data from Harcourt and Stewart 1987, Fig. 5; median individual's number of contests = 12 with support, 106 without it; $P < 0.05$, Sign Test).

First, the consanguinity rule: macaques support close relatives more frequently than they do distant ones even when the fact that relatives spend much time near one another is taken into account (Kurland 1977). Secondly, the risk rule: vervets and macaques tend to be wary of supporting animals against group members who are dominant to themselves, in the same way that they would be wary of competing with them in dyadic contests. In the free-ranging vervet (*Cercopithecus aethiops*) population on Barbados, for example, none of four mothers in either of two study groups ever supported their juvenile offspring against opponents dominant to themselves

(Horrocks and Hunte 1983). Similarly, in a captive group of macaques (*M. fascicularis*), relatives and mothers supported juveniles less frequently against group members dominant to themselves than against ones subordinate to them (Netto and van Hooff 1986).

However, whilst these two social parameters, of consanguinity with adversary and risk to self, are largely sufficient for decisions about whether or not to enter two-way contests, an alliance requires a third item of social information. Risk to the potential recipient if it is not supported has also to be part of the equation. Gorillas, for example, support immatures against dominant adults more often than they do against dominant immatures, who are less of a threat to the immature recipient than are adult opponents, being closer in size to it (Fig. 11.3). The same sort of decision is also made by rhesus monkeys when they respond to the screams given by immatures involved in contests (Gouzoules *et al.* 1984). These screams often elicit support from other group members, usually mothers, and indeed are thought to be given as a means of soliciting support. At least four acoustically distinct screams are given by immatures, the frequency of which varies with the identity of the opponent and the nature of the aggression. Thus, for example, 'noisy' screams are given when the opponent is higher ranking than the immature and the aggression involves physical contact, in other words when the immature is in some danger.

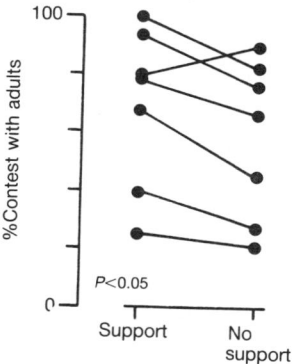

Fig. 11.3 Individuals judge level of risk to kin as well as to selves. Immature gorillas were more often supported when in contest with dominant adults (who are much larger than them) than with dominant immatures (who are only slightly larger). Values are the percentage of each immature's contests with and without support that were against dominant adults (rather than dominant immatures). (Data from Harcourt and Stewart 1987, Fig. 3; median individual's number of contests = 10 with support, 220 without it; $P < 0.05$, Wilcoxon matched pairs, signed ranks Test).

Alliances in contests and social intelligence 139

'Pulsed' screams, by contrast, are given when the opponent is a maternal relative, i.e. a comparatively benign opponent, and do not vary in frequency with the nature of the aggression. Gouzoules et al. demonstrated that mothers responded differentially to playbacks of the different screams, reacting more strongly and quickly to 'noisy' than to 'pulsed' screams.

It could be argued that the gorilla and macaque supporters are merely judging degree of fear shown by the victim. However, most contests in the gorilla groups involved only mild threat, with little or no apparent fearful response by the victim (Harcourt and Stewart 1987). For the macaques, Gouzoules et al. also argued that degree of fear was not involved. First, the screams were not a graded series; instead, they change categorically in concordance with the context. Secondly, the behaviour of the immature did not vary with the type of scream that it was giving; it was as likely to hold its ground, run away, or attack whether it was giving a 'noisy', a 'tonal', or a 'pulsed' scream.

Of course, the separate sets of information, i.e. consanguinity, risk to self, and risk to kin, are not acted upon independently of one another: the rules of alliance formation, as of dyadic competition, are contingent. Gorillas, for example, accept a higher level of risk when supporting close kin than distant kin: they supported close kin more often than they did distant kin, and in addition more often supported close kin against opponents dominant to themselves (Fig. 11.4).

So far, I have argued that alliances are more complex social interactions than are dyadic contests and, therefore, require greater social intelligence

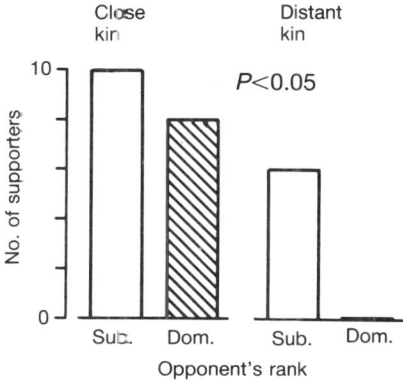

Fig. 11.4 Alliance rules are contingent. Gorillas were more likely to support close relatives ($r \geq 0.05$) than distant relatives ($r < 0.05$) against opponents dominant to themselves (hatched histogram), i.e. to risk retaliatory aggression. Data are the number of individuals who gave support in each context ($P < 0.05$, Sign Test).

for their successful conclusion. However, I imagine that in decisions of whether to give support or not to vulnerable kin, non-primates behave in a similar fashion to primates by balancing consanguinity, risk to self, and risk to kin. Yet, while relatively simple rules might typify almost all alliances among non-primates, they characterize only a fraction of them among primates.

Alliances have several functions

Alliances in non-primate groups almost entirely involve protection of vulnerable kin, excepting cases of males allying in competition for females (see later). In primate groups, by contrast, animals support offspring who are doing the threatening as well as those who are threatened; they support dominant as well as subordinate kin, adults as well as immatures, and adult males as well as adult females; in some circumstances, even non-kin ally with one another. For example, vervet mothers supported threatening offspring five times more frequently than they did threatened ones, and supported dominant offspring five times more frequently than they did subordinate ones (Horrocks and Hunte 1983); and the seven parous females in a captive group of macaques supported non-kin in over a third of all incidents of support (de Waal 1977). As a result, dominant or aggressive contestants, especially if they are non-kin, are sometimes allied with more often than are subordinate or threatened ones (*Cercopithecus*: Cheney 1983; *Macaca*: Colvin 1983; Netto and van Hooff 1986; *Papio*: Cheney 1977; *Theropithecus*: Bramblett 1970). Such a variety of participants, of potential recipients, and opponents, implies that supporters process far more information about group members and the context of the contest than if allying only to protect vulnerable kin.

It is highly unlikely that the same advantage is gained by a young immature as, say, a dominant female when it is supported in a fight. Therefore, at another level of analysis, the variety of participants also implies a variety of functions to the alliance, that is, benefits to the partners on which selection can act (Chapais 1983a; Datta 1983c; Vogel 1985; Harcourt 1987). Alliances thus serve not only to protect vulnerable kin, but also, for example, to facilitate their access to resources and to promote their long-term competitive ability in relation to other group members, i.e. their dominance rank. In addition, the supporter itself probably benefits directly from alliances by, for instance, being allowed to feed on the resource which its support enabled the recipient to win (Silk 1982), or by being supported later by the animals that it had previously supported (Packer 1977; Seyfarth and Cheney 1984). The range of beneficial consequences then itself indicates the range of information that needs to be incorporated into competitive decisions.

Clearly, the group members whom it is best to support will vary according to the outcome sought. That primates might indeed be making such distinctions is indicated by de Waal's (1977) work on macaques. He found that kin are supported both when victims of threat and when threatening, but non-kin only when threatening (Fig. 11.5): animals support relatives both to protect them and as a means of gaining access to a resource, but support non-relatives only in order to gain access.

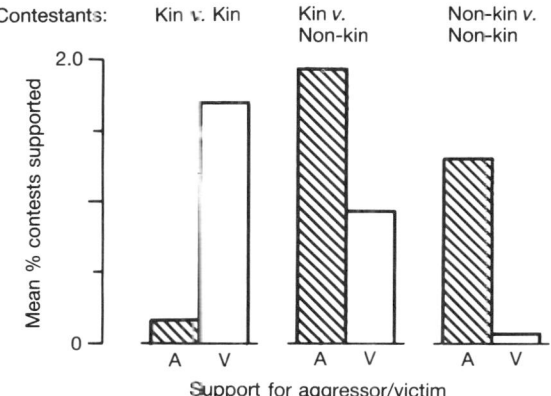

Fig. 11.5 Allies change with the function of the alliance. Relatives are frequently supported as victims (function = protection) and as aggressors (function = access to resources), but non-relatives mostly as aggressors. Values are the mean percentage of contests in a captive macaque group in which the first named contestant was supported as victim or aggressor, with original data sets separated in three-way comparison, kin/non-kin, aggressor/victim, dominant/subordinate. Mean values remove, as far as possible, differences in frequency of the above combinations as a confounding variable between the three displayed situations. (Data from de Waal 1977, Fig. 11).

The additional allies—non-kin as well as kin—and reasons for forming alliances—access as well as protection—require additional rules of behaviour. Furthermore, rules of alliance formation could be in conflict. Animals might have to decide between supporting an aggressive dominant non-relative in order to acquire a resource, or a threatened, subordinate relative in order to protect the relative. Even with the same beneficial outcome, say access, individuals might still face conflicting decisions. For instance, kin might make more reliable partners, because they are likely to co-operate, yet a dominant group member might be a better ally if it does co-operate (Cheney 1983). If primates do make such decisions, then allies cannot be chosen solely on the basis of consanguinity, or solely on the basis of competitive ability; decisions need to be made that combine and compare both sorts of information.

No report on non-primates begins to indicate that they are making these sorts of decisions and, therefore, processing the same required bulk of information about their fellow group members as are primates. By contrast, primates commonly do so, as shown by data on the distribution of both grooming and support among adult female vervet monkeys in the wild (Seyfarth 1980). [Grooming is a friendly behaviour, performed most often between mothers and offspring. While not an alliance in the sense used in this paper, it is regarded, like alliances, as functioning to establish, maintain and reinforce a relationship. It is effectively a solicitation of support, but with a time lag; animals groomed now might be readier to give support later (Cheney 1977; Seyfarth 1977, 1980; Fairbanks 1980; Silk 1982; Seyfarth and Cheney 1984).] Were animals reacting only to consanguinity as a determinant of the distribution of their grooming and support among group members, then in species where daughters remained in the group of their birth and rank was 'inherited', that is, daughters eventually ranked adjacent to their mother, females should groom and support those close in rank to themselves more than those distant in rank, since closely ranked animals are also closely related. Vervets are such a species. Nevertheless, females in the bottom half of a vervet group's linear dominance hierarchy groomed and allied with those in the top half as often as with each other (Fig. 11.6). This even distribution is not a consequence of random choice of partners, because low ranking females groomed high ranking ones more than twice as long as the dominant females groomed them. In other words, enough information has been processed about the

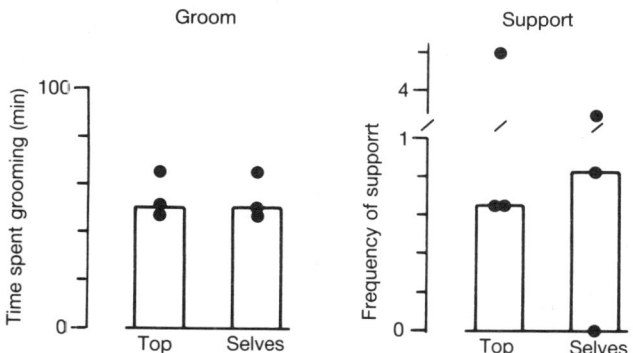

Fig. 11.6 Animals ally with a number of different partners and choices can be conflicting. Vervet females in the bottom half of the hierarchy groom and ally with females in the top half (TOP), to whom they are probably unrelated, as frequently as with one another (SELVES), i.e. with probable relatives. Values are each of the three group's mean individual's minutes spent grooming, and frequency of giving support in contests. (Data from Seyfarth 1980, Fig. 4; $n = 7$–8 females per group.)

advantages of alliances with individuals of high rank for dominant non-kin to be chosen as allies as often as are subordinate kin.

In sum, the decisions made are not simply dichotomous support/not support ones, but complex choices about who to support, when, under what circumstances, and with what beneficial consequence to be achieved.

Successful alliances depend on the ability to seize opportunities as they arise

If one ability characterizes intelligent action, surely it is the ability to change behaviour according to circumstances. Here I want to illustrate the point by describing two studies that indicate primates' abilities to grab chances in forming alliances. The studies indicate a situation of long static periods interspersed with rapid change in which individuals take quick tactical advantage of the behaviour and interactions of others to further their own competitive ability.

Ehardt and Bernstein (1986) observed a series of overthrows of the alpha family in a captive colony of rhesus macaques. Members of low ranking families combined forces and so viciously attacked the alpha family that animals were killed and the family deposed. The important point is that the low ranking family appeared to initiate the attacks suddenly, as the situation became temporarily favourable to them. For example, in two of the overthrows, internal fighting within a family apparently precipitated a combined attack on it by members of other famlies, as if they seized the opportunity that the familial discord offered them to advance their own cause.

Nearly 20 years ago, Marsden (1968) conducted a series of experiments to investigate the mechanism of acquisition of dominance rank in a captive macaque group. His procedure was to change the composition of the group, which had the effect of changing the sets of partners with which the group members could ally. Chapais (1985) extended Marsden's work, and obtained very similar results. Both of them observed a young daughter (who would normally be expected to remain subordinate to her mother, and to support and be supported by her in contests) apparently take advantage of a temporary change in the availability of allies, and begin to attack her mother and eventually to outrank her. Chapais' observations were far more detailed and precise than were Marsden's. The first ranking female's two young daughters were disputing rank with the second ranking female, B. Chapais saw B's daughter apparently react to overtures of friendship (i.e. grooming) from A's daughters by starting to threaten her own mother. Both B, who had seen the grooming between A's daughters and her own, and her daughter acted as if they expected A's daughters to support the attacks, which they did. In other words, B's daughter apparently took tactical advantage of the unlooked for friendship from A's daughters, and used it to further her own ends, namely high dominance rank. She achieved these to the extent of managing to dominate her own mother. The same goal was, Chapais suspected, behind A's daughters' friendliness to B's daughter.

Chapais went on to test this idea in subsequent removal and return experiments, which indicated, as did Marsden's, that the macaques were making moment to moment calculated decisions that were dependent on whom they could rely on for support, a calculation that itself depended on the changing behaviours and relative ranks of the other group members. In Chapais' group, the alpha female's older daughter eventually dominated the second ranking female, at which point the frequent attacks on her ceased, and her daughter once more dropped in rank below her mother. Chapais' conclusion about the end stages of the affair demonstrates the complexity of interpretation necessary if the sequence of events is to be understood. To paraphrase Chapais, once A's daughter was, as a result of previous supported attacks on B, undisputedly dominant to B, she had no more to gain from harrassing her. Therefore, she had no more to gain from forming further alliances with either her younger sister, or B's daughter, against B. As a result, B's daughter could not take advantage of the alpha family's attacks on her mother to ally with them against her. Thus, she reverted to being subordinate to her mother.

In sum, primates are consummate social tacticians.

Successful alliances require abilities not needed in two-animal contests

The main differences so far discussed between the intelligence required for two-animal contests and that necessary for successful support in contests is quantitative: more information needs to be processed in contests that involve alliances, but the type of information is probably very similar, even the same, in the two situations. In this and the next section I will review evidence to suggest that when forming alliances, primates require and use abilities that are not apparent in the conduct of dyadic contests. In dyadic contests an individual need estimate only the potential competitor's abilities, and its decisions about the contest require comparison only between the potential competitor and itself. Neither of these statements holds for alliances in primate groups.

Alliances in contests and social intelligence

Successful alliances require the ability to discriminate among others without reference to self

'By being capable of ranking [peers in a hierarchy], humans introduce a problem not possible with nonhuman primates' (Zivin 1977). The implication is that while primates are capable of 'self/not self' comparisons, or 'very similar/less similar to self' ones, only humans can discriminate among others without reference to self. That is almost certainly wrong (Seyfarth 1981, 1983). The ability to discriminate without reference to self is shown by macaques when giving support to and soliciting it from the more dominant of two contestants, whatever their rank in relation to the supporter's (also see Chapter 6).

Animals differ not only in their need for support, but also in their ability to give it. Dominant animals, for instance, are more effective supporters than are subordinate ones (Cheney 1977; Datta 1983b), and more dangerous adversaries. Therefore, when choosing partners with whom to ally in order to gain access to a resource, individuals should prefer dominant group members, and at the same time avoid forming alliances against them. Adult female rhesus macaques (*M. mulatta*) appear to follow precisely these rules when in contests among unrelated adult females. They support not just females dominant to themselves against those subordinate, but also the more dominant of two subordinate females (Fig. 11.7). It is the latter choice of recipient that implies the ability to rank individuals independently of reference to self. The females are following a 'divide and rule' tactic, it seems, whose effect is to reinforce their own status over

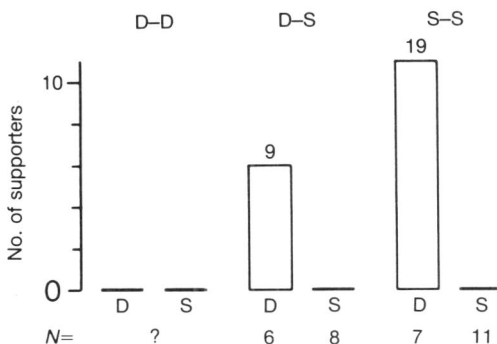

Fig. 11.7 Animals discriminate among group members without reference to self. Adult female rhesus macaques support the more dominant of two unrelated contestants when not only one contestant (D–S), but both (S–S), are subordinate to themselves. Contests between animals dominant to the potential supporter (D–D) will have been as common as S–S contests. Numbers above histograms show number of incidents of support. (Data from Chapais 1983b, Table 10.6; total incidents of support = 28.)

those subordinate to them, without antagonising females dominant to them (Chapais 1983b). Similarly, in 86 per cent of cases when one female bonnet macaque (*M. radiata*) inhibited grooming between other females, the interferer prevented grooming between two animals both lower ranking than itself. It appeared to be preventing the establishment of alliances between subordinates, who might combine to dominate it (Silk 1982).

The ability to distinguish among dominant group members is demonstrated by the fact that when soliciting support during contests, as opposed to giving it, animals direct their overtures to the most dominant among the dominant females (Fig. 11.8). Also, when grooming other females, females prefer higher ranking group members: 12 of 17 females in three wild vervet groups, who had two or more females dominant to themselves in their group, groomed at least one in the hierarchy above them more than they did the female ranked adjacently above them (Fig. 11.9). Furthermore, females competed to groom others in proportion to the rank of the groomed female (Fig. 11.10), a distribution that is difficult to explain except by the assumption that monkeys can rank others in a hierarchy.

The suggestion has been made that if an individual discriminates between two animals, both of whom are dominant (or subordinate) to it, the discrimination is made without reference to self. Of course, it could be the case that in the same way that more or less distant kin might be differentiated by degree of familiarity or similarity to self (Bateson 1980), so more or less

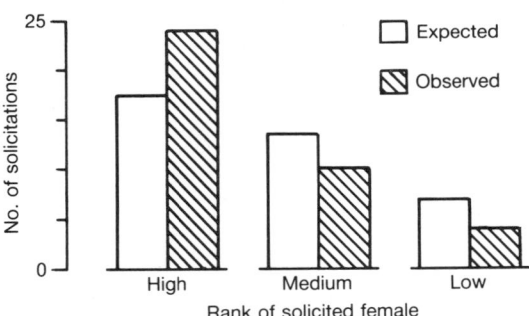

Fig. 11.8 Animals discriminate among group members without reference to self. When soliciting support from adult females dominant to their opponent, wild adolescent baboons solicit it most from the most dominant females. (Data from Walters 1980, p. 80; $n = 38$ solicitations.)

dominant group members might be separated by degree of dominance. If so, a common measure of dominance is needed. That this is unlikely is suggested by the observation that in hardly any study has a direct correlation been found between distance in rank among contestants and, for instance, the frequency of aggression or supplanting (see, e.g. Tables 4, 7, 8 in Sade 1967).

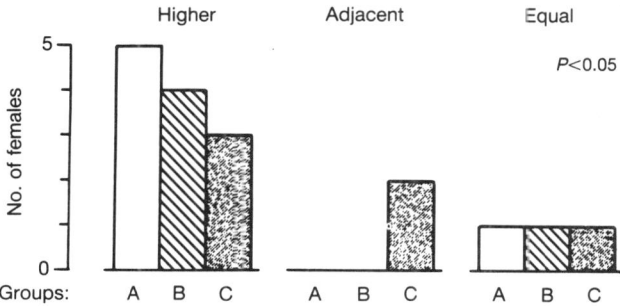

Fig. 11.9 Animals discriminate among group members without reference to self. Females groom (i.e. solicit support from) very high ranking females. Data are number of females in each of three vervet groups who groomed a female at least two ranks above them in the dominance hierarchy more often than they groomed the one adjacently above them (HIGHER), groomed the adjacently ranked one more (ADJACENT), and groomed the adjacent equally as long as at least one higher ranking (EQUAL). (Data from Seyfarth 1980, Fig. 4; $P < 0.05$, Sign Test.)

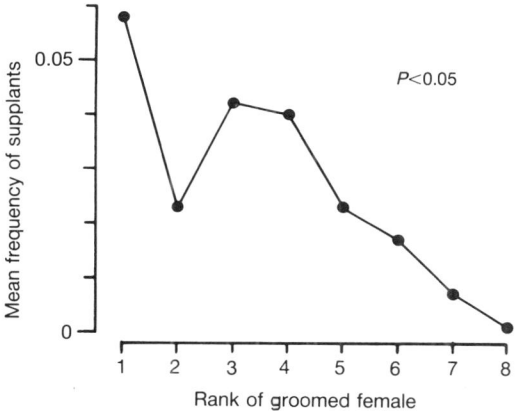

Fig. 11.10 Animals discriminate among group members without reference to self. Grooming of high ranking vervet females is interrupted by others more often than is grooming of low ranking females. Values are the mean (of three groups') rate per minute grooming at which the grooming of females of different rank was interrupted. (Data from Seyfarth 1980, Fig. 1; $r_s = 0.90$; $P < 0.05$.)

Decisions about alliances incorporate information about group members not immediately involved in the contest

So far, I have considered how the identity of only the contestants might influence decisions about whether or not to give support. However, the presence and identity of each contestant's allies needs also to be taken into account. A contestant must balance not only its own competitive ability with the opponent's (Maynard Smith and Parker 1976; Parker 1984), but also the competitive ability of the opponent's supporters with that of its own supporters and, in addition, the availability of the opponent's supporters compared to its own.

An indication that primates are using such extra information about potential allies when making decisions is the observation that offspring of high-ranking rhesus macaque mothers are more likely to threaten, and less likely to be threatened by, members of low-ranking families in the presence of their high-ranking mother (Fig. 11.11).

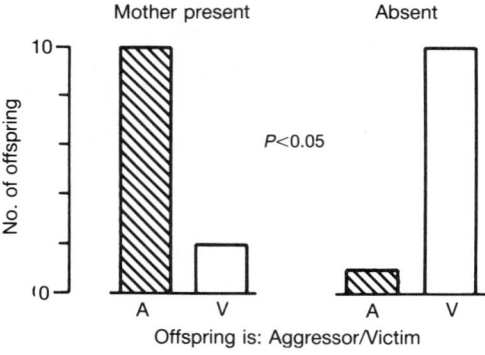

Fig. 11.11 Animals take into account the presence of potential supporters when threatening others. Data are number of offspring of high-ranking macaque mothers who threatened (A = aggressor) or were threatened by (V = victim) individuals dominant to them but subordinate to their mother in the presence (within 2 m) and absence (> 2m) of their mother. (Data from Datta 1983a, p. 99, 101; $P < 0.05$, Sign Test.)

A second instance concerns the observation that vervet females will ally with one another against dominant males, but very rarely against dominant females (Cheney 1983). Various factors can explain the distinction (Cheney 1983), one being the possibility of retaliatory alliances. If vervets are like macaques (Chapais 1983c), the male opponent is unlikely to be supported by others, whereas the female opponent is; moreover, a dominant female's main supporters are themselves likely to be dominant to the partners in the original alliance. The result is that while any two females can defeat a dominant male, they cannot win a contest with a dominant female, because her relatives will probably come to her aid. Thus the chances of the allies defeating the dominant opponent depend not so much on the relative competitive ability of the opponent (Maynard Smith and Parker 1976), but rather on whether or not the opponent is supported and on the competitive ability of the supporters.

Thirdly, and very importantly for the argument that information processing abilities might be involved, animals tend to support not just dominant females, but the immature offspring of dominant females (Fig. 11.12). The significance of this fact is that the support of the offspring is often given at a time when the offspring are subordinate to the supporter, and thus useless themselves as allies. It must be supposed, therefore, that the targeted ally is the dominant high-ranking mother, not her immature offspring (Cheney 1977).

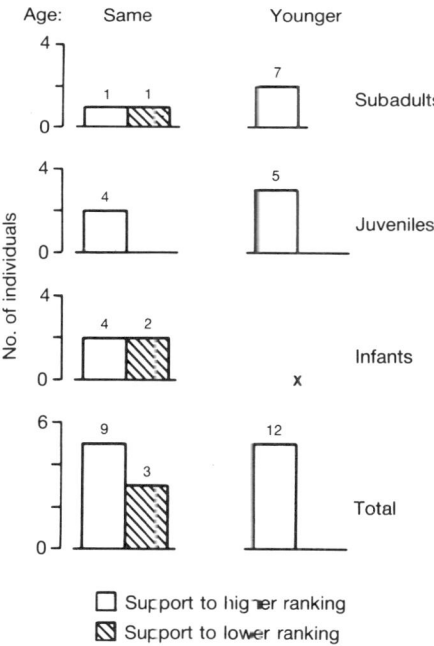

Fig. 11.12 Animals support the offspring of potential supporters, rather than the supporters themselves. Immatures support the immature offspring of mothers higher ranking than their own (open histogram) more often than they do the offspring of lower ranking mothers (hatched histogram), even when the recipients are in an age class younger than their own (YOUNGER) and hence subordinate to them. Numbers above histograms are number of incidents of support in each context. (Data from Cheney 1977, Figs. 1b, 2b.)

Finally, Cheney and Seyafarth's (1986; see Chapter 6) study of redirected aggression similarly shows that account seems to be taken of group members not directly involved in the contest, but who are potential allies of the contestants. Animals attacked the close relatives of those that attacked them or, if they were over three years old, the relatives of those who attacked their own relatives.

Conclusion

I have argued (a) that social interactions that involve alliances are more complex than those that do not, and (b) that on current evidence primates both more often use alliances in competition than do non-primates, and form more complex alliances. At present, these are in the nature of possibilities rather than proven hypotheses. For example, I have argued that the ability to choose among different categories of group members requires intelligence if the subsequent alliances are to be successful in the long run. Yet is the vervet monkey female who decides to choose as its ally a high ranking female instead of its own low ranking sister processing more information or more complex information, or processing the information in a more complex way than the starling who decides to fly to a distant good food source instead of a nearby poor one (Tinbergen 1981), or the flycatcher who exploits the more profitable perch (McNamara and Houston 1985), or the shrew who usually chooses the reliable food source over the unreliable, but sometimes good one (Barnard et al. 1985)? I am certain that she is, but that has yet to be rigorously demonstrated.

To separate the possibilities, experimental testing of individuals' abilities rapidly to process complex social information is needed, as is far more detailed study of the process of alliance formation. Take, for instance, the observation that low ranking animals tend to avoid attacking the immature offspring of high ranking mothers in the presence of the mother. Is this because they have already experienced retaliation from the mother (simple association learning)?; or is it because, without any prior interaction with the high ranking mother or its offspring, the low ranking animals are aware from previous observation of the high ranking mother's dominance over their own mother, that even if their own mother supported them, they would still be in danger from the high ranking mother (intelligent calculation of odds)? The only way to separate the two possibilities is detailed observation in an experimental situation of the events leading up to the interaction.

The ability of animals to choose between and manipulate objects has a long history of study. By comparison, experimental testing of social intelligence is almost non-existent, although social companions are as much objects in the environment to be manipulated as are differently weighted objects (Mason 1978; Dasser 1985; Cheney and Seyfarth, Chapter 19). For instance, monkeys can display transitive reasoning in assessing the relative weights of objects (McGonigle and Chalmers 1977); has anyone attempted to test whether they use transitive reasoning in their assessment of dominance hierarchies, and subsequent courting of allies? As far as I know, they have not. Perhaps the closest that anyone has come are Cheney and Seyfarth (1986) in their study of redirected aggression (see Chapter 6). Their work seems to indicate a process of transitive reasoning in vervet monkeys that, put anthropomorphically, runs along the lines of, 'If A is to its kin as I am to mine, then A's attack on my relative is equivalent to my subsequent attack on A's relative'.

I have little doubt that primates are socially very intelligent, and that future observational, experimental, and mathematical analysis of alliances will demonstrate that the behaviour is far more complex than any performed in dyadic contests without alliances. I am less sure that primates use alliances in more complex ways than do non-primate mammals. At present, the suggestion that they do is based more on lack of reporting for non-primates than on solid observational or theoretical analysis. I suspect that the lack of reporting is based in part on the background of the observers. Many primatologists have sociological and ethological training, and are, therefore, ready to recognize and note complexities of social interaction. Ornithologists too are ethologists. However, non-primatological mammalogists are far more often ecologists, and neither they nor their studies are designed to analyse social behaviour.

In addition to this methodological consideration, primates and non-primate mammals live under different social and ecological conditions. Alliances occur only under certain circumstances, which might be more often fulfilled in primate groups. Allies will be useful where individuals compete for resources by contesting them ('interference', or 'defence', competition). Now, the advantage of contesting resources is determined partly by the subject's own relative competitive ability, but also to a large extent by the quality, abundance, and distribution of the resources (for a review, see Davies and Houston 1984). Only where resources are of high quality and not very abundant (and hence worth defending) and somewhat clumped (hence defensible) will defence competition, and therefore support in that competition, be a viable option (Wrangham 1980, 1982).

It is significant that the best established accounts of support in contests among non-primates concern male lions allying with each other in competition for females (Bygott *et al.* 1979; Packer and Pusey 1982): lionesses are high quality, relatively rare, clumped resources. An additional predisposition to co-operation is divisibility of resources (Wrangham 1980, 1982), and a lioness might be a divisible resource if her litter can be fathered by more than one male.

Alliances ought also to be common where opportunities exist for repeatable, calculated exchange of support, and punishment of cheating (Trivers 1971; Axelrod and Hamilton 1981), namely in stable social groups of relatives. Such groups are characteristic of many primate species, but less so, it seems, of non-primates. Nevertheless, some predictions can be made on the basis of the possible ecological and social correlates of the use of support in contests as a competitive strategy. I suspect that alliances might be found to be common in some of the social omnivores, such as coatimundies (*Nasua*), peccaries (*Tayassu*), and carnivores, such as lions and hyaenas (*Crocuta*). For these animals, foods are often of high quality and clumped, and the females in a group are familiar relatives who remain together. It might be significant that coatimundies apparently support non–relatives (Russell 1983) and that hyaena females 'inherit' their dominance rank (Frank 1986), a phenomenon that in Old World primates results largely from consistent support in contests (Cheney 1977; Berman 1980; Datta 1983*a*, *b*, 1986; Horrocks and Hunte 1983; Harcourt and Stewart 1987).

The fact that the subject of this book, social intellect, is confined to primates is unfortunate, but inevitable. It results from the vast disparity in the distribution of evidence, and from the pervasive assumption that primates are more intelligent than non-primates. Whether or not that assumption is correct, we will progress little further in our understanding of the evolution and operation of social intelligence unless we institute far more broadly based comparative studies. Alliances will be seen, I have suggested, only under defined social and environmental conditions. We therefore require both naturalistic observation of non-primates whose social organization resembles that of primates, and also experimental manipulation of the environment, for example, of the distribution of resources, in order that the behaviour of non-primates can be recorded in the same environment as induces alliances in primates.

Acknowledgements

I thank the National Geographic Society and the H. F. Guggenheim Foundation for funding the field work that stimulated my interest in the use of co-operation as a competitive strategy; and Drs Dorothy Cheney, Robin Dunbar, Robert Seyfarth, and Kelly Stewart for very helpful criticism of the manuscript.

Are primates mind readers?

12

A group of young chimpanzees in a 1-acre field: leadership and communication*
EMIL W. MENZEL

The title succinctly describes the set-up, for in this innovative series of studies Menzel simply recorded the patterns of exploration and association among his chimps, in response to objects such as food which he positioned in their enclosure. In the part quoted here, he shows the great sensitivity of chimpanzees to the superior knowledge of a normally disregarded individual. In the process, he documents how an intelligent chimpanzee can thereby manipulate others in a deceitful way, improving its feeding rate.

Introduction

The procedure of the present experiments was a variation of the delayed-response test (Fletcher 1965). The sole operational differences between the present experiments and the previous ones were the following. (a) The 'goal' that we provided was hidden from sight when the chimps were turned loose, instead of being left visible for all to see (Menzel and Davenport 1962; Menzel and Draper 1965). (b) Only a single individual, rather than the whole group, was given any cue as to the nature and location of the hidden object. That is, we gave one individual a cue and the other chimps had to get their cues from him.

It is inconceivable that the chimps had never before experienced comparable conditions 'naturally' in their own group life, and it is doubtful that they attached any more significance to these changes in procedure than to any other such change. However, the new procedure permitted us to experimentally assess the extent to which a single individual is capable of controlling the nature and direction of group activity ('leadership': Gibbs

---

* Extracted from *Behavior of Nonhuman Primates* (ed. A. M. Schrier and F. Stollnitz), Academic Press, New York, 1974.

1969) and how he does this (which is, in part, a question of 'communication'). Neither the degree to which individuals differed in their ability to lead group travel nor (especially) the efficiency with which the chimps communicated about the environment was surmised in advance from our own previous observations or from the contemporary literature on those topics (see, for example, Mason's 1970 review).

Can a single individual determine a group move?

On a given trial, the entire group of six chimps was locked into the release cage as usual, so that they could not see into the field. Next, we hid six pieces of fruit 60–150 ft from the release cage under leaves or grass, behind a tree, etc., varying these locations semi-randomly across the field by selecting a new pair of x, y co-ordinates on each trial, except trials following one on which the animals failed to get the food. Then, on alternating trials either no animal was shown the food (control condition for olfaction and unknown sources of cues), or one animal was removed from the group, carried to the food, shown it without being allowed to touch it, and then returned to the group. About 2 minutes later the group was released. The chimp that had seen food on a given trial was operationally designated the 'leader' for that trial; and those who had not seen the food were designated 'followers'. Bandit and Belle were chosen as leaders because they were at this point the only animals that could be carried off by us without screaming and struggling to get back to the group, and because Bandit seemed the least likely candidate as a 'group leader' in foraging. Experimental and control conditions were alternated from trial to trial because we anticipated that the group would return to previously rewarded places on the control condition, and we wished to assess the temporal course of 'extinction' here across a series of trials. Bandit was tested first for several days, and then we switched to Belle.

The group as a whole headed straight for the food on the experimental trials, but on control trials they first ran back to the place where food had been on the previous trial, and then searched more or less at random, sometimes actually stepping on the food without detecting it (which suggests that cues from odour, inadvertent behaviours on the part of experimenters, etc., were of negligible importance).

The food was found within 2½ minutes in 73 per cent of the 55 experimental trials (with little difference in performance for the different leaders), and in one of 46 control trials. After a few days the chimps stopped running on the control trials, and sat around as if waiting for someone else to make the first move; or they got involved in play. On none of the trials on which the leaders failed to get the group to the food did the leaders ever search in a false direction; rather, the failures occurred because the group simply did not run. It is most likely that leader 'errors' reflected something other than memory failure, and a major possibility was satiation, for if we added a novel object to the pile, the group moved out and the leader grabbed the object and left the food to the others.

Communication of food quality

In a later experiment we initially assumed that the chimps would not be able to communicate to each other the relative value of two desirable, but hidden goals. Because we were so surprised with chimps' actual performance, we wished to cross-check the results. Here, the same two leaders were shown two different goals that at times varied in quality (high-preference fruits versus low-preference vegetables). The question was once more: can the animals pool the information received by the two leaders, and proceed first to the generally preferred goal?

Unfortunately, this test had to be terminated prematurely due to causes unrelated to the test[1]. However, the results were completely consistent with the last test. If both leaders were shown vegetables, or if both were shown fruit, or if the better leader (Belle) had been shown fruit and the poorer leader (Bandit) had been shown vegetables, the group majority—usually including Bandit—went with Belle to her goal. However, if the poorer leader was shown fruit and the better leader saw vegetables, then the poorer leader won at least half the time. Excluding ties, the group majority thus went to fruit versus vegetables on 21/27 trials, irrespective of who had seen which goal.

The critical results from both this experiment and the last one were clearly significant. It should also be mentioned that more detailed analyses that took into account all of the spatial data and did not simply lump the reactions of all six animals on a given trial into a single 'majority response', were even more overwhelmingly indicative of communication of some sort.

The question of intent and an observation on 'lying'

It might be objected that the chimpanzee group movements, individual locomotion, and gestures cannot possibly be considered 'real' communication unless they are used intentionally. Do the leaders who walk out toward food or tap their companions on the shoulder respond to their own behaviour as do the companions? Do the animals hold such signs in

[1] Bandit performed a 'playful' flying tackle on the author and laid him up for 2 weeks with a sprained back.

common? If not, the behaviour cannot be said to function in the same fashion as linguistic signs (G. Mead, cited by Morris 1955).

As far as I know, the only way in which the question of intent can be assessed with animals is to examine how the leader's actions vary as a function of the consequences they produce in the actions of the followers.

Locomotion and visual orientation, as well as tapping and some molecular signals, are clearly 'intentional' by this criterion. The present chimpanzees seldom started out 'independently' toward food without glancing back at their companions. Even a fast running start from the release cage would stop cold if no one followed, and there was a continuous feedback between leaders and followers until the group was within relatively close range of the goal, the leader usually adjusting his rate so as to just barely keep ahead. The details of these social interactions have been described more fully (Menzel 1971). Probably, all vertebrates that move in packs or close-knit groups adjust their movements to each other, but the extent to which the behaviour of most species can also be interpreted as a cue of objects is still an open question.[2]

Probably, the clearest evidence that the chimpanzees knew what effect their own behaviour was having on others (and varied it accordingly) occurred in some interactions between Rock (the most dominant chimp as well as a relative stranger) and Belle over a period of several months. If tested when Rock was not present, Belle invariably led the group to food and nearly everybody got some. In tests conducted when Rock was present, however, Belle became increasingly slower in her approach to the food. The reason was not hard to detect. As soon as Belle uncovered the food, Rock raced over, kicked or bit her, and took it all.

Belle accordingly stopped uncovering the food if Rock was close. She sat on it until Rock left. Rock, however, soon learned this, and when she sat in one place for more than a few seconds, he came over, shoved her aside, searched her sitting place, and got the food.

Belle next stopped going all the way. Rock, however, countered by steadily expanding the area of his search through the grass near where Belle had sat.

Eventually, Belle sat farther and farther away, waiting until Rock looked in the opposite direction before she moved toward the food at

---

[2] Sam Goldberger and Roberta Black of the State University of New York at Stony Brook tested 'Red', a golden retriever dog, with the same 'pointing' technique the author used. They found that with only a small amount of training, Red would go in any direction pointed out in a 360° outdoor field, even after several minutes' delay between the pointing cue and opportunity to respond; even without having seen the goal object (a stick) thrown out by the experimenters; and even when he was walked around in a circle during the delay period to break his motor set. Control tests ruled out olfactory and other immediate sensory cues. If anything, the accuracy of Red's travel with respect to the hidden objects was greater than that of these chimpanzees. It would not surprise me if pack–hunting carnivores communicate the direction and nature of their prey to each other as many hunters claim.

all—and Rock, in turn, seemed to look away until Belle started to move somewhere. On some occasions, Rock started to wander off, only to wheel around suddenly precisely as Belle was about to uncover the food.

Often Rock found even carefully hidden food that was 30 ft or more from Belle, and he orientated repeatedly at Belle and adjusted his place of search appropriately if she showed any signs of moving or orientating in a given direction. If Rock got very close to the food, Belle invariably gave the game away by a 'nervous' increase in movement. However, on a few trials she actually started off a trial by leading the group in the opposite direction from food, and then, while Rock was engaged in his search, she doubled back rapidly and got some food. In other trials when we hid an extra piece of food about 10 ft away from the large pile, Belle led Rock to the single piece, and while he took it she raced for the pile. When Rock started to ignore the single piece of food to keep his watch on Belle, Belle had temper 'tantrums'.

Goodall (1971) reports some similar observations on wild chimpanzees that she interprets as intentional 'misleading' (see Chapter 16), or deception, and the present observations not only support this as a possibility in chimpanzees, but do so under conditions where the location of the goal and the social conditions were varied systematically and we could usually predict ahead of time when the behaviour in question would appear. It seems quite conceivable from our other data that the reason Goodall's followers could be fooled is that the leader's behaviour had in effect 'told' them that something more important lay in the opposite direction. They were not necessarily following merely for the sake of following.

13

'Does the chimpanzee have a theory of mind?' revisited
DAVID PREMACK

Over many years, Premack has shown an uncanny skill in using the tools of behaviourism to pose apparently mentalistic questions to captive chimpanzees. For this chapter, we asked him to assess the results of perhaps his most ambitious such attempt: can chimpanzees be said to understand the minds of others?

In 1978, we raised the question 'Does the chimpanzee have a theory of mind?' (Premack and Woodruff 1978), by which we meant, does the ape do what humans do: attribute states of mind to the other one, and use these states to predict and explain the behaviour of the other one? For example, does the ape wonder, while looking quizzically at another individual, What does he really *want*? What does he *believe*? What are his *intentions*? Notice that in raising this question we do not ask, Is the ape an intentional system? Does it have wants, beliefs, hopes, plans, etc.? We take it for granted that the ape is an intentional system and ask, instead, 'Does the ape think that other apes are intentional systems?' In other words, does the ape attempt to account for ape behaviour in the same way that humans attempt to account for human behaviour?

In even so much as raising this question, we already imply that language or speech is not a necessary condition for having a theory of mind. There is a widespread tendency to see language as the source of virtually all the 'interesting' properties of the human mind (e.g. Schwartz 1980; Bickerton 1987). We do not subscribe to this view. Many important competences, such as that of social attribution are, we believe, competences in their own right, not secondary properties derived from a competence in language. This is not to say that language cannot influence these competences. Language can amplify competences in several ways, the most obvious being that of giving the individual a greater awareness of the competence, thereby enabling him to use it in more powerful ways.

How shall we decide whether or not the ape has a theory of mind? An important part of the 1978 paper was the description of experiments that could deal with this question. In a later section, I shall discuss some of these experiments as well as report others that have been done more recently. However, there is an overriding point that takes precedence over the individual experiments. This concerns the need for experiments. This need is less widely recognized than one might suppose (given the overall progress of science and its obvious dependence on experiment). There are many who appear to believe that the present question (and others like it) can be answered simply by making observations in the field or even the zoo. An astute observer will somehow see answers to these questions in the behaviour of the animals. On this view, not only is there no need for experiments, probably they should be avoided. Experiments are harmful, or at least misleading, for they create unnatural situations in which what animals do is neither representative nor revealing.

In this regard, it is of interest to compare the progress that has been made in two different fields—one with a clear experimental tradition, another with no such tradition—since the publication of 'Theory of Mind'. In primatology, a basically nonexperimental field, the last 10 years have yielded a handsome collection of anecdotes. Anecdotes were once imperilled by disuse, but the reaction against the sterilities of American Behaviourism has reopened closed gates, returning them to favour. In Child Development, the last 10 years have yielded a handsome collection of experiments. It is of interest to compare what we have been taught by the anecdotes in one case, and by the experiments in the other.

The Primatologists have taken as their point of entry apparent cases of deception (see Chapters 15–17). 'Deception' in apes and monkeys has been reported in diverse settings—food, sex, the handling of another one's infant, etc. (Whiten and Byrne 1988a.) This diversity is itself important. It distinguishes deception in the ape and monkey from that reported in birds. A plover, for example, will fly from its nest, leading a potential intruder away from its fledglings; but it will not employ a comparable tactic in leading a competitor away from food, a receptive mate, a piece of potential nesting material, etc. In the plover, 'deception' is an innate disposition restricted to protection of the young. Like other innate dispositions it can be modified by learning. For example, the bird can learn to distinguish the pseudo-intruder (who merely circumnavigates the nest) from the serious one (who heads directly for the nest), no longer bothering even to 'deceive' the former, while steadfastly continuing to 'deceive' the latter (Ristau, in press). However, this does not change the fact that the bird's 'deception' is an inflexible device that cannot be applied to any target except protection of the young. The bird is analogous to a human who could tell lies only about pilfering fudge; he could not tell lies about dirtying the carpet, breaking the lamp, taking money from his mother's purse, or about lying itself. We would look closely at such a 'person', wondering whether it was child or robot. Deception in the ape and monkey is not so restricted; for this reason alone, the primate's behaviour is of interest with respect to the question of 'theory of mind', while the bird's is not.

One of the frequently reported cases of primate deception concerns the sexual behaviour of the male chimpanzee (Goodall 1986). Although orgasm in the male is ordinarily accompanied by a distinctive cry, when copulating with a favourite female the male sometimes suppresses his sexual cry. In this way he avoids having to share the female with others. For if other males were drawn to the scene, they would discover the receptivity of the female and, given the promiscuity of the chimpanzee, the male would lose his exclusive rights.

In calling this observation an anecdote, I do not mean to question the reliability of the report or even, for that matter, the general accuracy of the account as to the circumstance in which the act occurs. Although, in fact, these are serious problems surrounding the use of anecdotes, for the sake of discussion, let us assume that none of them apply. Let us assume that suppression of sexual cries does not occur except as sexual rivals are near by—the base level of the act is zero—and that the agreement among judges as to the occurrence of the act is respectably high. Given observations of this character, what more could one ask? Where is the need for experiment?

The need for experiments arises with regard to the interpretation we place on the act. In general, an individual can engage in social behaviour for either of two quite different reasons. In one case, he acts so as to affect what the other individual does; in the other case, so as to affect what the other individual believes. Only in the second case is the behaviour relevant to and evidence for a theory of mind; for only in the second case does the actor attribute a state of mind to the other one. How can we distinguish between the two cases?

We can picture the hypothetical situation in which the behaviour itself will tell the difference. For instance, suppose I do not want you to fish in my pond. I could bring this about in either of two quite different ways. I could attempt to influence your belief concerning the presence or absence of fish in the pond. I could assure you that there are no fish there, i.e. I could do so provided you spoke. However, speech or language is not a necessity. If you lacked language, I could substitute perceptual evidence: stir up the water and point out the absence of fish, or have you sit with me for hours, watching while I failed to catch anything. These observations, arranged specifically for your benefit, are likely to influence your belief about the lack of fish in the pond. In an alternative approach, I could ignore your belief and deal with your behaviour directly. Whenever I caught you fishing in the pond, I could punish you. Punishment will reduce the likelihood of your fishing in the pond and, if severe enough, eliminate it altogether. Now if we found two species, one of which specialized in the first approach (always arranged perceptual evidence for the benefit of the 'listener'), the other of which specialized in the alternative approach (did nothing but punish the 'listener'), we could reasonably entertain rather different hypotheses about the mentation of the two species. Only in the former would it seem reasonable to ascribe a belief in belief. Unfortunately, cases of this hypothetical kind—where distinctive behaviours invite distinctive interpretations—are hard to come by; actual social behaviour will not have this helpful character. Most acts that are carried out to affect what another one believes will not look any different from acts that are carried out to affect what another one *does*.

Consider again the ape who suppresses his sexual cry. Does he do this to affect the beliefs of rival males, or merely to affect what they do? Were human adults to behave in a comparable manner we would not hesitate to assign a rich interpretation. We would say that the actor knows that if he does not suppress his cry, his rivals will hear him; if they do hear him they will know what he is doing; and if they know what he is doing, they will want to do the same. This series of assumptions credits the actor with at least the following: he understands:

(1) the distinction between seeing and hearing, recognizing that sound can be effective where vision is not;
(2) that belief or knowledge depends upon perception;
(3) that desire (want) depends upon knowledge or belief.

Which of the assumptions that we grant in the human case can be maintained for the chimpanzee?

It may be thought that, at this point, we could advance our cause by turning to other anecdotes, using one to disambiguate the other. Regrettably, this is not the case. All anecdotes are about equally lacking in resolving power so far as the present issue is concerned. Each of them can be interpreted frugally (he acts so as to affect what the other one does) or in the more profligate way. To resolve this issue we must turn to experiments.

Consider, for example, our first two assumptions:

(1) the actor understands the difference between seeing and hearing;
(2) he recognizes that knowledge depends upon perception.

Fortunately, both of these assumptions have been tested with young children. By the age of 2½, children appreciate the difference between seeing and hearing, recognizing the different conditions on which they depend (Flavell 1978). Do they take the next step, and understand that knowledge or belief depends upon perception? This is evidently a more demanding task for, though by the age of 3 children already attribute knowing/believing to the other one, they do not understand that knowledge depends upon perception. That is, they use the concepts of know and believe, but not in the adult manner, for they do not know how these states come about. They do not have a causal theory of knowledge.

When children are allowed (or not allowed) to see what is in a box, or are told or not told what is in the box, and then subsequently asked, 'Do you know what's in the box?' they answer correctly. However, if they are shown the same conditions for another child—either he is or is not permitted to look, either he is or is not told—and asked the comparable question, 'Does he/she know what's in the box?' they answer incorrectly. In his own case, the child apparently answers simply by determining whether or not he can answer the question, 'What is in the box', but in using this internal knowledge he does not understand how the knowledge came about—that it depends upon his having either seen or been told what's in the box. Hence, he cannot answer the question in the case of the other child for, of course, he has no access to that child's internal knowledge. Not until the child is over 4 years old does he develop a causal theory of knowledge (Wimmer *et al.* in press). Now what bearing does the child's performance have on the claims we can make for the chimpanzee?

A considerable bearing, for in 20 years of comparing chimpanzees with children we have only one case in which a chimpanzee passed a test that was failed by 3½-year-old children (Woodruff *et al.* 1979). Barring the one exception, a good rule of thumb has proved to be: if the child of 3½ years cannot do it neither can the chimpanzee. In this case, moreover, we also have two tests that were done on chimpanzees. The results of these tests tend to agree with those from the children's tests.

In the first of these tests, my former colleague Guy Woodruff put four juvenile apes into a circumstance in which they needed the assistance of the trainer to obtain food from a locked container. The trainer (carrying the key on a chain around his neck) willingly followed the animals across the ½-acre compound to the locked container provided only that the ape did not dash off, but took care to 'lead' him, i.e. turned around, glancing back from time to time, waiting for him to catch up if he fell too far behind. Evidently, all four animals found this a natural form of communication, for they were immediately successful, regularly ending up at the container with the trainer only a few steps behind. He promptly unlocked the container, giving the animal the long-awaited banana. Then one day we blindfolded the trainer.

Now the blindfolded trainer did not follow the 'beckoning' animal, but being unable to see, remained seated on the ground. Three of the four animals, finding that they had headed into the field without being followed, returned, set out again, returned; in the end, they more or less dragged the trainer across the field, pulling him by the key chain around his neck (it was a slow trip because, being unable to see, he moved cautiously). Only one of the four apes responded appropriately to the blindfold. Rather than set out like the others, returning when she found that she had not been followed—and only then remove the blindfold—she removed it the moment she saw it on the trainer's face. Moreover, the only blindfold she removed was one placed over the eyes; blindfolds placed over the nose and mouth or around the head were not disturbed. Evidently, the 4-year-old African-born Jessie understood seeing, or at least she understood that the behaviour she wanted from the trainer, his following her, depended upon his having unobscured eyes. Whether she had a more complete theory of seeing—understood the role of light, position, attention, etc.—is something that remains to be studied.

Notice the sense in which this test demands less than the one given the children. The ape is asked only to understand the relation between two observable conditions: uncovered eyes (or seeing) and normal following which, of course, is a behavioural act; whereas the children were asked to understand the relationship between one observable and one unobservable condition: uncovered eyes (or seeing) and knowledge (a state of mind). It should be easier to grasp a relationship between two observables than one between an observable and an unobservable; we predict that the child will pass the ape's test at an age when it still fails its own.

Although I cannot see where the food is hid, I can see which of you can see where it is

In a second case, four 6–7-year-old domestic-born chimpanzees were given a test more nearly comparable in its demands to the one given the children. Where the first test required only an understanding of the relation between 'seeing' and behaviour, this test required an understanding of the relation between 'seeing' and *knowing*. In brief, the animal was shown that, while it could not see which of two containers was baited, one of the two trainers standing nearby could. This trainer could see because she had an unobstructed view of the containers, whereas the other trainer (like the animal itself) could not see because her view of the containers was blocked by an intervening object. Before choosing between the two containers, the animal was allowed to seek the advice of one of the trainers. If the animal understood the conditions on which seeing depends, and recognized the advantage that seeing has for knowing, it should seek the advice of the seeing trainer. Moreover, it should not only seek her advice, but should follow it, always choosing the container to which she pointed.

The test involved three steps, the first two of which were pretraining. In the first step, the animal was simply familiarized with two opaque containers that were placed before it; food was placed in one of them on each trial and the animal was allowed to choose between them. In the second step, the baiting was obscured from the animal's view by placing an upright box before the animal and inserting the containers in the box. Although the animal could see the trainer place the containers in the box, take out a piece of fruit, kneel down behind the box, rise up with the fruit no longer in hand, it could not see the actual baiting. When the box was removed, leaving behind the containers, the animal was again allowed to choose; it chose willingly but now, of course, at chance level. In the third and essential step, a screen was placed perpendicular to the animal's cage (sometimes on the right, sometimes to the left), and on each trial, two trainers took up positions on opposiet sides of the containers. One stood on the side without the screen so that she could readily see the containers, whereas the other stood on the far side of the screen so that she could not see the containers. (The position of the screen, trainers, and of the baited container was, of course, varied from trial to trial in a counterbalanced fashion so that each condition occurred equally often on both sides.) The animal could readily see which trainer on a given trial had an unimpeded and which an impeded view of the baiting, and it was given an opportunity to exploit this knowledge. The box was removed revealing the two containers, and the screen was taken away; the two trainers then came forward and stood behind the containers equidistant from them. The animal chose between the two trainers by pulling either of two strings, one of which was attached to each trainer. The trainer chosen stepped forward and tapped one container or the other. The trainer who had witnessed the baiting always tapped the wrong container. Ordinarily, of course, the unwitnessing trainer would have responded at chance level, but we downgraded his advice in this manner to help the animal see the advantage of choosing the witnessing trainer.

All four animals readily adopted the practice of pulling the trainer forward by the string and, to a degree that varied over animals, followed the trainer's advice, choosing the container that she indicated. Three of the four animals solved the problem in the sense of choosing significantly more often than chance the trainer who had witnessed the baiting. These animals chose correctly essentially from the beginning: if we divide the 24 trials given them into blocks of four, we find that they performed as well on the first block as on the last. However, one of the three animals who chose the correct trainer did not consistently follow her advice; she followed the trainer's tapping only 15 out of 24 trials ($P > 0.05$) and thus could not be said to have genuinely solved the problem. A fourth animal neither chose the correct trainer above chance level nor followed her advice (a further possibility, that the animal choose at chance level and yet follow the trainer's advice was not found). Hence, only two of the four animals solved the problem fully, both choosing the trainer who witnessed the baiting and following her advice; two others failed, either fully or in part.

These results suggest that two of the four juveniles understand the conditions on which seeing depends and recognize the advantage that seeing has for knowing, i.e. for obtaining the kind of information that the animal wants. This is a strong interpretation to place on the results of one test, of course, and we need controls to eliminate weaker alternatives. For instance, we must deal with the possibility that the animal simply has a preference for the trainer who sees the baiting — and will always choose this trainer whether or not it needs the information the trainer possesses.

Suppose the animal has observed the baiting and therefore knows which container is baited; nevertheless, when offered a choice between trainers who did and did not witness the baiting, it still chooses the witnessing trainer. In fact, these are the results we obtained with the two animals who were fully successful on the earlier test. When we eliminated the box that enclosed the containers — thus restoring the baiting to the animal's view — and repeated the experiment, two of the animals continued to choose the trainer who witnessed the baiting. What does this mean?

It need not be taken to mean that the animal acted out of 'blind' preference for a 'seeing' trainer, and has no comprehension of the relation between seeing and knowing. In a choice between trainers who do and do not know, why not choose the knowing one? (The animal might even realize that it has difficulty avoiding a trainer's advice, and therefore should not choose the ignorant trainer.) The point is, choosing between the trainers cost the animal nothing. We must arrange an experiment in which this choice is costly, and then determine whether the animal will elect to pay for the right to seek advice only when its view of the baiting is blocked and it does not know which container to choose.

Although, as things stand, we cannot definitively interpret the results of the present tests, for the sake of discussion I shall proceed as though they can be interpreted and will treat the outcome for the two successful animals as defending the strong claim. Then the implications of these results for the interpretation of the sexual anecdote obviously depend on the animal; for the successful animals they are of one kind, for the unsuccessful another. Consider first the implications in the case of the unsuccessful animals.

The rich interpretation we would give in the case of comparable sexual behaviour by a human adult, 'The individual suppresses his cry because he knows that if he is overheard his rivals will *know* what he is doing', etc., is unwarranted. For a creature who does not understand the relationship between perception and knowledge, the most we can say is: 'The individual suppresses his cry because he knows that if he does not do so his rivals will rush in and destroy his alliance', etc. This claim entails a relationship between two pieces of behaviour, suppression of cry by the one individual and interference by others, and has no implications for a 'theory of mind'. The appreciation of the causal relationship between the two pieces of behaviour does not entail any attribution, neither of want, belief, or any other state of mind. Moreover, we could give a still weaker account, one that did not credit the ape with causal understanding of any kind.

How could an animal, without attributing any state of mind to the other one, learn the correlation between his sex cry and interference by other animals, and thus learn to suppress his cry? The standard answer of the tough-minded learning account is well known. The sex cry is the last act in a sequence that began with courtship (presentation by the female) and culminated in copulation. The arrival of the other males and their taking over of the female is aversive. Aversive events, by definition, suppress the occurrence of preceding acts, especially the act that is last or most temporally proximal to the aversive event.

If this account were correct, then close observation might reveal that the sex cry was not the only act suppressed. Suppose that the ape, while copulating, clung with one hand to an overhead branch (as is sometimes shown in videotapes of apes copulating); interference by rivals could lead to suppression of this act too. That is, if 'last' acts are especially prone to suppression by aversive events, grasping-the-branch-with-right-hand might be no less subject to suppression than the sex cry itself. If the data actually bore this out, we would be obliged to take seriously the traditional learning account. On the other hand, if the sex cry is uniquely prone to suppression, the learning account would be in trouble. For learning can explain the suppression of an act only on the grounds of its temporal proximity to the aversive event; all 'last' acts must be equally prone to suppression.

A cognitive treatment of the learning may be more compelling. On this view, the sex cry is not like other acts but has special status (after all, clutching a branch will not give one away to other animals). The animal recognizes the unique connection between the sex cry and the likelihood that others will find him out. When he hears someone else's sex cry—and begins to run towards its source—he runs with a definite expectation of what he will find: animals are copulating (he may even be able to picture which animals depending on how intimately he knows the social group). Knowing that other animals' sex cries bring him running, he infers that his sex cry will have the same effect on them. Consequently, when he copulates (and does not want to be disturbed by rivals) he suppresses his sex cry (more so than other acts).

Notice, incidentally, that the two accounts are not mutually exclusive. It is often assumed that if cognitive variables apply, conditioning or learning ones do not, and vice versa. This view fails to recognize that we may find, in lower species, only the primitive system, the conditioning one, but find, in more evolved species, both systems. In our own species, for example, we find both, cognition having to do largely with voluntary behaviour, conditioning with involuntary (Premack 1986). Finally, notice that neither the conditioning nor the cognitive account are attributional; neither requires granting the animal a theory of mind. In neither account does the animal attribute states of mind—knowing, wanting, or the like—to the other one.

Consider now the implications in the case of the successful animals. They evidently understand that knowing depends on seeing. This conclusion is supported by an earlier test we gave Sarah—an unconventional match-to-match problem in which there were two samples (rather than the usual one) and two alternatives, each of which matched one of the samples. If given only one choice per problem, which alternative should she choose? That was the question Sarah faced. We provided her with a basis on which to decide by arranging that the trainer (giving her the problem) stare at one of the two samples. When given this problem, Sarah promptly climbed up the mesh of the cage, improving her view of the trainer's gaze, climbed down and chose the alternative corresponding to the sample at which the trainer was looking. This is a weaker test whose results are at best suggestive, but when combined with the results of the other test, strengthen the claim that at least some chimpanzees understand the importance of looking—and even perhaps seeing. Ideally, we should establish how well-formed or complete is the animal's theory of seeing. Does it understand that seeing depends on a specifiable physical relation between the viewer and object viewed, the eyes being open, a certain amount of light, a minimum duration of exposure, even perhaps attention, and that all of this may vary with the complexity of the item viewed? These are answerable questions though we have not yet addressed them.

Let us assume that animals which understand the relationship between knowing and seeing, also understand the parallel relationship in the case of hearing. This would warrant making the strong claim, viz. 'If I am overheard, they will know what I am doing', etc. and we might generalize this strong claim, applying it to all animals who are observed to suppress their cries. It would be desirable to corroborate this claim, however.

A simple experiment would do the job: allow animals to copulate with favourite mates both when rivals are and are not present. The claim would collapse, of course, if we found sex cries to be suppressed whether rivals were present or not.

That is, what would collapse is the assumption that the suppression is of a rational kind, based on the analogical argument I raised earlier. 'I interfere with others when I hear their sex cries, they will do the same to me if they hear mine. Therefore, I shall suppress mine'. Suppression produced by reasoning of this kind should be quite unemotional, I would think, and therefore readily confinable to just the right occasion. It would not spill over onto occasions when rivals were not present. Moreover, on the assumption that copulation-with-cry is more enjoyable than without cry (suppression presumably would detract from the pleasure) if there were no countermanding force (no rival in the vicinity) we should expect to hear the sex cries. If, however, we find suppression even though rivals are not present, we might consider that the suppression was not the exclusive product of analogical reasoning, but was based at least in part on painful learning. On some occasion the animal must have failed to suppress its cries and paid the price, losing its exclusive prerogatives. We may be advised to adopt this view in any case—even without the hypothetical results of our Gedankenexperiment—for analogical reasoning is not a standard practice of the chimpanzee. We find it only in the language-trained chimpanzee (Premack 1984; Matsuzawa and Premack unpublished data). Hence, even apes who understand the relation between knowledge and perception may nonetheless be unable to anticipate the results of their sex cries. They may have to suffer one or more painful experiences before learning to suppress the cry, and only come to understand the relationship between the two events after the fact.

Consider the analogy we looked at earlier. 'I interfere with him on the occasion of his sex cries, he will do the same to me', etc. The analogy as written is strictly behavioural or non-attributional: it makes no reference to the other one's mind. It could just as well be written in an attributional form, however. For example, 'When I hear sex cries, I expect to find some one copulating; when others hear my sex cries they will expect to find me copulating', etc. Having given thought to his own expectations, the individual now attributes expectations to the other one—giving the analogy an attributional form. To ask which is correct, the attributional or non-attributional, is, of course, to ask the question we are trying to answer.

Perhaps we could gain some leverage on this question by combining the results of field and laboratory. Our current laboratory finding is that some adolescent apes appear to understand the relationship between knowledge and perception. Suppose, in addition, that unless the animal is at least pre-adolescent—6 or 7 years old—we do not find suppression. Suppression not only of sex cries, of course, which naturally would be confined to the older animal, but suppression of any kind so long as it qualified as deceit or deception.

If pre-adolescent animals neither showed deceit nor understood the relationship between perception and knowledge, we might suppose that a causal theory of knowledge was a prerequisite for the practice of deception; but this is almost certainly false. Children show both deceit and pretend play (Leslie 1987) well before reaching an age at which they have a causal theory of knowledge. And we have found, or actually produced, 'deceitful' suppression in apes only *c*. 3–4 years old, far too young for a causal theory of knowledge (Woodruff and Premack 1981). Each of four young apes was put together with a trainer in the following way: the ape knew which of two containers was baited, but could not reach either of them, whereas the trainer did not know which container was baited, but could reach both of them. To obtain food the ape had to 'inform' the trainer where the food was, either in an involuntary way (by glancing at the baited container, bodily orientation, changes in rocking behaviour, etc.) or voluntary way (by pointing with their foot or hand). Actually, the ape's problem was a bit more complicated. They confronted not one trainer but two, one who was benign and gave them the food whenever he chose the correct container, another who was hostile and kept the food for himself whenever he chose the correct container. In time, all four animals learned to suppress their responding to the hostile trainer (while at the same time continuing to 'inform' the benign one). Moreover, the oldest of the four animals went beyond deceit by suppression, the false negative. She actively directed the hostile trainer to the incorrect container, false positive, pointing to the incorrect container with her foot, pointing being of special interest because it is a form of behaviour that is not seen in the wild. Actually, her 'lying' to the hostile trainer involved both false positives and false negatives, for even as she pointed to the wrong container she also suppressed the looking she would normally have directed at the right container.

All of these developments took many trials (largely because the animals were very young), leading critics to say, 'this is not interesting, it's just conditioning, anecdotes are what are interesting'. However, how many 'trials' go into producing the anecdotes that are reported from the field? Since this is rarely known, readers are led to indulge their ignorance and to draw romantic conclusions. The behaviour took place in the field; therefore, it is natural and could not be based on conditioning. How does that follow? There appear to be readers who believe that unless an experimenter rings a bell the phenomenon is not conditioning. Though conditioning is studied in the laboratory, it is not a laboratory phenomenon. Most conditioning, quite obviously, occurs in the real world (and does not require an experimenter). Conversely, cognition is not exclusively a field phenomenon; it can take place quite nicely in the laboratory. Indeed, in the case of chimpanzees, advanced cognition would appear to be largely a laboratory phenomenon. For only the chimpanzee who has been specially trained—exposed to the culture of a species more evolved than itself—shows analogical reasoning. Here is an interesting irony. Most of the deceit described in anecdotal reports from the fields is probably based on simple learning; whereas advanced cognition, such as analogical reasoning, is confined to the laboratory.

Belief is special

So far we have proceeded as though all states of mind were of comparable complexity—entailed the same psychological processes—such that if an individual instantiated or attributed one state of mind it would instantiate or attribute all of them. This is almost certainly false. Some states of mind are more complex than others. Belief, for example, is a more complex state than is perception or desire. Though the distinctions we require here are controversial and suffer from too close contact with the vernacular, we can make some progress simply by dividing all states of mind into two groups, simple and not so simple. Simple states are those produced by processes that are hard-wired, automatic or reflex-like, and encapsulated (for enlightening discussion see Pylyshyn 1980; Fodor 1983). In perception, for example, the prototypic simple state, the proximal stimulus leads more or less inexorably to the end state, with little or any input from 'outside' sources. While perception is the prototypic simple state, we may add others: first, certain basic motivational states; and secondly, somewhat more controversially, expectancy, a state that is produced by conditioning or simple learning. These three states—seeing, wanting, expecting—have in common a restricted and automatic production process that is independent of language both at the level of input to the system and of internal representation. We might call all such states sensory—even though this entails stretching 'sensory' a bit—to underline their simplicity and independence of conceptual processes.

Complex states, of which belief is the prototype, are of course everything that simple states are not. Belief is not automatic, encapsulated, or hard-wired; moreover, it definitely depends on language, most certainly at the level of internal representation though often also at the level of input to the system. Belief is not a completely unitary entity; we can distinguish at least two forms. In its simplest form, belief concerns the reliability of sensory states. Ordinarily, sensory states lead directly to action, but an individual (or species) may reach a stage of development where it calls these states into question. 'Do I really see X?' 'Do I really want X?' 'Are my expectations of X well founded?' The discovery that one's sensory states are subject to unreliability has a social parallel: the discovery that the information told one by a second party is subject to unreliability. Both cases lead to the same outcome. The action to which these states would normally lead is deflected, and the states themselves are examined in the light of available evidence. The examination can be quite simple, as in the case of a child who is just discovering the disparity between seem and real (e.g. Flavell *et al.* forthcoming), or quite elaborate as in the case of a sophisticated adult. However, even when simple the process is not hard-wired, automatic, or encapsulated, and the mental representation on which it depends has a discursive or language-like form. Normally, the process culminates in a decision either to accept or reject the information that one is given—by one's own sensory states or by another party—and this decision is belief in its simplest form.

Have any non-human species reached a stage of development such that they do not invariably act directly on the information given them, but sometimes 'call to question' the information itself? We answer 'no' unhesitatingly for at least some species, e.g. we do not believe that bees debate the reliability of the information encoded in the dances of their conspecifics. The bee does not, we think, distinguish between conspecifics whose dance it believes and others whose dance it doubts or rejects. So much for the bee. By the time we reach the chimpanzee we are no longer certain. There are field reports claiming that the adult ape responds differently to the food grunts of mature and immature conspecifics. Given the food grunts of another adult, the ape is said to climb directly up the tree without so much as a glance; but given the food grunt of a juvenile, the animal is said to pause, to peer up into the branches, having a check for itself before possibly wasting its time climbing a tree for which the message is unreliable (Nishida 1987). As in the case of all anecdotes, however, we do not know what interpretation to put on this behaviour. Is it the result of simple learning, a discrimination based on good experience with old conspecifics, mixed experience with young conspecifics? 'Mixed' experience can produce exactly the kind of vacillation that is described for the one stimulus. If so, this is no more evidence for the informed investigative process culminating in belief (simple form) than is any other case of vacillation. That is, not every species that vacillates does so because it has discovered the unreliability of sensory states or social messages. We recognize a by now familiar bind. There is no experimental work on this issue and, therefore, there can be no sound judgement.

Perhaps we should fall back on our rule of thumb: capacities that do not appear in the 3½-year-old child will not be found in the ape. Children are about 4 years old when they begin to distinguish between 'what X looks like and what X really is', thus offering at least some evidence for the discovery of the unreliability of sensory states. Taken in the light of our informal rule, these data imply that the discovery of sensory unreliability—and thus of belief even in its weak form—lies beyond the chimpanzee, requiring a stage of development the species does not reach. Of course, the conclusion is risky or inadvisable. The children's data are based inescapably on a test that makes specific demands as to level of consciousness or degree of knowing. If we relax those requirements we may obtain positive evidence at an earlier age. There is no substitute for testing the animal itself (not at least until our theories of test demands become more than the informal suggestions which they presently are).

Belief has a more advanced form, one that goes beyond questioning the reliability of information. In its advanced form belief is the decision to accept as an explanation (of one phenomenon or another) conditions that do not depend on sensory states. Familiar examples (in our culture) include: the earth is round, the soul is immortal, species evolve, germs cause disease, God is omniscient, the mind is in the brain, and so forth. Although some of these beliefs can be more easily driven back to sensory states than others, none of them really depend on sensory states. The vast majority of people who believe, for example, in the germ theory of disease and in the immortality of the soul have never seen either germ or soul; nor are they waiting to see them, suspending their belief in one or the other until such time as they do. Belief of this kind has two essential prerequisites. First, a high level of language competence not only for the sake of internal representation, but also for the communication of the information that is typically the source of the belief. Secondly, a conscious high level use of explanation based on causal theory, i.e. the belief that one event causes another. Indeed, the belief that events are caused is itself an example of the present form of belief: it differs from more ordinary beliefs only in that it is somewhat more species-specific (the genetic-experiential mix varies from one belief to another) and it is obviously foundational, a belief that makes other beliefs possible. Both these requirements and particularly their combination make it unlikely that we shall find belief of this form in non-human species, though a definitive judgement awaits the experimental analysis. We may find interesting precursors.

A special variant of the advanced form of belief is to be found in the personality traits we assign to one another. Although this case does not introduce any new properties—it is a classical case of belief pretending to depend more on sensory states than it actually does—it has the virtue of enabling us to contrast belief with expectancy, and thus of showing how much stronger the one state is than the other. In humans, a social encounter can produce either belief, expectancy or both. Thus, Bill, having encountered John, may be led to expect that if he does X, John will do Y. Alternatively, Bill's encounter may lead him to believe that John is a particular kind of person and, therefore, likely to do not only Y in situation X, but a diversity of things in indeterminately many situations, some of which Bill can specify, others of which he can only dimly sense. In the human, these are not mutually exclusive states; he may have expectancies ('narrow predictions') based on conditioning, beliefs ('broad predictions') based on more cognitive processes. The question is whether the chimpanzee resembles the human in having both, or is unlike the human in being restricted to expectancy. One way to find out is to examine the chimpanzee's anticipations of another one's behaviour. Are they broad or narrow? Dependent on specific situations or largely independent of situation?

We tested this, in a sense, inadvertently, when we accidentally allowed the four young apes who participated in the benign/hostile trainer experiment to encounter the hostile trainer outside the experimental room. He was passing through the hall, still wearing the costume identifying him as hostile trainer (he had removed his mask, but not his neck-to-floor white gown) when the animals chanced to be taken out of the test room. Those of us who beheld the encounter were taken aback by the animals' acceptance of the hostile trainer, the readiness with which they either climbed upon him or took his hand and were led off to the compound. The observation taught us that chimpanzee view of liar (the hostile trainer regularly pointed them to the unbaited container) is not necessarily the human view. Humans are inclined to believe that someone who would intentionally misdirect them, who would lie to them, is a bad person generally and should be avoided or dealt with warily. The young chimpanzees appeared to have a much narrower view; their negative expectations concerning the hostile trainer did not appear to extend outside the test room. Perhaps this is only because of their extreme youth; three-year-old children might show a similar 'innocence' (if so, I should be greatly interested in the possible developmental transition from expectancy to belief in the child). Sarah, when tested in a manner comparable to that of the young animals, reacted far more aggressively, hurling objects at the hostile trainer from below the mesh of her cage at a speed so dangerous we terminated the experiment. But this aggression took place in the test room itself during the experiment proper; we have no idea how she might have reacted to him outside the test space. Once again, to resolve the issue we need experiments not serendipitous comparisons or anecdotes.

The chimpanzee's theory of mind

The immediate implication of the preceding section for present purposes is this: if the chimpanzee does have a theory of mind, it will be weaker than the human one. We have seen that the states of mind the chimpanzee is most likely to instantiate are the sensory ones—seeing, wanting, expecting. Belief is more doubtful, the advanced form especially, but even the weaker one concerning the reliability of information. Now a species will not attribute to others states of mind that it does not instantiate itself. Hence, the only states of mind the chimpanzee may attribute—if it attributes any at all—may be the simple ones—seeing, wanting, expecting.

Anecdotes aside, the evidence we have for evaluating theory of mind in the ape (or any animal) is painfully thin. We have two experimental paradigms, one previously reported (that I will summarize briefly in a moment), a new one that I will describe in a later section. Both have their limitations, the former being weak, designed to avoid false negatives (but at the possible expense of false positives), the latter stronger and designed for the opposite reasons.

In the original paradigm, the animal was shown videotapes of a human actor in a cage experiencing 'problems' of one kind or another. In one series, he encountered food that was inaccessible in various ways, and in another series, equipment that was deficient in one respect or another. The videotape depicting the actor in the throes of the problem was put on hold, and the animal was given either two or three photographic alternatives, one of which constituted a solution to the problem. For example, the actor was shown struggling to reach bananas overhead, and the alternatives consisted of the actor stepping onto a chair in one case and reaching to the side with a stick in another. The animal's task in each series was to pick one of the photographs and place it in a designated location.

Sarah performed well above chance, picking the correct alternative on the first trial on nearly all problems (10/12 correct on one series, 12/14 on another). How does behaviour of this kind support the claim that the ape attributes state of mind to the actor? Notice that the selection of solutions presupposes problems, and a problem is not physically instantiated by the videotape. A videotape is merely a sequence of events, e.g. an actor jumping up and down in a cage with bananas overhead. That the actor is not merely jumping up and down, but 'wants' the bananas and is 'trying' to get them, etc., is an interpretation of the actor's behaviour, one the depends on assigning certain states of mind to the actor; for a problem is nothing so much as an individual who is seen as having certain states of mind that are unfulfilled.

Three-year-old children tested in the same manner (with both ape videotapes and others adapted for them, e.g. a sibling attempting to reach cookies on top of the refrigerator) did not merely fail, but failed in a way that tends to corroborate the present interpretation. Rather than choose solutions, they picked alternatives physically resembling some presumably salient item in the videotape, e.g. a yellow bird or flower matching the colour of the bananas. This suggests that they did not so much read or interpret the videotape as react to it at the physical level, as a sequence of events.

The paradigm could be strengthened by some further steps. In one case, the actor would be shown reacting in different ways to the same object, e.g. not always trying to obtain the bananas, but in a second case, trying to destroy them, in a third case surprised or taken aback by their presence, etc. The alternatives given the animal would be designed to show that it understood the actor's attitude. For example, the animal could choose the actor throwing stones, when the actor wanted to destroy the bananas; stepping up onto a chair when he desired them, etc. Correct choices of this kind would emphasize that the animal was not merely picking photographs depicting what it would itself do in similar circumstances.

It is sometimes suggested that Sarah's choices merely represent her choice of the correct 'next act' in a familiar sequence. The problem with this interpretation concerns the concept of 'next'. 'Next' is not the simple idea that it may seem to be, as could be shown by the following procedure. We start by familiarizing Sarah with a sequence in which the trainer jumps up and down persistently attempting to reach the bananas; at some point, he pauses, hitches up his pants, and then resumes jumping. After thoroughly familiarizing Sarah with this videotape, we then test her by presenting the tape, stopping it exactly where the actor's next act is hitching up his pants. We offer as alternatives: the actor stepping up onto a chair, reaching out with a stick, and hitching up his pants. What will Sarah choose? The animal will pick, I assume, not next acts, but *relevant* next acts where relevance is defined by the intention the animal attributes to the actor. To obtain inaccessible bananas, stepping up onto a chair is a relevant act; hitching up one's pants is not. A sophisticated individual such as human adult or even perhaps an older child could, I assume, be taught to pick specifically next acts no matter how irrelevant, but it would be a difficult idea to teach a chimpanzee.

New paradigm

The new paradigm was designed to avoid the inferential indirection of the old one, to obtain more direct evidence (direct, even though one can't speak to the animal). Sarah was the subject again, and again she was tested in home cage. We mounted a cabinet on the wall immediately in front of her cage in a position fully visible to Sarah. The cabinet was divided into left and right halves, and painted white and black, respectively. The left side was stocked exclusively with good things, the right with bad. Good consisted of pastries that a favourite trainer, Bonnie, shared with Sarah during a daily tea time. Bad consisted of rubber snakes, putrid rotting rubber and a cup of faeces; items so untoward that Bonnie did not merely feign dislike of them, but signalled her disgust by quite genuine gestures including wearing rubber gloves when handling the material.

Both for the intermittent restocking of the cabinet and the daily tea time, Bonnie needed Sarah's co-operation. For she could not open the cabinet door alone. The lock on the door was controlled by Sarah; only when she pushed a button inside her cage was the trainer able to open the door.

Sarah's latency on opening the door (pushing the button) once Bonnie entered her cage area was almost immediately stable. After three of the daily visits by Bonnie, Sarah's latency fell to an average of about 7 seconds and so remained for the rest of the test.

On day 18 we introduced the first experimental variation. A 'villain' (masked and concealed in a gown) entered Sarah's room, forced the cabinet door with a crowbar, took out all the material, placing them on the floor before Sarah, and then replaced them, totally reversing their position in the cabinet. Videotapes confirmed the 'villain's' report: Sarah responded hostily, throwing things out under the mesh of her cage at the 'villain'. Bonnie arrived 15 minutes after the 'villain' had left, at her usual time.

What was Sarah's response? Did she greet Bonnie in an unusual manner? More important, did she hesitate in opening the cabinet door, i.e. in pushing her button? One might have thought she would because she was in a position to know, first, that the location of the goods was now reversed, and secondly, that Bonnie did not know this (because, of course, the 'villain' had made the change in Bonnie's absence). Any one knowing this would see that Bonnie was in jeopardy: she might put her hand where it did not belong! And an individual taking this into account might be expected to show some hesitation in opening the door.

However, Sarah did not show any change, either in general demeanour or in latency on pushing the button (she did only one aberrant thing: after Bonnie had already opened the door, Sarah pushed the button a second time), and she showed the same equanimity following four subsequent intrusions by the 'villain'. Either Sarah did not know the pertinent facts, knew but did not care, or both.

Negative outcomes seldom lend themselves to diagnosis. Sarah could have failed for any of dozens of reasons, but one possibility is more interesting than the others. To react appropriately on this test, i.e. to show evidence for theory of mind, Sarah must have had separate representations of what she knows and of what she knows the trainer knows; this was not required by the first paradigm. What Sarah knew and what the actor knew did not differ. For example, both Sarah and the actor could see that the bananas were out of reach, the cord was not plugged in, the flame was out, etc. There was no disparity in the knowledge of the two parties and, therefore, no need for separate representations, or strictly speaking for a representation of a representation. Sarah could have represented the actor's knowledge or perception simply as being the same as her own; she was not obliged to represent her knowledge on the one hand, and her knowledge of the actor's knowledge on the other. On two earlier occasions we applied the old paradigm to test material different from the original material in that it did require that Sarah represent separately her knowledge and her knowledge of the other's (different) knowledge; she failed both tests, perhaps not accidentally (Premack and Premack 1983).

There is, on the whole, only suggestive evidence for theory of mind in the chimpanzee (or any non-human). Moreover, even the positive evidence leaves open the possibility of a theory of mind weaker than the human one. First, should the ape attribute any states at all, these are certain to be a small subset of those the human attributes. Secondly, the ape may be incapable of attributing different states of knowledge than its own. We may wish to distinguish three degrees of theory of mind: (a) species that make no attributions of any kind, presumably the case for the vast majority of species; (b) species whose attributions are unlimited in any respect except perhaps for number of embeddings (e.g. John thought that Mary believed that Bill thought that . . .), presumably the case for humans (by the time they are 4 years old) (e.g. Wimmer and Perner 1983); and (c) species that make attributions but attributions that are limited in a number of respects, possibly the case for the chimpanzee.

14

The intentional stance in theory and practice
DANIEL C. DENNETT

In the first part* of this chapter, one of today's most exciting philosophers elegantly charts the increasing 'levels' (embedding) of intentionality which may in principle underly primate vocal behaviour, and suggests a simple method for picking out the real level. Then, in the second part, he calls his own bluff. He visits the very primatologists on whose data he theorized, and discusses the difficulties in executing his simple test in practice.

Section 1: Intentional systems in cognitive ethology

I. The problem

The field of cognitive ethology provides a rich source of material for the philosophical analysis of meaning and mentality, and even holds out some tempting prospects for philosophers to contribute fairly directly to the development of the concepts and methods of another field. As a philosopher, an outsider with only a cursory introduction to the field of ethology, I find that the new ethologists, having cast off the straitjacket of behaviourism and kicked off its weighted overshoes, are looking about somewhat insecurely for something presentable to wear. They are seeking a theoretical vocabulary that is powerfully descriptive of the data they are uncovering and at the same time a theoretically fruitful method of framing hypotheses that will eventually lead to information-processing models of the nervous systems of the creatures they are studying (see Roitblat 1982). It is a long way from the observation of the behaviour of, say, primates in the wild to the validation of neurophysiological models of their brain activity, and finding a sound interim way of speaking is not a trivial task. Since the methodological and conceptual problems confronting the ethologists appear to me to bear

* Section 1 extracted from *Behavioral and Brain Sciences*, **3**, 343–350 (1983).

striking resemblances to problems I and other philosophers have been grappling with recently, I am tempted to butt in and offer, first, a swift analysis of the problem; secondly, a proposal for dealing with it (which I call intentional system theory); and thirdly, an analysis of the continuity of intentional system theory with the theoretical strategy or attitude in evolutionary theory often called adaptationism.

The methodology of philosophy, such as it is, includes as one of its most popular (and often genuinely fruitful) strategies the description and examination of entirely imaginary situations, elaborate thought experiments that isolate for scrutiny the presumably critical features in some conceptual domain. In *Word and Object*, W. V. O. Quine (1960), gave us an extended examination of the evidential and theoretical tasks facing the 'radical translator', the imaginary anthropologist-linguist who walks into an entirely alien community—with no string of interpreters or bilingual guides—and who must figure out, using whatever scientific methods are available, the language of the natives. Out of this thought experiment came Quine's thesis of the 'indeterminacy of radical translation', the claim that it must always be possible in principle to produce nontrivially different translation manuals, equally well supported by all the evidence, for any language. One of the most controversial features of Quine's position over the years has been his uncompromisingly behaviourist scruples about how to characterize the task facing the radical translator. What happens to the task of radical translation when you give up the commitment to a behaviouristic outlook and terminology? What are the prospects for fixing on a unique translation of a language (or a unique interpretation of the 'mental states' of a being) if one permits oneself the vocabulary and methods of 'cognitivism'? The question could be explored via other thought experiments, and has been in some regards (Bennett 1976; Dennett 1971: Lewis 1974), but the real-world researches of Seyfarth *et al.* (1980a) with vervet monkeys in Africa will serve us better on this occasion. Vervet monkeys form societies of sorts and have a language of sorts, and of course, there are no bilingual interpreters to give a boost to the radical translators of Vervetese. This is what they find:

Vervet monkeys give different alarm calls to different predators. Recordings of the alarms played back when predators were absent caused the monkeys to run into the trees for leopard alarms, look up for eagle alarms, and look down for snake alarms. Adults call primarily to leopards, martial eagles, and pythons, but infants give leopard alarms to various mammals, eagle alarms to many birds, and snake alarms to various snakelike objects. Predator classification improves with age and experience. (Seyfarth *et al.* 1980, p. 801.)

This abstract is couched, you will note, in almost pure Behaviourese—the language of *Science* even if it is no longer exclusively the language of science. It is just informative enough to be tantalizing. How much of a language, one wants to know, do the vervets really have? Do they really communicate? Do they mean what they say? Just what interpretation can we put on these activities? What, if anything, do these data tell us about the cognitive capacities of vervet monkeys? In what ways are they—must they be—like human cognitive capacities, and in what ways and to what degree are vervets more intelligent than other species by virtue of these 'linguistic' talents? These loaded questions—the most natural ones to ask under the circumstances—do not fall squarely within the domain of any science, but whether or not they are the right questions for the scientist to ask, they are surely the questions that we all, as fascinated human beings learning of this apparent similarity of the vervets to us, want answered.

The cognitivist would like to succumb to the temptation to use ordinary mentalistic language more or less at face value, and to respond directly to such questions as: What do the monkeys know? What do they want, and understand, and mean? At the same time, the primary point of the cognitivists' research is not to satisfy the layman's curiosity about the relative IQ, as it were, of his simian cousins, but to chart the cognitive talents of these animals on the way to charting the cognitive *processes* that explain these talents. Could the everyday language of belief, desire, expectation, recognition, understanding, and the like also serve as the suitably rigorous abstract language in which to describe cognitive competences?

I will argue that the answer is yes. Yes, if we are careful about what we are doing and saying when we use ordinary words like 'believe' and 'want', and if we understand the assumptions and implications of the strategy we must adopt when we use these words.

The decision to conduct one's science in terms of beliefs, desires, and other 'mentalistic' notions, the decision to adopt 'the intentional stance', as I call it (Dennett 1971, 1976, 1978*a*, 1981*a*, *b*, *c*), is not an unusual sort of decision in science. The basic strategy of which this is a special case is familiar: changing levels of explanation and description in order to gain access to greater predictive power or generality—purchased, typically, at the cost of submerging detail and courting trivialization on the one hand and easy falsification on the other. When biologists studying some species choose to call something in that species' environment food and leave it at that, they ignore the tricky details of the chemistry and physics of nutrition, the biology of mastication, digestion, excretion, and the rest. Even supposing many of the details of this finer-grained biology are still ill-understood, the decision to leap ahead, in anticipation of fine-grained biology, and rely on the well-behavedness of the concept of food at the level of the theory appropriate to it, is likely to meet approval from the most conservative risk takers.

The decision to adopt the intentional stance is riskier. It banks on the soundness of some as yet imprecisely described concept of information—not the concept legitimized by Shannon–Weaver information theory (Shannon 1949), but rather the concept of what is often called semantic information. (A more or less standard way of introducing the still imperfectly understood distinction between these two concepts of information is to say that Shannon–Weaver theory measures the capacity of information-transmission and information-storage vehicles, but is mute about the contents of those channels and vehicles, which will be the topic of the still-to-be-formulated theory of semantic information. See Dretske 1981 [and multiple book review in *BBS* 6(1) 1983] for an attempt to bridge the gap between the two concepts.) Information, in the semantic view, is a perfectly real, but very abstract commodity, the storage, transmission, and transformation of which is informally—but quite sure-footedly—recounted in ordinary talk in terms of beliefs and desires and the other states and acts philosophers call intentional.

II. *Intentional system theory*

Intentionality, in philosophical jargon, is—in a word—aboutness. Some of the things, states, and events in the world have the interesting property of being about other things, states, and events; figuratively, they point to other things. This arrow of reference or aboutness has been subjected to intense philosophical scrutiny and has engendered much controversy. For our purposes, we can gingerly pluck two points from this boiling cauldron, oversimplifying them and ignoring important issues tangential to our concerns.

First, we can mark the presence of intentionality—aboutness—as the topic of our discussions by marking the presence of a peculiar logical feature of all such discussion. Sentences attributing intentional states or events to systems use idioms that exhibit *referential opacity*: they introduce clauses in which the normal, permissive, substitution rule does not hold: This rule is simply the logical codification of the maxim that a rose by any other name would smell as sweet. If you have a true sentence, so runs the rule, and you alter it by replacing a term in it by another, different term that still refers to exactly the same thing or things, the new sentence will also be true. Ditto for false sentences—merely changing the means of picking out the objects the sentence is about cannot turn a falsehood into a truth. For instance, suppose Bill is the oldest child in class; then if it is true that

(1) Mary is sitting next to Bill,
then, substituting 'the oldest child in class' for 'Bill', we get
(2) Mary is sitting next to the oldest child in class,
which must be true if the other sentence is.

A sentence with an *intentional idiom* in it, however, contains a clause in which such substitution can turn truth into falsehood and vice versa. (This phenomenon is called referential opacity because the terms in such clauses are shielded or insulated by a barrier to logical analysis, which normally 'sees through' the terms to the world the terms are about.) For example, Sir Walter Scott wrote *Waverly*, and Bertrand Russell (1905) assures us

(3) George IV wondered whether Scott was the author of *Waverly*,

but it seems unlikely indeed that

(4) George IV wondered whether Scott was Scott.

(As Russell remarks, 'An interest in the law of identity can hardly be attributed to the first gentleman of Europe; 1905, p. 485.) To give another example, suppose we decide it is true that

(5) Burgess fears that the creature rustling in the bush is a python,

and suppose that in fact the creature in the bush is Robert Seyfarth. We will not want to draw the conclusion that

(6) Burgess fears that Robert Seyfarth is a python.

Well, in one sense we do, you say, and in one sense we also want to insist that, oddly enough, King George was wondering whether Scott was Scott. But that's not how he put it to himself—and that's not how Burgess conceived of the creature in the bush, either—that is, as Seyfarth. It's the sense of conceiving as, seeing as, thinking of as that the intentional idioms focus on.

One more example. Suppose you think your next-door neighbour would make someone a good husband and suppose, unbeknownst to you, he's the Mad Strangler. Although in one very strained sense you could be said to believe that the Mad Strangler would make someone a good husband, in another more natural sense you don't, for there is another—very bizarre and unlikely—belief that you surely don't have which could better be called the belief that the Mad Strangler would make a good husband.

It is this resistance to substitution, the insistence that for some purposes how you call a rose a rose makes all the difference, that makes the intentional idioms ideally suited for talking about the ways in which information is represented in the heads of people—and other animals.

So the first point about intentionality is just that we can rely on a marked set of idioms to have this special feature of being sensitive to the *means of reference* used in the clauses they introduce. The most familiar such idioms are 'believes that', 'knows that', 'expects (that)', 'wants (it to be the case that)', 'recognizes (that)', 'understands (that)'.

In short, the 'mentalistic' vocabulary shunned by behaviourists and celebrated by cognitivists is quite well picked out by the logical test for referential opacity.

The second point to pluck from the cauldron is somewhat controversial, although it has many adherents who have arrived at roughly the same conclusion by various routes: the use of intentional idioms carries a presupposition or assumption of *rationality* in the creature or system to which the intentional states are attributed. What this amounts to will become clearer if we now turn to the intentional stance in relation to the vervet monkeys.

III. Vervet monkeys as intentional systems

To adopt the intentional stance toward these monkeys is to decide—tentatively, of course—to attempt to characterize, predict, and explain their behaviour by using intentional idioms, such as 'believes' and 'wants', a practice that assumes or presupposes the rationality of the vervets. A vervet monkey is, we will say, an *intentional system*, a thing whose behaviour is predictable by attributing beliefs and desires (and, of course, rationality) to it. Which beliefs and desires? Here there are many hypotheses available, and they are testable in virtue of the rationality requirement. First, let us note that there are different grades of intentional systems.

A *first-order* intentional system has beliefs and desires (etc.), but no beliefs and desires about beliefs and desires. Thus, all the attributions we make to a merely first-order intentional system have the logical form of:

(7) x *believes that* p;
(8) y *wants that* q;

where 'p' and 'q' are clauses that themselves contain no intentional idioms. A *second-order* intentional system is more sophisticated; it has beliefs and desires (and no doubt other intentional states) about beliefs and desires (and other intentional states)—both those of others and its own. For instance,

(9) x *wants* y to *believe* that x is hungry;
(10) x *believes* y *expects* x to jump left;
(11) x *fears* that y *will discover* that x has a food cache.

A *third-order* intentional system is one that is capable of such states as

(12) x *wants* y to *believe* that x *believes* he is all alone.

A *fourth-order* system might want you to think it understood you to be requesting that it leave. How high can we human beings go? 'In principle', forever, no doubt, but in fact I suspect that you wonder whether I realize how hard it is for you to be sure that you understand whether I mean to be saying that you can recognize that I can believe you to want me to explain that most of us can keep track of only about five or six orders, under the best of circumstances. (See Cargile (1970) for an elegant and sober exploration of this phenomenon.)

How good are vervet monkeys? Are they really capable of third-order or higher-order intentionality? The question is interesting on several fronts. First, these orders ascend what is intuitively a scale of intelligence: higher-order attributions strike us as much more sophisticated, much more human, requiring much more intelligence. There are some plausible diagnoses of this intuition. Grice (1957, 1969) and other philosophers (see especially Bennett 1976) have developed an elaborate and painstakingly argued case for the view that genuine *communication*, speech acts in the strong, human sense of the word, depend on at least three orders of intentionality in both speaker and audience.

Not all interactions between organisms are communicative. When I swat a fly I am not communicating with it, nor am I if I open the window to let it fly away. Does a sheep dog, though, communicate with the sheep it herds? Does a beaver communicate by slapping its tail, and do bees communicate by doing their famous dances? Do human infants communicate with their parents? At what point can one be sure one is really communicating with an infant? The presence of specific linguistic tokens seems neither sufficient nor necessary. (I can use English commands to get my dog to do things, but that is at best a pale form of communication compared to the mere raised eyebrow by which I can let someone know he should change the topic of our conversation.) Grice's theory provides a better framework for answering these questions. It defines intuitively plausible and formally powerful criteria for communication that involve, at a minimum, the correct attribution to communicators of such third-order intentional states as

(13) Utterer *intends* Audience to *recognize* that Utterer *intends* Audience to produce response *r*.

So one reason for being interested in the intentional interpretation of the vervets is that it promises to answer—or at least help answer—the questions: Is this behaviour really linguistic? Are they really communicating? Another reason is that higher-orderedness is a conspicuous mark of the attributions speculated about in the sociobiological literature about such interactive traits as reciprocal altruism. It has even been speculated (by Trivers 1971), that the increasing complexity of mental representation required for the maintenance of systems of reciprocal altruism (and other complex social relations) led, in evolution, to a sort of brain-power arms race. Humphrey (Chapter 2) arrives at similar conclusions by a different and in some regards less speculative route. There may then be a number of routes to the conclusion that higher-orderedness of intentional characterization is a deep mark—and not just a reliable symptom—of intelligence.

(I do not mean to suggest that these orders provide a uniform scale of any sort. As several critics have remarked to me, the first iteration—to a second-order intentional system—is the crucial step of the recursion; once one has the principle of embedding in one's repertoire, the complexity of what one can then in some sense entertain seems plausibly more a limitation of memory, or attention span, or 'cognitive workspace' than a fundamental measure of system sophistication. Thanks to 'chunking' and other, artificial, aids to memory, there seems to be no interesting difference between, say, a fourth-order and a fifth-order intentional system; but see Cargile 1970 for further reflections on the natural limits of iteration.)

However, back to the empirical question of how good the vervet monkeys are. For simplicity's sake, we can restrict our attention to a single apparently communicative act by a particular vervet, Tom, who, let us suppose, gives a leopard alarm call in the presence of another vervet, Sam. We can now compose a set of competing intentional interpretations of this behaviour, ordered from high to low, from romantic to killjoy. Here is a (relatively) romantic hypothesis (with some variations to test in the final clause):

4th-order: Tom *wants* Sam to *recognize* that Tom *wants* Sam to *believe* that
    there is a leopard
    there is a carnivore
    there is a four-legged animal
    there is a live animal bigger than a breadbox

A less exciting hypothesis to confirm would be this third-order version (there could be others):

3rd-order: Tom *wants* Sam to *believe* that Tom *wants* Sam to run into the trees.

Note that this particular third-order case differs from the fourth-order case in changing the speech act category: on this reading the leopard call is an imperative (a request or command) not a declarative (informing Sam of the leopard). The important difference between imperative and declarative interpretations (see Bennett 1976, pp. 41, 51) of utterances can be captured—and then telltale behavioural differences can be explored—at any level of description above the second order—at which, *ex hypothesi*, there is no intention to utter a speech act of either variety. Even at the second order, however, a related distinction in effect-desired-in-the-Audience is expressed, and is in principle behaviourally detectable, in the following variations:

2nd-order: Tom *wants* Sam to *believe* that
    there is a leopard
    he should run into the trees

This differs from the previous two in not supposing Tom's act involves ('in Tom's mind') any recognition by Sam of his (Tom's) own role in the situation. If Tom could accomplish his end equally well by growling like a leopard, or just somehow attracting Sam's attention to the leopard without Sam's recognizing Tom's intervention, this would be only a second-order case. (Cf. I *want* you to *believe* I am not in my office; so I sit very quietly and don't answer your knock. That is not communicating.)

1st-order: Tom *wants* to cause Sam to run into the trees (and he has this noise-making trick that produces that effect; he uses the trick to induce a certain response in Sam).

On this reading the leopard cry belongs in the same general category with coming up behind someone and saying 'Boo!'. Not only does its intended effect not depend on the victim's recognition of the perpetrator's intention; the perpetrator does not need to have any conception at all of the victim's mind: Making loud noises behind certain things just makes them jump.

0-order: Tom (like other vervet monkeys) is prone to three flavours of anxiety or arousal: leopard anxiety, eagle anxiety, and snake anxiety.[1] Each has its characteristic symptomatic vocalization. The effects on others of these vocalizations have a happy trend, but it is all just tropism, in both utterer and audience.

We have reached the killjoy bottom of the barrel: an account that attributes no mentality, no intelligence, no communication, no intentionality at all to the vervet. Other accounts at the various levels are possible, and some may be more plausible; I chose these candidates for simplicity and vividness. Lloyd Morgan's canon of parsimony enjoins us to settle on the most killjoy, least romantic hypothesis that will account systematically for the observed and observable behaviour, and for a long time the behaviourist creed that the curves could be made to fit the data well at the lowest level prevented the exploration of the case that can be made for higher-order, higher-level systematizations of the behaviour of such animals. The claim that, in principle, a lowest-order story can always be told of any animal behaviour (an entirely physiological story, or even an abstemiously behaviouristic story of unimaginable complexity) is no longer interesting. It is like claiming that in principle the concept of food can be ignored by biologists—or the concept of cell or gene for that matter—or like claiming that in principle a purely electronic-level story can be told of any computer behaviour. Today we are interested in asking what gains in perspicuity, in predictive power, in generalization, might accrue if we

---

[1] We can probe the boundaries of the stimulus-equivalence class for this response by substituting for the 'normal' leopard such different 'stimuli' as dogs, hyenas, lions, stuffed leopards, caged leopards, leopards dyed green, firecrackers, shovels, motorcyclists. Whether these independent tests are tests of *anxiety specificity* or of the *meaning* of one-word sentences of Vervetese depends on whether our tests for the other components of our *n*th-order attribution, the nested intentional operators, come out positive.

adopt a higher-level hypothesis that takes a risky step into intentional characterization.

The question is empirical. The tactic of adopting the intentional stance is not a matter of replacing empirical investigations with aprioristic ('armchair') investigations, but of using the stance to suggest which brute empirical questions to put to nature. We can test the competing hypothesis by exploiting the rationality assumption of the intentional stance. We can start at either end of the spectrum; either casting about for the depressing sorts of evidence that will demote a creature from a high-order interpretation, or hunting for the delighting sorts of evidence that promote creatures to higher-order interpretations (cf. Bennett 1976). We are delighted to learn, for instance, that lone male vervet monkeys, travelling between bands (and hence out of the hearing—so far as they know—of other vervets) will, on seeing a leopard, silently seek refuge in the trees. So much for the killjoy hypothesis about leopard-anxiety yelps. (No hypothesis succumbs quite so easily, of course. *Ad hoc* modifications can save any hypothesis, and it is an easy matter to dream up some simple 'context' switches for leopard-anxiety yelp mechanisms to save the zero-order hypothesis for another day.) At the other end of the spectrum, the mere fact that vervet monkeys apparently have so few different things they can say holds out little prospect for discovering any real theoretical utility for such a fancy hypothesis as our fourth-order candidate. It is only in contexts or societies in which one must rule out (or in) such possibilities as irony, metaphor, storytelling, and illustration ('second-intention' uses of words, as philosophers would say)[2] that we must avail ourselves of such high-powered interpretations. The evidence is not yet in, but one would have to be romantic indeed to have high expectations here. Still, there are encouraging anecdotes.

Seyfarth reports (personal communication) an incident in which one band of vervets was losing ground in a territorial skirmish with another band. One of the losing-side monkeys, temporarily out of the fray, seemed to get a bright idea: it suddenly issued a leopard alarm (in the absence of any leopards), leading all the vervets to take up the cry and head for the trees—creating a truce and regaining the ground his side had been losing (see Chapter 16). The intuitive sense we all have that this is possibly (barring killjoy reinterpretation) an incident of great cleverness is amenable to a detailed diagnosis in terms of intentional systems. If this act is not just a lucky coincidence, then the act is truly devious, for it is not simply a case of the vervet uttering an imperative 'get into the trees' in the expectation that all the vervets will obey, since the vervet (being rational—our predictive lever) should not expect a rival band to honour *his*

---

[2] See Quine (1960) pp. 48–49, on second-intention cases as the 'bane of theoretical linguistics'.

imperative. So either the leopard call is considered by the vervets to be informative—a warning, not a command—and hence the utterer's credibility, but not authority, is enough to explain the effect, or our utterer is more devious still: he *wants* the rivals to *think* they are *overhearing* a command *intended* (of course) only for his own folk, and so on. Could a vervet possibly have that keen a sense of the situation? These dizzying heights of sophistication are strictly implied by the higher-order interpretation taken with its inevitable presupposition of rationality. Only a creature capable of appreciating these points could properly be said to have those beliefs and desires and intentions.

Another observation of the vervets brings out this role of the rationality assumption even more clearly. When I first learned that Seyfarth's methods involved hiding speakers in the brush and playing recorded alarm calls, I viewed the very success of the method as a seriously demoting datum, for if the monkeys really were Gricean in their sophistication, when playing their audience roles they should be perplexed, unmoved, somehow disrupted by disembodied calls issuing from no known utterer. If they were oblivious to this problem, they were no Griceans. Just as a genuine Communicator typically checks the Audience periodically for signs that it is getting the drift of the communication, a genuine Audience typically checks out the Communicator periodically for signs that the drift it is getting is the drift being delivered.

To my delight, however, I learned from Seyfarth that great care had been taken in the use of the speakers to prevent this sort of case from arising. Vervets can readily recognize the particular calls of their band—thus they recognize Sam's leopard call as Sam's, not Tom's. Wanting to give the recordings the best chance of 'working', the experimenters took great care to play, say, Sam's call only when Sam was neither clearly in view and close-mouthed or otherwise occupied, nor 'known' by the others to be far away. Only if Sam could be 'supposed' by the audience to be actually present and uttering the call (though hidden from their view), only if the audience could believe that the noisemaker in the bush was Sam, would the experimenters play Sam's call. While this remarkable patience and caution are to be applauded as scrupulous method, one wonders whether they were truly necessary. If a 'sloppier' scheduling of playbacks produced just as 'good' results, this would in itself be a very important *demoting* datum. Such a test should be attempted; if the monkeys are baffled and unmoved by recorded calls except under the scrupulously maintained circumstances, the necessity of those circumstances would strongly support the claim that Tom, say, does believe that the noisemaker in the bush is Sam, that vervet monkeys are not only capable of believing such things, but must believe such things for the observed reaction to occur.

The rationality assumption thus provides a way of taking the various hypotheses seriously—seriously enough to test. We expect at the outset that there are bound to be grounds for the verdict that vervet monkeys are believers only in some attenuated way (compared to us human believers). The rationality assumption helps us look for, and measure, the signs of attenuation. We frame conditionals such as

(14) if x believed that p, and *if x was rational*, then since 'p' implies 'q' x would (have to) believe that q.

This leads to the further attribution to x of belief that q,[3] which, coupled with some plausible attribution of desire, leads to a prediction of behaviour, which can be tested by observation or experiment.[4]

Once one gets the knack of using the rationality assumption for leverage, it is easy to generate further telling behaviours to look for in the wild or to provoke in experiments. For instance, if anything as sophisticated as a third- or fourth-order analysis is correct, then it ought to be possible, by devious (and morally dubious!) use of the hidden speakers to create a 'boy who cried wolf'.[5] If a single vervet is picked out and 'framed' as the utterer of false alarms, the others, being rational, should begin to lower their trust in him, which *ought* to manifest itself in a variety of ways. Can a 'credibility gap' be created for a vervet monkey? Would the potentially nasty results (remember what happened in the fable) be justified by the interest such a positive result would have? (Editors' Note: this has now been done, see Chapter 6.)

IV. How to use anecdotal evidence: the Sherlock Holmes method

One of the recognized Catch 22s of cognitive ethology is the vexing problem of anecdotal evidence. On the one hand, as a good scientist, the ethologist knows how misleading and, officially, unusable anecdotes are, and yet on the other hand they are often so telling! The trouble with the canons of scientific evidence here is that they virtually rule out the description of anything but the oft-repeated, oft-observed, stereotypic behaviour of a species, and this is just the sort of behaviour that reveals no particular intelligence at all—all this behaviour can be more or less plausibly explained as the effects of some humdrum combination of

---

[3] 'I shall always treasure the visual memory of a very angry philosopher, trying to convince an audience that "if you believe that A and you believe that if A then B then you *must* believe that B". I don't really know whether he had the moral power to coerce anyone to believe that B, but a failure to comply does make it quite difficult to use the word "belief", and that is worth shouting about' (Kahneman 1982).

[4] The unseen normality of the rationality assumption in any attribution or belief is revealed by noting that (14), which explicitly assumes rationality, is virtually synonymous with (plays the same role as) the conditional beginning: if x *really* believed that p, then since 'p' implies 'q' . . .

[5] I owe this suggestion to Susan Carey, in conversation.

'instinct' or tropism and conditioned response. It is the novel bits of behaviour, the acts that couldn't plausibly be accounted for in terms of prior conditioning or training or habit, that speak eloquently of intelligence; but if their very novelty and unrepeatability make them anecdotal and, hence, inadmissible evidence, how can one proceed to develop the cognitive case for the intelligence of one's target species?

Just such a problem has bedevilled Premack and Woodruff (1978), for instance, in their attempts to demonstrate that chimps 'have a theory of mind'; their scrupulous efforts to force their chimps into nonanecdotal, repeatable behaviour that manifests the intelligence they believe them to have engenders the frustrating side effect of providing prolonged training histories for the behaviourists to point to in developing their rival, conditioning hypotheses as putative explanations of the observed behaviour. [See the commentaries and replies in: 'Cognition and Consciousness in Nonhuman Species' (*BBS* 1(4) 1978; see also Premack: 'The Codes of Man and Beasts' *BBS* 6(1) 1983; and Chapter 13].

We can see the way out of this quandary if we pause to ask ourselves how we establish our own higher-order intentionality to the satisfaction of all but the most doctrinaire behaviourists. We can concede to the behaviourists that any single short stretch of human behaviour can be given a relatively plausible and not obviously *ad hoc* demoting explanation, but as we pile anecdote upon anecdote, apparent novelty upon apparent novelty, we build up for each acquaintance such a biography of apparent cleverness that the claim that it is all just lucky coincidence—or the result of hitherto undetected 'training'—becomes the more extravagant hypothesis. This accretion of unrepeatable detail can be abetted by using the intentional stance to provoke one-shot circumstances that will be particularly telling. The intentional stance is in effect an engine for generating or designing anecdotal circumstances—ruses, traps, and other intentionalistic litmus tests—and predicting their outcomes.

This tricky tactic has long been celebrated in literature. The idea is as old as Odysseus testing his swineherd's loyalty by concealing his identity from him and offering him temptations. Sherlock Holmes was a master of more intricate intentional experiments, so I shall call this the *Sherlock Holmes method*. Cherniak (1981) draws our attention to a nice case:

In 'A Scandal in Bohemia', Sherlock Holmes' opponent has hidden a very important photograph in a room, and Holmes wants to find out where it is. Holmes has Watson throw a smoke bomb into the room and yell 'fire' when Holmes' opponent is in the next room, while Holmes watches. Then, as one would expect, the opponent runs into the room and takes the photograph from where it was hidden. Not everyone would have devised such an ingenious plan for manipulating an opponent's behaviour; but once the conditions are described, it seems very easy to predict the opponent's actions. (p. 161.)

In this instance Holmes simultaneously learns the location of the photograph and confirms a rather elaborate intentional profile of his opponent, Irene Adler, who is revealed to *want* the photograph; to *believe* it to be located where she goes to get it; to *believe* that the person who yelled 'fire' *believed* there was a fire (note that if she believed the yeller wanted to deceive her, she would take entirely different action); to *want* to retrieve the photograph without letting anyone *know* she was doing this, and so on.

A variation on this theme is an intentional tactic beloved of mystery writers: provoking the telltale move. All the suspects are gathered in the drawing room, and the detective knows (and he alone knows) that the guilty party (and only the guilty party) *believes* that an incriminating cuff link is under the gateleg table. Of course the culprit *wants* no one else to *believe* this, or to *discover* the cuff link, and *believes* that in due course it will be discovered unless he takes covert action. The detective arranges for a 'power failure'; after a few seconds of darkness the lights are switched on and the guilty party is, of course, the chap on his hands and knees under the gateleg table. What else on earth could conceivably explain this novel and bizarre behaviour in such a distinguished gentleman.?[6]

Similar stratagems can be designed to test the various hypotheses about the beliefs and desires of vervet monkeys and other creatures. These stratagems have the virtue of provoking novel but interpretable behaviour, of *generating anecdotes* under controlled (and hence scientifically admissible) conditions. Thus, the Sherlock Holmes method offers a significant increase in investigative power over behaviourist methods. This comes out dramatically if we compare the actual and contemplated research on vervet monkey communication with the efforts of Quine's imagined behaviouristic field linguist. According to Quine, a necessary preliminary to any real

---

[6] It is a particular gift of the playwright to devise circumstances in which behaviour—verbal and otherwise—speaks loudly and clearly about the intentional profiles ('motivation', beliefs, misunderstandings, and so forth) of the characters, but sometimes these circumstances grow too convoluted for ready comprehension; a very slight shift in circumstance can make all the difference between utterly inscrutable behaviour, and lucid self-relevation. The notorious 'get thee to a nunnery' speech of Hamlet to Ophelia is a classic case in point. Hamlet's lines are utterly bewildering until we hit upon the fact (obscured in Shakespeare's minimal stage directions) that while Hamlet is speaking to Ophelia, he *believes* not only that Claudius and Polonius are listening behind the arras, but that they *believe* he doesn't *suspect* that they are. What makes this scene particularly apt for our purposes is the fact that it portrays an intentional experiment: Claudius and Polonius, using Ophelia as decoy and prop, are attempting to provoke a particularly telling behaviour from Hamlet in order thereby to discover just what his beliefs and intentions are; they are foiled by their failure to design the experiment well enough to exclude from Hamlet's intentional profile the belief that he is being observed, and the desire to create false beliefs in his observers. See, for example, Dover Wilson (1951). A similar difficulty can bedevil ethologists: 'Brief observations of avocet and stilt behavior can be misleading. Underestimating the bird's sharp eyesight, early naturalists believed their presence was undetected and misinterpreted distraction behavior as courtship' (Sordahl 1981, p. 45).

progress by the linguist is the tentative isolation and identification of native words (or speech acts) for 'Yes' and 'No', so that the linguist can enter into a tedious round of 'query-and-assent'—putting native sentences to co-operative natives under varying conditions and checking for patterns in their yes and no responses (Quine 1960, Chap. 2). Nothing just like Quine's game of query-and-assent can be played by ethologists studying animals, but a vestige of this minimalist research strategy is evident in the patient explorations of 'stimulus substitution' for animal vocalizations—to the exclusion, typically, of more manipulative (if less intrusive) experiments (see note 1). So long as one is resolutely behaviouristic, however, one must miss the evidential value of such behaviour as the lone vervet quietly taking to the trees when a 'leopard stimulus' is presented. However, without a goodly amount of such telling behaviour, no mountain of data on what Quine calls the 'stimulus meaning' of utterances will reveal that they are communicative acts, rather than merely audible manifestations of peculiar sensitivities. Quine of course realizes this, and tacitly presupposes that his radical translator has already informally satisfied himself (no doubt by using the powerful, but everyday, Sherlock Holmes method) of the richly communicative nature of the natives' behaviour.

Of course the power of the Sherlock Holmes method cuts both ways; failure to perform up to expectations is often a strongly demoting datum.[7] Woodruff and Premack (1979) have tried to show that chimpanzees in their lab can be full-fledged *deceivers*. Consider Sadie, one of four chimps used in this experiment. In Sadie's sight, food is placed in one of two closed boxes she cannot reach. Then either a 'co-operative' or a 'competitive' trainer enters, and Sadie has learned she must point to one of the boxes in hopes of getting the food. The competitive trainer, if he discovers the food, will take it all himself and leave. The co-operative trainer shares the food with Sadie. Just giving Sadie enough experience with the circumstances to assure her appreciation of these contingencies involves training sessions that give the behaviourist plenty of grist for the 'mere reinforcement' mill. (In order to render the identities of the trainers sufficiently distinct, there was strict adherence to special costumes and rituals; the competitive trainer always wore sunglasses and a bandit's mask, for instance. Does the mask then become established as a simple 'eliciting stimulus' for the tricky behaviour?)

[7] I do not wish to be interpreted as *concluding* in this paper that vervet monkeys, or laboratory chimpanzees, or any non-human animals have *already been shown* to be higher-order intentional systems. Once the Sherlock Holmes method is applied with imagination and rigour, it may very well yield results that will disappoint the romantics. I am arguing in favour of a method of raising empirical questions, and explaining the method by showing what the answers *might be* (and why); I am not giving those answers in advance of the research.

Still, setting behaviourists' redescriptions aside, will Sadie rise to the occasion and do the 'right' thing? Will she try to deceive the competitive trainer (and only the competitive trainer) by *pointing to the wrong box*? Yes, but suspicions abound about the interpretation.[8] How could we strengthen it? Well if Sadie really intends to deceive the trainer, she must (being rational) start with the belief that the trainer does not already know where the food is. Suppose, then, we introduce all the chimps in an entirely different context to transparent plastic boxes; they should come to know that since they—and anyone else—can see through them, anyone can see, and hence come to know, what is in them. Then on a one-trial, novel behavioural test, we can introduce a plastic box and an opaque box one day, and place the food in the plastic box. The competitive trainer then enters, and lets Sadie see him looking right at the plastic box. If Sadie still points to the opaque box, she reveals, sadly, that she really doesn't have a grasp of the sophisticated ideas involved in deception. Of course, this experiment is still imperfectly designed. For one thing, Sadie might point to the opaque box out of despair, seeing no better option. To improve the experiment, an option should be introduced that would appear better to her only if the first option was hopeless, as in this case. Moreover, shouldn't Sadie be puzzled by the competitive trainer's curious behaviour? Shouldn't it bother her that the competitive trainer, on finding no food where she points, just sits in the corner and 'sulks' instead of checking out the other box? Shouldn't she be puzzled to discover that her trick keeps working? She should wonder. Can the competitive trainer be that stupid? Furthermore, better-designed experiments with Sadie—and other creatures—are called for.[9]

Not wanting to feed the galling stereotype of the philosopher as an armchair answerer of empirical questions, I will nevertheless succumb to the temptation to make a few predictions. It will turn out on further exploration that vervet monkeys (and chimps and dolphins, and all other

---

[8] It is all too easy to stop too soon in our intentional interpretation of a presumably 'lower' creature. There was once a village idiot who, whenever he was offered a choice between a dime and a nickel, unhesitatingly took the nickel—to the laughter and derision of the onlookers. One day someone asked him if he could be so stupid as to continue choosing the nickel after hearing all that laughter. Replied the idiot, 'Do you think that if I ever took the dime they'd ever offer me another choice?'

The curiously unmotivated rituals that attended the training of the chimps as reported in Woodruff and Premack (1979) might well have baffled the chimps for similar reasons. Can a chimp wonder why these human beings don't just eat the food that is in their control? If so, such a wonder could overwhelm the chimps' opportunities to understand the circumstance in the sense the researchers were hoping. If not, then this very limit in their understanding of such agents and predicaments undercuts somewhat the attribution of such a sophisticated higher-order state as the desire to deceive.

[9] This commentary on Premack's chimpanzees grew out of discussion at the Dahlem conference on animal intelligence with Sue Savage-Rumbaugh, whose chimps, Austin and Sherman, themselves exhibit apparently communicative behaviour (Savage-Rumbaugh 1978) that cries out for analysis and experimentation via the Sherlock Holmes methods.

higher nonhuman animals) exhibit mixed and confusing symptoms of higher-order intentionality. They will pass some higher-order tests and fail others; they will in some regards reveal themselves to be alert to third-order sophistications, while disappointing us with their failure to grasp some apparently even simpler second-order points. No crisp, 'rigorous' set of intentional hypotheses of any order will be clearly confirmed. The reason I am willing to make this prediction is not that I think I have special insight into vervet monkeys or other species, but just that I have noted, as anyone can, that much the same is true of us human beings. We are not ourselves unproblematic exemplars of third-, fourth-, or fifth-order intentional systems. And we have the tremendous advantage of being voluble language users, beings that can be plunked down at a desk and given lengthy questionnaires to answer, and the like. Our very capacity to engage in linguistic interactions of this sort seriously distorts our profile as intentional systems, by producing illusions of much more definition in our operative systems of mental representation than we actually have (Dennett 1978a, Chaps. 3, 16; Dennett 1981b). I expect the results of the effort at intentional interpretation of monkeys, like the results of intentional interpretation of small children, to be riddled with the sorts of gaps and foggy places that are inevitable in the interpretation of systems that are, after all, only imperfectly rational (see Dennett 1981a, c; 1982).

Still, the results, for all their gaps and vagueness, will be valuable. How and why? The intentional stance profile or characterization of an animal— or for that matter, an inanimate system—can be viewed as what engineers would call a set of specs— specifications for a device with a certain overall information-processing *competence*. An intentional system profile says, roughly, *what information* must be receivable, usable, rememberable, transmittable by the system. It alludes to the ways in which things in the surrounding world must be represented—but only in terms of distinctions drawn or drawable, discriminations makable—and not at all in terms of the actual machinery for doing this work (cf. Johnston 1981 on 'task descriptions'.) These intentional specs, then, set a design task for the next sort of theorist, the representation-system designer.[10] This division of labour is already familiar in certain circles within artificial intelligence (AI): what I have called the intentional stance is what Newell (1982) calls 'the knowledge level'. Oddly enough, the very defects and gaps and surd places in the intentional profile of a less than ideally rational animal, far from creating problems for the system designer, point to the shortcuts and stopgaps Mother Nature has relied upon to design the biological system; they hence make the system designer's job easier.

[10] In the terms I develop in 'Three Kinds of Intentional Psychology (Dennett 1981b), intentional system theory specifies a semantic engine which must then be realized – mimicked, in approximation, by a syntactic engine designed by the subpersonal cognitive psychologist.

Suppose, for example, that we adopt the intentional stance towards bees, and note with wonder that they seem to *know* that dead bees are a hygiene problem in a hive. when a bee dies its sisters *recognize* that it has died, and, *believing* that dead bees are a health hazard, and *wanting*, rationally enough, to avoid health hazards, they *decide* they must remove the dead bee immediately. Thereupon, they do just that. Now if that fancy an intentional story were confirmed, the bee system designer would be faced with an enormously difficult job. Happily for the designer (if sadly for bee romantics), it turns out that a much lower order explanation suffices: dead bees secrete oleic acid; the smell of oleic acid turns on the 'remove-it' subroutine in the other bees; put a dab of oleic acid on a live, healthy bee, and it will be dragged, kicking and screaming, out of the hive (Gould and Gould 1982; Wilson *et al.* 1958).

Someone in artificial intelligence, learning that, might well say: 'Ah how familiar! I know just how to design systems that behave like that. Shortcuts like that are my stock in trade'. In fact there is an eerie resemblance between many of the discoveries of cognitive ethologists working with lower animals and the sorts of prowess mixed with stupidity one encounters in the typical products of AI. For instance, Roger Schank (1976) tells of a 'bug' in TALESPIN, a story-writing program written by James Meehan in Schank's lab at Yale, which produced the following story: 'Henry Ant was thirsty. He walked over to the river bank where his good friend Bill Bird was sitting. Henry slipped and fell in the river. Gravity drowned'. Why did 'gravity drown'? (!) Because the program used a usually reliable shortcut of treating gravity as an unmentioned *agent* that is always around pulling things down, and since gravity (unlike Henry in the tale) had no friends (!), there was no one to pull it to safety when it was in the river pulling Henry down.

Several years ago, in 'Why Not the Whole Iguana?' (Dennett 1978*b*) I suggested that people in AI could make better progress by switching from the modelling of human microcompetences (playing chess, answering questions about baseball, writing nursery stories, etc.) to the whole competences of much simpler animals. At the time I suggested it might be wise for people in AI just to *invent* imaginary simple creatures and solve the whole-mind problem for them. I am now tempted to think that truth is apt to be both more fruitful, and, surprisingly, more tractable, than fiction. I suspect that if some of the bee and spider people were to join forces with some of the AI people, it would be a mutually enriching partnership.

Section 2: Out of the Armchair*

In June of 1983, I had a brief introduction to ethological field work, observing Seyfarth and Cheney observing the vervet monkeys in Kenya. Once I got in the field and saw, first hand, some of the obstacles to performing the sorts of experiments I had recommended, I found some good news and some bad news. The bad news was that the Sherlock Holmes method, in its classical guise, had very limited applicability to the vervet monkeys—and by extrapolation, to other 'lower' animals. The good news was that by adopting the intentional stance one can generate some plausible and indirectly testable hypotheses about why this should be so, and thereby learn something important about the nature of the selection pressures that probably have shaped the vervets' communication systems.

A vocalization that Seyfarth and Cheney were studying during my visit had been dubbed the Moving Into the Open (or MIO) grunt. Shortly before a monkey in a bush moves out into the open, it often gives a MIO grunt. Other monkeys in the bush will often repeat it—spectrographic analysis has not (yet) revealed a clear mark of difference between the initial grunt and this response. If no such echo is made, the original grunter will often stay in the bush for 5 or 10 minutes and then repeat the MIO. Often, when the MIO is echoed by one or more other monkeys, the original grunter will thereupon move cautiously into the open.

What does the MIO grunt mean? We listed the possible translations to see which we could eliminate or support on the basis of evidence already at hand. I started with what seemed to be the most straightforward and obvious possibility:

'I'm going.'
'I read you. You're going.'

What, however, would be the use of saying this? Vervets are in fact a taciturn lot, who keep silent most of the time, and are not given to anything that looks like passing the time of day by making obvious remarks. Then could it be a request for permission to leave?

'May I go, please?'
'Yes, you have my permission to go.'

This hypothesis could be knocked out if higher ranking vervets ever originated the MIO in the presence of their subordinates. In fact, higher ranking vervets do tend to move into the open first, so it doesn't seem that MIO is a request for permission. Could it be a command, then?

'Follow me!'
'Aye, Aye, Cap'n.'

Not very plausible, Cheney thought.

---

* Excerpted from Dennett (in press, *a*, *b*).

Why waste words with such an order when it would seem to *go without saying* in vervet society that low-ranking animals follow the lead of their superiors? For instance, you would think that there would be a vocalization meaning 'May I?' to be said by a monkey when approaching a dominant in hopes of grooming it. And you'd expect there to be two responses: 'You may' and 'You may not', but there is no sign of any such vocalization. Apparently, such interchanges would not be useful enough to be worth the effort. There are gestures and facial expressions which may serve this purpose, but no audible signals.

Perhaps, Cheney mused, the MIO grunt served simply to acknowledge and share the fear:
 'I'm really scared.'
 'Yes. Me too.'
Another interesting possibility was that the grunt helped with co-ordination of the group's movements:
 'Ready for me to go?'
 'Ready whenever you are.'
A monkey that gives the echo is apt to be the next to leave. Or perhaps even better:
 'Coast clear?'
 'Coast is clear. We're covering you.'
The behaviour so far observed is compatible with this reading, which would give the MIO grunt a robust purpose, orienting the monkeys to a task of co-operative vigilance. The responding monkeys do watch the leave-taker and look in the right directions to be keeping an eye out. 'Suppose then, that this is our best candidate hypothesis,' I said. 'Can we think of anything to look for that would particularly shed light on it?' Among males, competition overshadows co-operation more than among females. Would a male bother giving the MIO if its only company in a bush was another male? Seyfarth had a better idea: suppose a male originated the MIO grunt; would a rival male be devious enough to give a dangerously misleading MIO response when he saw that the originator was about to step into trouble? The likelihood of ever getting any good evidence of this is minuscule, for you would have to observe a case in which the Originator didn't see and Responder did see a nearby predator *and* Responder saw that Originator didn't see the predator. (Otherwise Responder would just waste his credibility and incur the wrath and mistrust of Originator for no gain.) Such a coincidence of conditions must be extremely rare. This was an ideal opportunity, it seemed, for a Sherlock Holmes ploy.

Seyfarth suggested that perhaps we could spring a trap with something like a stuffed python that we could very slyly and surreptitiously reveal to just one of two males who seemed about to venture out of a bush. The technical problems would clearly be nasty, and at best it would be a long shot, but with luck we might just manage to lure a liar into our trap. However, on further reflection, the technical problems looked virtually insurmountable. How would we establish that the 'liar' had actually seen (and been taken in by) the 'predator', and wasn't just innocently and sincerely reporting that the coast was clear? I found myself tempted (as often before in our discussions) to indulge in a fantasy: 'If only I were small enough to dress up in a vervet suit, or if only we could introduce a trained vervet, or a robot or puppet vervet who could . . . ' and slowly it dawned on me that this recurring escape from reality had a point: there is really no substitute, in the radical translation business, for going in and talking with the natives. You can test more hypotheses in half an hour of attempted chitchat than you can in a month of observation and unobtrusive manipulation. However, to take advantage of this you have to become obtrusive; you—or your puppet—have to enter into communicative encounters with the natives, if only in order to go around pointing to things and asking 'Gavagai?' in an attempt to figure out what 'Gavagai' means. Similarly, in your typical mystery story caper, some crucial part of the setting up of the 'Sherlock Holmes method' trap is—*must be*—accomplished by imparting some (mis)information verbally. Manoeuvring your subjects into the right frame of mind—and knowing you've succeeded—without the luxurious efficiency of words can prove to be arduous at best, and often next to impossible.

In particular, it is often next to impossible in the field to establish that particular monkeys have been shielded from a particular bit of information. Since many of the theoretically most interesting hypotheses depend on just such circumstances, it is often very tempting to think of moving the monkeys into a lab, where a monkey can by physically removed from the group and given opportunities to acquire information that the others don't have and that the test monkey knows they don't have. Just such experiments are being done, by Seyfarth and Cheney with a group of captive vervets in California, and by other researchers with chimpanzees. The early results are tantalizing, but equivocal (of course), and perhaps the lab environment, with its isolation booths, will be just the tool we need to open up the monkeys' minds, but my hunch is that being isolated in that way is such an unusual predicament for vervet monkeys that they will prove to be unprepared by evolution to take advantage of it.

The most important thing I think I learned from actually watching the vervets is that they live in a world in which secrets are virtually impossible. Unlike orangutans, who are solitary, and get together only to mate and when mothers are rearing offspring; unlike chimps, who have a fluid social organization in which individuals come and go, seeing each other fairly often, but also venturing out on their own a large proportion of the time, vervets live in the open in close proximity to the other members of their groups, and have no solitary projects of any scope. Thus, it is a rare occasion indeed when one vervet is in a position to learn something that it alone knows and knows that it alone knows. (The knowledge of the others' ignorance, and of the possibility of maintaining it, is critical. Even when one monkey is the first to see a predator or a rival group, and knows it, it is almost never in a position to be sure the others won't very soon make the same discovery.) However, without such occasions in abundance, there is little to impart to others. Moreover, without frequent opportunities to recognize that one knows something that the others don't know, devious reasons for or against imparting information cannot even exist—let alone be recognized and acted upon. I can think of no way of describing this critical simplicity in the Umwelt of the vervets, this missing ingredient, that does not avail itself explicitly or implicitly of higher-order intentional idioms.

In sum, the vervets couldn't really make use of most of the features of a human language, for their world—or you might even say their lifestyle—is too simple. Their communicative needs are few, but intense, and their communicative opportunities are limited. Like honeymooners who have not been out of each other's sight for days, they find themselves with not much to say to each other (or to decide to withhold). If they couldn't make use of a fancy, human-like language, we can be quite sure that evolution hasn't provided them with one. Of course, if evolution provided them with an elaborate language in which to communicate, the language itself would radically change their world, and permit them to create and pass secrets as profusely as we do. And then they could go on to use their language, as we use ours, in hundreds of diverting and marginally 'useful' ways. Without the original information-gradients needed to prime the evolutionary pump, however, such a language couldn't get established.

So we can be quite sure that the MIO grunt, for instance, is not crisply and properly translated by any familiar human interchange. It can't be a (pure, perfect) command or request or question or exclamation because it isn't part of a system that is elaborate enough to make room for such sophisticated distinctions. When you say 'Wanna go for a walk?' to your dog and he jumps up with a lively bark and expectant wag of the tail, this is not really a question and answer. There are only a few ways of 'replying' that are available to the dog. It can't do anything tantamount to saying 'I'd rather wait till sundown', or 'Not if you're going to cross the highway', or even 'No thanks'. Your utterance is a question in English, but a sort of melted-together mixture of question, command, exclamation, and mere harbinger (you've made some of those going-out-noises again) to your dog (Bennett 1976, 1983). The vervets' MIO grunt is no doubt a similar mixture, but while that means we shouldn't get our hopes too high about learning Vervetese and finding out all about monkey life by having conversations with the vervets, it doesn't at all rule out the utility of these somewhat fanciful translation hypotheses as ways of interpreting—and uncovering—the actual informational roles or functions of these vocalizations. When you think of the MIO as 'Coast clear?' your attention is directed to a variety of testable hypotheses about further relationships and dependencies that ought to be discoverable if that is what MIO means—or even just 'sort of' means.

Deception

15

Tactical deception of familiar individuals in baboons*
RICHARD W. BYRNE and ANDREW WHITEN

The topic of deception merits a number of chapters, for it is a particularly sensitive yardstick for the depth of Machiavellian intelligence a species can display. In this chapter we introduce the concept of Tactical Deception and detail the concrete examples out of which it arose.

In this paper we detail several samples of baboon behaviour, the significance of which is summarized in our title. 'Deception' is accepted as a well-established phenomenon in animals and plants (Wickler 1968; Hailman 1977; Krebs and Dawkins 1984). By its very nature, it involves the use of an activity or body form which occurs also as an 'honest' counterpart, generally of higher frequency. However, the relationships between honest and deceitful versions, and indeed between deceiver and victim, vary widely.

The term 'tactical' is used here to emphasize a contrast between the very short-term changes occurring in a single animal between deceitful and honest versions of behaviour, and those involved in common types of deception described in the literature. Dawkins and Krebs (1978) note that 'interspecific deceit is so well known as to pass almost without comment', but that 'ethologists have failed to find many unequivocal cases of successful intraspecific deceit'. Mimicry and camouflage account for much of the former, where the deceptive and honest versions of anatomy and behaviour typically occur in different species and therefore cannot, of course, be flexibly alternated by one individual. Even in the less well documented intraspecific case, examples of honest and deceptive versions of behaviour tend to be the alternative strategies adopted by different individuals, as in the fairly common ruse of some males who mimic females, such that other males are misled (e.g. scorpion flies: Thornhill 1979). It is with such long-term 'strategies' that we contrast short-term 'tactics' in which the deception uses elements from an honest counterpart in the individual's repertoire.

---

* Reprinted from *Animal Behaviour* 33, 669–673 (1985).

Finally, our title emphasizes that our data refer to interactions between familiar individuals exhibiting long-term social relationships. Any deception must be frequency-dependent: in particular, 'misleading signals must occur only rarely in relation to correct ones' (Wiley 1983). In the context of a baboon group, the problem is particularly severe:

> In close-knit groups of animals, where individuals recognise one another and interact with each other over extended periods, the long-term penalties which may arise if deceit is discovered may more than outweigh any short-term gain it makes possible (Slater 1983).

Cheney and Seyfarth (1980), in field experiments, have confirmed that primates discriminate not only their relationships with other individuals, but also those existing between other pairs of individuals (Chapter 6).

Tactical deception in such intimate circumstances may be under-reported in animal behaviour because, as predicted by the last paragraph, it occurs, if at all, at low frequency in any one group and with some subtlety; the data collected on it are, therefore, 'anecdotal' and for any one study the researchers are justifiably cautious or coy about publishing it. This is a vicious circle.

Accordingly, we first present a provisional definition of such 'intimate tactical deception': (i) acts from the normal repertoire of an individual, (ii) used at low frequency, and in contexts different from those in which it uses the high frequency (honest) version of the act, (iii) such that another, familiar individual, (iv) is likely to misinterpret what the acts signify, (v) to the advantage of the actor.

Secondly, Table 15.1 presents candidates for this category of behaviour, observed during an 18-month field study by ourselves and P. Henzi on two groups of baboons.

The necessary third step is classification, and some possible bases for this are indicated in the table. Any attempt at a rigid taxonomy would be premature at this stage; however, when a fuller corpus has been assembled, at least the following issues will need to be addressed.

Firstly, phylogenetic distribution: does intimate tactical deception occur in all taxa in which the necessary complexity of social groupings exist, or are there other limitations on whether animals are capable of showing such sophisticated behaviour? Chimpanzees are described as masters of social sensitivity and deception (de Waal 1982; Kummer and Goodall 1985), but at present one cannot be sure whether their uniqueness lies in their mental abilities or in the ease with which human observers perceive subtle and complex behaviour in a species so closely related to themselves. Indeed, the relationship of these behaviours to lying and other intimate tactical deception in man needs to be clarified: see the illuminating discussion by Dennett (1983), which also contains an anecdote from vervet monkeys, similar to our type 2.

Table 15.1 Examples of intimate tactical deception

Background	Examples	'Honest' context from which act borrowed	Costs to directly manipulated individual X	Cost to third parties	Modality	Use of individual as 'social' tool to manipulate others
Type 1: here, the actor is a juvenile who uses screaming, normally associated with attack or threat by another individual, to gain access to underground food items which normally requires extensive digging or loosening of turf from hard ground	(a) (16 Sept 1983) Adult female ML is digging, probably to obtain a deep growing corm. Young juvenile PA approaches to 2 m and looks at her, then scans round; no other baboons are in view. PA looks back at ML and screams. Adult female SP runs into view towards them, then chases ML over a slight cliff and out of sight. SP, who is PA's mother, normally defends him from attack. When both females are out of sight PA walks forward and continues digging in ML's hole (b) (8 Sept 1983) Adult female ML is feeding where a patch of turf has been loosened. Young juvenile PA tentatively approaches and although ML makes no threat, PA screams. As happens in the normal context, JG, the only adult male of the group, runs towards them and ML retreats, leaving PA to feed on food source. Two minutes later this sequence is repeated. Five minutes later the same sequence recurs. JG again acting as if the vocalization signified an attack by ML on PA, but this time he chases ML for 20 m, jumps upon her and lunges as though to bite her (c) (19 Sept 1983) Adult female SP and her young juvenile PA are feeding about 60 m from JG, the adult male. PA screams and JG run toward them, but I (RB) cannot see the detail of the interaction until I can get closer to them, when PA is feeding in a loosened patch of turf, SP is walking rapidly away and JG is watching her. The actors' positions are just as in (b), except that PA's mother SP is departing instead of female ML	Threat or attack from another individual	Support in feeding competition	Loss of food	Vocalization	Yes

Table 15.1 Continued

Background	Examples	'Honest' context from which act borrowed	Costs to directly manipulated individual X	Cost to third parties	Modality	Use of individual as 'social' tool to manipulate others
Type 2: here, exaggerated gestures of 'looking', normally used when a potential predator or a baboon troop has been sighted by one individual, are used by juveniles to avoid attack by an adult male	(a) (23 May 1983) Subadult male ME attacks one of the young juveniles who screams repeatedly, pursuing ME while screaming (this is common when aid has just been successful solicited). Adult male HL and several other adults run over the hill into view, giving aggressive pantgrunt calls; ME, seeing them coming, stands on hindlegs and stares into the distance across the valley. HL and the other newcomers stop and look in this direction; they do not threaten or attack ME. No predator or baboon troop can be seen through 10 × 40 binoculars (b) (8 Dec 1983) Adult male DV is chasing older juvenile male SZ aggressively. Seeing me (RB), SZ leaps into the air and stares at me—an immense, comical over-reaction. DV stops chasing SZ and also stares at me, then begins eating. However, for the next 30 s, SZ remains standing on hindlegs staring at me, casting quick glances at DV who is still eating	First appearance of baboon troop or possible predator	Attack by X diverted	Loss of support in agonism None	Looking	No

Tactical deception of familiar individuals in baboons

Table 15.1 *Continued*

		Recruitment of support in attack on animal dominant to agent	Attack by X diverted	Attack by X	Looking	(In passive sense)
Type 3: these examples concern the use of aid-solicitation, normally used to recruit support from a third party against a dominant animal, to divert aggression towards a subordinate and away from the agent	(a) (15 July 1983) Adult male JG displaces female PK from her feeding patch. She responds by immediately enlisting JG, by characteristic rapid flicks of gaze from potential ally to target, to attack juvenile PA. PA (who is not PK's offspring) was merely feeding 2 m away, but is threatened and chased by JG. PK meanwhile continues feeding on her patch (b) (23 Aug 1983) Female PK is attacked by adult male JG, and responds by enlisting JG against juvenile PA, who was once more uninvolved. However, this time JG does not threaten PA, though he halts his attack on PK					
Type 4: at the time of this example, the younger male HL was in the process of taking over the troop from the older male DV: HL was dominant and monopolized oestrous adult females, yet DV still led and determined troop movements, especially in the range periphery	(6 July 1983) The adult male DV is displaced by the adult male HL on what appears to be a rich source of underground food items in loose turf. HL allows some other adult individuals to feed here, but DV rises and moves off briskly. The group are at the edge of their home range in an area they rarely visit and where DV has led them, sc within 3 min, all six individuals at the turf patch, including HL, move away following the direction in which DV departed. Two minutes later DV swings around in a circle and returns to feed by himself on the turf patch	Troop leadership	Loss of food	Loss of food, in those individuals tolerated by X at patch	Locomotion	No

Secondly, evolution: in all types of intimate tactical deception discussed above there is a clear benefit to the individual who is the agent of the deceptions. However, there are costs to other individuals, some of whom are relatives of the agent. For instance, in type 1 the mother and the putative father of the agent are both deceived and used as social tools, and in example (c) the actor's mother apparently loses food. The constraints imposed by kin selection on patterns of intimate tactical deception need to be explored.

Thirdly, ontogeny: with present data, one cannot know whether intimate tactical deceptions represent innovations by particular animals or simply rare items in the repertoire. Since, even in our limited data set, developmental explanations could range from shaping by reinforcement to 'intelligent' switching of solution methods between different problems, it becomes important to establish the antecedents of intimate tactical deception by detailed longitudinal study (see Byrne and Whiten 1987).

With the aim of assembling a data base from which some of these questions may be tackled, we would be delighted to hear of candidate examples of intimate tactical deception in any animal species (see Chapters 16–18; Byrne and Whiten 1988).

16

The manipulation of attention in primate tactical deception
ANDREW WHITEN and RICHARD W. BYRNE

We have recently collated records of tactical deception contributed by many primatologists working on different species. In this survey the predominance of manipulation of others' attention was striking and here we treat this topic in detail, evaluating evidence that deception is sometimes sophisticated enough to imply that the agent can mentally represent others' mental states.

Introduction: studying tactical deception in primates

Skilful deception of one individual by another is an obvious topic to focus on in the investigation of Machiavellian intelligence and social expertise, for by its very nature it seems to involve the actor being one step ahead of the unfortunate dupe in the games which occur in primate society. Deception is, therefore, a major topic addressed in this book. Here, we examine one aspect of primate deception, as revealed in a wide-ranging survey which involved data collected by many primatologists on many different primates. A phenomenon which stands out in the overall pattern of this data is the extent to which primate deception is concerned with the perception and manipulation of the attention of others, and it is on this phenomenon we focus here. It is a particularly apt topic for the present volume because it allows us to examine evidence for the contention that primates act as 'natural psychologists'. We must start, however, with an explanation of the nature of our survey.

In the beginning, we were not studying deception as such, but conducting a project essentially concerned with the behavioural ecology of chacma baboons. During this, we both observed several instances of different types of deception, and although the occurrences were few, we entered detailed records in our field notes. We found this behaviour to resemble a series of published observations, which were meagre, scattered, and almost entirely limited to chimpanzees (Goodall 1971; Menzel 1974;

Kummer 1982; de Waal 1982; Ettlinger 1983). In marked contrast, informal discussions with other primatologists repeatedly elicited a response like, 'oh yes, I've seen that sort of thing, it doesn't happen very often and I expect I miss it quite often too, but . . . ', and we were then treated to an account. For good reasons behavioural scientists are wary of trying to publish a handful of such anecdotes, but there is a danger that real natural phenomena will as a result never be recognized. In Byrne and Whiten (Chapter 15; see also Quiatt 1984) we discussed why this might be so in the case of the rare and subtle behaviours to be expected in primate tactical deception, and (taking a deep breath!) we published our own modest corpus of records.

Our longer-term aim was to gather a much wider corpus of data of this sort, which we saw as the only way to establish the reality and nature of such elusive behaviour. For this reason, we were careful to offer as clear a definition as possible of 'tactical deception' and to explain just what appeared to be special about such behaviour as we had observed it in primates, by contrast with descriptions in the literature on other groups of animals (e.g. Wickler 1968; Thornhill 1979; Mitchell and Thompson 1986). We shall assume from here on that the reader has assimilated what we include in the scope of 'tactical deception' (Chapter 15). Our next step was to distribute a brief questionnaire to 115 primatologists asking for records (including null records) of similar behaviour. Rather than pass judgement ourselves on the adequacy of the evidence for tactical deception in each submission, we simply collated all of them in a catalogue of 'candidate records' of tactical deception (Whiten and Byrne 1986).

The goal of assembling such a corpus of data is to see if any repeated patterns emerge, and the pattern which we first discerned amounts to a classification distinguishing different functional consequences of deceptive tactics (Whiten and Byrne 1988). The classification involves five major functional classes divided into thirteen subclasses, and it offers an overview of the range and scope of tactical deception observed to date (Table 16.1).

The manipulation of attention

When we took stock of the overall pattern presented by our classification of the corpus, we were struck by something which, although it was not something we were looking for initially, stands out in the picture as a whole: much of tactical deception in primates is concerned with the manipulation of the attention of other individuals. The most obvious point is simply that seven of the thirteen subclasses (see Table 16.1) fall under the heading of either *concealment* or *distraction*, the two major classes in which the deception is essentially about manipulation of the *target's*

Table 16.1 *Types of tactical deception*

Records of tactical deception have been sorted into five major functional classes, divided in turn into thirteen subclasses (Whiten and Byrne 1988). Illustrative records are quoted in the text. The *agent* is the individual performing the deceptive act and the *target* is the individual who poses the problem the *agent's* behaviour deals with.

Concealment

The *agent's* behaviour functions to conceal something from the *target*.

Hiding from view. The *agent* hides an object, or a part or whole of itself, by screening it from the *target's* view.
Acoustic concealment. The *agent* acts quietly, such that the *target's* attention is not attracted.
Inhibition of attending. The *agent* avoids looking at a desirable object when such looking would lead one or more *targets* to notice it.

Distraction

The *agent's* behaviour functions to distract the *target's* attention away from some locus at which it is directed, to a second locus.

Distract by looking away. The *agent* distracts the *target's* attention from one locus by looking away at another locus in such a way that the *target* also looks there.
Distract by looking and vocalization. The *agent* distracts the *target's* attention by looking away at another locus and vocalizing in such a way that the *target* looks there or at least loses the original focus of attention.
Distract by leading away. The *agent* leads the *target* away from the first locus to another one, allowing the *agent* to return to the first location free of competition.
Distract with intimate behaviour. The *agent* shifts the *target's* attention to some part or extension of its own body, which is highlighted posturally or gesturally.

Creating an image

The *agent's* behaviour functions to create an impression which, rather than merely affecting the *target's* attention as above, causes the *target* to misinterpret the behaviour's significance for itself in other ways.

Present neutral image. The image is non-threatening just in the sense that it is of little or no significance to the *target*.
Present affiliative image. The image is not merely neutral, but is affiliative.

Table 16.1 *(continued)*

Manipulation of target using social tool

The *agent* manipulates one individual, the *tool*, so as to affect the *target* to the *agent's* advantage.

Deceive tool about agent's involvement with target. The *agent's* behaviour misleads the *tool* about the significance of the involvement of the *agent* with the *target*.
Deceive tool-1 about tool-2's involvement with target. The *agent's* behaviour misleads one *tool* about the significance of the involvement of a second *tool* with the *target*.
Deceive target about agent's involvement with tool. The *target* is deceived about the significance of the results of the *agent's* action on the *tool*.

Deflection of target to fall-guy

The function of this behaviour is to divert the *target* who poses a problem towards a passive victim, the *fall-guy*.

---

attention. In *concealment*, the behaviour of the '*agent*' (the individual practising the deception) functions to conceal something from the '*target*' (the individual who poses the problem which the '*agent's*' behaviour appears designed to deal with). In *distraction*, the *agent's* behaviour functions to shift the *target's* attention away from one locus in the environment, to a second locus; this is deceptive when the entity to which attention is normally drawn by the act is counterfeit or non-existent. Thus, in *concealment*, the misinterpretation of the *agent's* behaviour which is engineered in the *target* involves misinterpreting what is available for attention; in *distraction*, the *target's* misinterpretation is about what should be the highest priority for its attention.

It might be objected that this way of weighing the prevalence of attention-manipulation depends on just how the categories are lumped, or split. However, the same picture emerges if we simply count the number of records catalogued under each class: *concealment* and *distraction* together account for 50, whereas the other three classes include only 29. We must be cautious about this alternative measure of prevalence, too, because in some cases a record really does refer to a single behavioural incident, in others it is just an example of a behaviour which the informant reports to have seen a number of times — often an indefinite number. The point is that, in the present tentative state of this corpus of data, there appears to be a prevalence of both raw records and different forms (subclasses) of *concealment* and *distraction*. In addition, the other classes of deception often require the monitoring or manipulation of attention to underpin the

The manipulation of attention in primate tactical deception 215

further deceptive components which distinguish these categories from *concealment* and *distraction*, as we shall see below.

Primates as psychologists: representing the attention of others

In some records there is evidence that the *agent's* behaviour can be sufficiently finely tuned to the task of manipulating the *target's* attention that we are forced to credit the *agent* with the ability to mentally represent attentional states in the *target*. The idea of the primate as 'psychologist' or 'mind reader' (Premack and Woodruff 1978; Humphrey 1980; Krebs and Dawkins 1984; see also Chapter 5) thus starts to get purchase in specific records of behaviour. Consider the following instance of *hiding from view* (index numbers refer to our 1986 catalogue):

The unit was resting. An adult female spent 20 minutes in gradually shifting in a seated position over a distance of about 2 metres to a place behind a rock about 50 cm high where she began to groom the subadult follower of the unit—an interaction not tolerated by the adult male. As I was observing from a cliff slightly above the unit, I could judge that the adult male leader could, from his resting position, see the tail, back and crown of the female's head, but not her front, arms and face; the subadult male sat in a bent position while being groomed and was also invisible to the leader. The only aspect that made me doubt that the arrangement was accidental was the exceptionally slow, inch by inch shifting of the female. This had in fact caused me to focus on her behaviour so long before she had reached the final position (No. 26. Kummer, Hamadryas baboons).

Kummer's observation shows that the female carefully adjusted her behaviour to a goal-state in which the *target* (the adult male leader) was unable to attend to what her hands were doing. The implication is that she was acting so as to minimize the discrepancy between a mental representation of this goal state, and a representation of the *target's* current ability to attend to her hands. We suggest that this gives us one clear-minded approach to the otherwise slippery issue of what it really means to say a primate is acting as a natural psychologist. Natural psychologist is a justifiable description where an explanation must be invoked which involves the *agent's* brain minimizing discrepancies between representations of the ideal and actual mental state of the *target*.

Further illustrations of *hiding from view* which display the fine engineering of deceptive behaviour to the attentional states of others have been provided by de Waal (Nos. 66, 67, 99; 1982, 1985, chimpanzees) and the following (No. 100) sums up the expertise involved:

Often a low-ranking male will sit with his upper arm resting on his knee and his hand loosely hanging down so that a female in front of him can see his erect penis, but apes on the side cannot see it. This inconspicuous form of concealment occurs together with quick glances at dominant males. Needless to say, the subordinate always uses the hand on the body side which is turned towards the dominants.

Following Dennett's distinctions between different orders of intentionality (Chapter 14), we can discriminate orders of 'nesting' in mental representations about the world. A first-order representation would be one which coded simply for states of the physical world, like the location of a certain food, or the fact that an individual often behaves aggressively. A psychologist-primate, like the baboon and chimpanzees whose *hiding from view* has been described above, would be capable of second-order representations, coding for mental representations in other individuals, such as their actual or potential focus of attention. We shall consider third-order representations later; for the moment let us examine the scope for second-order representations in the different types of tactical deception we have distinguished.

The most general observation is that in all cases it is necessary for the *agent* to monitor the actual or potential attention of the *target* to the *agent*. However, although evidence like that reviewed above implies that it is within a monkey's competence for this always to be achieved through second-order representations, in many specific records it is impossible to discount alternative and simpler explanations. For instance, there are other records of *hiding from view* which involve the *agent* hiding totally from the *target* (mating surreptitiously being a common example), and then it may be that the *agent* is monitoring just the presence of the *target*. The importance of the records of partial *hiding from view* is thus precisely that they rule out the use of such first-order representations.

The position is different in the case of *distraction*, which is a matter of shifting the *target's* attention from one locus to another. One or both of these will not be the *agent*, so recognition of the full-face image of the *target* presented when the *target* looks at the *agent*, which could be used for the simple discrimination '*target* facing me/not facing me', will not be adequate for monitoring all the shifts in the *target's* attention which are involved in distraction. Thus evidence for second-order representation emerges in an example of *distract by leading away* which was observed in the course of experiments where just one of a group of chimpanzees (Belle) knew the location of a food source:

If Rock got very close to the food, Belle invariably gave the game away in a 'nervous' increase in movement. However on a few trials she actually started off a trial by leading the group in the opposite direction from the food, and then, while Rock was engaged in his search, she doubled back rapidly and got some food. (No. 95, Menzel 1974).

The manipulation of attention in primate tactical deception 217

That Belle raced back to the food at locus-1, only when 'Rock was engaged in his search', implies that what she was monitoring can only be appropriately described at the level of Rock's attentional state, which had to be directed towards locus-2 rather than locus-1 before Belle could make a dash for the latter. Consider another illustration, this time from the category *distract with intimate behaviour*:

One of the female baboons at Gilgil grew particularly fond of meat, although the males do most of the hunting. A male, one who does not willingly share, caught an antelope. The female edged up to him and groomed him until he lolled back under her attentions. She then snatched the antelope carcass and ran. (No. 62; Strum, in Jolly 1985, Olive baboons).

The important point here is that the female distracted the male by grooming him 'until he lolled back under her attentions'. As in the case of Belle, there is the implication that the *agent* is able to accurately represent the *target's* waning attention to locus-1 (here, the carcass), permitting the *agent's* access to it once the *target's* attention is adequately centred on locus-2 (here, his experience of being groomed).

So far we have focused heavily on the two classes of tactical deception inherently concerned with the manipulation of attention. However, in the other classes there is usually also a necessity to monitor the attention of others. Just as in the case of *distraction*, where the *agent* must monitor not only the *target's* attention to itself, but also attention to one or two other loci, the triadic classes—*manipulation of target using social tool, and deflection of target to fall-guy*—require monitoring of more than one object. However, here there is added complexity because the objects are other individuals. Even if such interactions were managed through only first-order representations, the difficulties of simultaneously monitoring the behaviour of several other individuals could, as we argued in Chapter 5, impose special demands on working memory. In the following record of *manipulation of target using social tool*, for example, the male KMO needs to monitor the attentional states of his allies who he uses as a *tool*, of the competitor male who is the *target*, and probably also of the female prize:

From a distance KMO screams at DMU, who is consorting an oestrous female. Adult males GBA, SIO and KTE join KMO and they converge on DMU. KMO hangs back slightly as they approach DMU. When DMU finally 'breaks' and leaves the female to chase his challengers, he chases GBA, the coalition member who's closest to him. KMO runs to the female and herds her in the opposite direction. The remaining coalition members do not interfere. This example is unusual, in that KMO initiated the coalition himself and then appeared to hang back deliberately until the consorting male chased someone else. (No. 29; Rasmussen, Yellow baboons)

In addition to such complexity, there is evidence of second-order representation of the *dupe's* attention in some triadic episodes, like the series in which we observed a juvenile, in competition for a food source, to scream as if threatened, thus eliciting attack on the competitor by an ally (Table 15.1, Chapter 15). The important observation here is that the behaviour was not exhibited when the *tool* could attend to the lack of a real attack. In all cases, it was done when the *tool* was available and capable of acoustic attention to the behaviour, but not visual attention to its deceitful aspects.

Evidence for second-order representations is sparse. We need not only more records, but also better quality ones: only observations which follow Kummer's model description of the fineness of social adjustment will readily serve as evidence for representation of others' mental states. However, we would suggest that the records quoted so far may well be the tip of an iceberg which for obvious reasons is largely submerged beneath current scientific horizons. Our analysis offers some support for second-order representations, but just as important, we hope, is the explicit specification of what data is required to test such hypotheses. In this spirit of exploration, let us venture yet further, and ask about evidence for higher order psychological representations.

Evidence for third-order representations

In a third-order representational system, the *agent* would represent the *target's* representation of the representations of someone else, which could be the *agent* (see Chapter 5). Candidates for such a possibility come most obviously from the categories *inhibition of attending*, *distract by looking away*, and *deflection of target to fall-guy*. In the case of the first of these, *inhibition of attending*, for example, the *agent's* behaviour apparently takes into account the ability of the *target* to exploit the *agent's* attention, so as to guide its own:

S's group travelling slowly between feeding sites in a relatively straight line along a narrow trail. Four other animals behind S in line. S looks up into *Hypericum* tree and spies a nearly obscured clump of *Loranthus vine*. Without looking at those behind her, she sits down by the side of the trail and begins to intently self-groom until the others have passed her and all are out of sight some 15 foot ahead. Only then did S stop 'self-grooming' to rapidly climb into the tree, break off the vine clump and descend with it to the trail to hastily feed on it before running to catch up with the group. (No. 2; Fossey, gorillas).

Examining this step by step, we should first note that the ability of the *target* to follow the *agent's* gaze is certainly a good candidate for *second-order* representation, for in doing so it is as if the *target* were able to see the world through the eyes of the *agent*. By saying 'as if' here, we are not implying that in humans this is in some sense what really happens, whereas in, say, gorillas, it just looks this way; it is quite plausible that the processes are similar—and second order—in both. Like all cases of second-order representation we have discussed, what is *actually* happening is that the *target* makes observations of the surface behaviour of the *agent*—its head and eye movements—relative to the physical world; its brain must then perform complex computations on the data so provided so as to infer the attentional focus of the other individual (see Butterworth and Cochran 1980, for an analysis of what is involved in human infants developing the capacity for 'joint' visual attention with their mothers).

Inhibition of attention does not in itself constitute evidence that the *agent* is representing the *target's* capacity for second order representation; the *agent* might have learned just not to look at something they want, when others are present. However, in the record which follows, the observer's detailed noting of the glance of the *agent* at the *target* suggests that the *agent* did, in fact, know that the *target* could use his own focus of attention in the way described above:

We took to hiding some fruits, one here, one there, up in the trees; youngsters such as Figan quickly learnt to search for these whilst the adult males were busy loading up at the boxes. One day, some time after the group had been fed, Figan suddenly spotted a banana that had been overlooked—but Goliath was resting directly underneath it. After no more than a quick glance from the fruit to Goliath, Figan moved away and sat on the other side of the tent so that he could no longer see the fruit. Fifteen minutes later, when Goliath got up and left, Figan, without a moment's hesitation, went over and collected the banana. Quite obviously he had sized up the whole situation: if he had climbed for the fruit earlier Goliath, almost certainly, would have snatched it away. If he had remained close to the banana he would probably have looked at it from time to time: chimps are very quick to notice and interpret the eye movements of their fellows, and Goliath would possibly, therefore, have seen the fruit himself. And so Figan had not only refrained from instantly gratifying his desire, but had also gone away so that he could not give the game away by looking at the banana (No. 86, Goodall 1971).

We should note in passing that records of *inhibition of attention* imply also that the *agent* is representing another aspect of the *target's* attentional repertoire. *Inhibition of attention* does not occur when the *target* is likely to see the food in any case; it has been recorded only with respect to 'secret' food resources which there is a good chance the *target* will not notice. It is thus also likely that the *agent* is representing this aspect of the *target's* attentional powers—the likelihood of the *target* noticing, or not noticing, some particular item in the environment. This is not a third-order representation, but it is a sophisticated second-order one on which rests the efficient deployment of the third-order ability we have just been discussing.

If some primates are capable of third-order representations, this could also underly the *distraction* tactic of *looking away*, for here the *agent*, instead of inhibiting attention, does just the reverse:

I have seen exaggerated gestures of looking . . . directed at either an observer or where an observer normally sits in laboratory conditions when a juvenile is trying to avoid an attack by a more dominant individual. In all of these cases the observers had done nothing unusual and were just sitting recording behaviour as they had been doing for months previously (No. 25; Chamove, stump-tailed macaques).

By staring at a focal point, the *agent* encourages and exploits just that second-order capacity of the *target* to tune into the *agent's* focus which was disabled in the case of *inhibition of attention*. Again, deception could in principle be achieved without the *agent* representing the *target's* second-order capacity, for it could simply have learned to stare into the distance when in a fix. In this type of deception, records of the *agent* glancing at the *target* (Table 15.1, Chapter 15) do not carry the same weight as in the case of Figan quoted above, for they may simply function to check that the *target* has stopped its aggressive chase. More convincing would be the *agent* stopping *looking away* as soon as the *target* ceases to attend to it. Even better would be any evidence that an individual, having learned either to *conceal* by *inhibiting attention*, or *distract by looking away*, then spontaneously generated whichever of these two tactics it had not previously exhibited, implying that it was putting the hypothesized third-order representation to different use. As yet, no such observations have been contributed. The same goes for the possibility of third-order representations in the case of *deflect target to fall-guy*, where the *agent* uses its gaze to direct the *target's* attention to the *fall-guy*.

It seems appropriate to finish this section with the best candidate for fourth-order representation of attention, which involves a nice example of counter-deception:

One chimp was alone in the feeding area and was going to be fed bananas. A metal box was opened from a distance. Just at the moment when the box was opened, another chimp approached at the border of the clearing. The first chimp quickly closed the metal box and walked away several metres, sat down and looked around as if nothing had happened. The second chimp left the feeding area again, but as soon as he was out of sight, he hid behind a tree and peered at the individual in the feeding area. As soon as that individual approached and opened the metal box again, the hiding individual approached, replaced the other and ate the bananas (No. 24; Plooij).

We shall leave as a test for the student reader the specification of the sense in which this might be a case of fourth-order representation, and what further observations would be required to confirm or deny such a hypothesis!

Phylogenetic differences and the evolution of attention manipulation

Do primates show especial sophistication in attention manipulation? Is the evidence for superiority of intellect in apes versus monkeys (Passingham 1982) and monkeys versus prosimians (Chapter 3) reflected in differences in the types of behaviour examined in this chapter? As we have already stressed, quantitative comparisons involving the numbers of records in any category is a risky business because of the problem of what constitutes a unit record. However, there are several points where, when we ask about the existence of species differences in the repertoire of tactical deception, it makes sense to discuss at least the presence versus absence of records of any particular category of deception, as well as what appear to be qualitiative differences, in the corpus of records as it stands at present.

At present it is the prosimian data which tells the simplest comparative story, for our two 'prosimian respondents' simply noted that they had never observed any tactical deception. If this is a true inability, it may explain why we have so few prosimian replies — negative data is always tedious to report. The absence in prosimians of both tactical deception and, consequently, the representation of others' attention which we have argued to be often implicated is, then, a working hypothesis which other prosimian researchers are invited to dispute!

By contrast, candidate records of tactical deception in cercopithecine monkeys have been contributed for every subclass in our taxonomy. Closer inspection of these, however, indicates that further distinctions are possible between how the monkeys behave, and the patterns of behaviour shown in the great ape examples of the same subclass of deception. In the category of *concealment*, there is no monkey example which matches those for gorillas and chimpanzees which have already been given as examples of *inhibition of attention*, where what is concealed is an object. In those ape examples, it was clear that the *agent* had seen the 'secret' food, and then refrained from drawing attention to it. In the monkey records, there is only clear *inhibition of attention* to the *target* itself. Thus,

subordinates sometimes ignore threats they must have seen. They sit frozen still, with a body position indicating readiness for flight, and look around, especially slightly upward, while avoiding turning the head in the dominant's direction (although they seem to monitor the dominant by means of peripheral vision) (No. 64; de Waal 1986, Rhesus macaques)

and

ignoring as a social strategy, used in many social situations . . . may be the most common form of deceptive behaviour (comment on No. 6; Altmann, Yellow baboons).

Such behaviour does not begin to argue for an explanation involving third-order representations, unlike that concerning the secret food sources in the chimpanzee and gorilla records discussed above. It is possible that monkeys inhibit attention to others whom they wish to avoid, whereas only apes inhibit attention to resources they want (and see also Chapter 19 for support for the implication that monkeys are poor at representing object knowledge). If this monkey-ape distinction were to be confirmed by further research, a simian capacity for ignoring could, nevertheless, have been an important preadaptation for evolution of the ape's more complex ability; which, we may add, has apparently not been described in any other species. Other species clearly show *hiding from view* and *acoustic concealment*, as in the case of hunting cats, for example; but it is not so clear that they can manage such partial hiding as we have highlighted as evidence for second-order representations.

Turning to *distraction*, perhaps the closest non-primate example to those described here is that in which ant-shrikes were reported to give false alarm calls which distracted other members of a mixed species flock, allowing the ant-shrikes to steal their food (Munn 1986). This meets our definition of tactical deception and seems to be a clear case of distraction. However, evidence of second-order representations is not easily forthcoming in the case of broadcast signals and none is apparent in Munn's account.

Within the primate distraction corpus, there is again one category in which the ape records are more impressive than the simian ones. In the chimpanzee cases of *distract by leading away*, the repeated and skilful use of the tactic implies that the *agent* is indeed pursuing the goal of distracting the *target*:

In other trials when we hid an extra piece of food about 10 ft away from the large pile, Belle led Rock to the single piece, and while he took it she raced for the pile. When Rock started to ignore the single piece of food to keep his watch on Belle, Belle had temper tantrums (No. 96; Menzel 1974).

By contrast, the monkey examples are either isolated incidents, or involve a less clearly defined effort to lead (see Table 15.1, Chapter 15 for an example), such that we can have little confidence in dismissing the explanation of opportunism, exhibited once the *target* has left the resource.

Surprisingly, perhaps, there are other cases where, according to our corpus of records, chimpanzees do not show a subclass of deception for which we do have monkey records. Thus, the only chimpanzee case of *distract with intimate behaviour* was an isolated and odd record of stone-throwing. Even more surprisingly, the inherently triadic classes are so far bereft of chimpanzee records—strange for a species with such a reputation for triadic manipulation! An apparent absence of records of

The manipulation of attention in primate tactical deception

such behaviour in non-primates is consistent with a lack of such triadic interactions as are involved in alliances (see Chapters 5 and 11), but this can hardly function as an explanation in the case of chimpanzees. Perhaps then these forms of deception are only common at intermediate levels of social intellect; although chimpanzees may be clever enough to perform the acts, their companions may be too clever to be fooled! (Are chimpanzees third-order, rather than just second-order psychologists?) This possibility is plausible in the case of *distract with intimate behaviour*, where it is inherently difficult for the *agent's* deception not to be found out. In the monkey examples of *manipulation of target using social tool* (and also in *deflection of target to fall-guy*), the *target* seems extraordinarily gullible. Future contributions to our catalogue will show whether chimpanzees, and indeed other great apes, are too socially intelligent to be duped by these forms of tactical deception.

17
Deception and social manipulation in symbol-using apes
SUE SAVAGE-RUMBAUGH and
KELLY McDONALD

In humans, deception can take the elaborate, linguistic form of lying. This chapter shows that symbol-using apes use this facility to indulge in particularly intelligent deceit, but it also details many other forms of deception—often no less impressive—which become apparent in the intimate ape–human relationships which ape language research programmes can generate.

What is deception?

Deception can take any number of forms in the human species, but most often, when we speak of being deceived, we mean that we have been lied to. That is, someone has represented something inaccurately. However, when misrepresentation occurs we treat it as deception only so long as it appears that intent was involved. We deem it important that the deceiver realizes things are not as he or she suggests. It must also be the case that the erroneous information is put forth with the explicit purpose of making us behave differently than we would if we knew the real state of affairs. If the agent is not attempting to make us believe X is true, when in fact it is not, then we rarely apply the term 'deception'. Thus, someone may tell us how to go to Main Street, but if we find their directions inadequate we do not assert that we were deceived unless we believe that they had something to gain by preventing us from finding Main Street.

Deception cannot involve the misrepresentation of real and present events, for such events are too easily verified and the deceiver readily embarrassed. Deception must involve the characterization of things that are not present or self-evident so that the deceived party has no immediate means of invalidating the information.

Among non-human primates, what sort of things are likely to be topics of communication when the information to be presented is not self-evident? It seems that the most probable candidate is information about future actions. Primates might indicate that they are about to do Y when in fact they really plan to do X. Whiten and Byrne's category of concealment (Chapter 16) corresponds to this type of deception. Take for example, Fossey's female gorilla who stopped travelling and acted as though she was going to groom herself. Instead, she obtained an obscure and luscious clump of *Loranthus* vine once the others had passed by without noticing it.

Other examples offered by Whiten and Byrne are less clear. For instance, in 'ignoring' the target, one may ask: Is the agent really presenting a false intent? A mother who ignores the milk-begging gestures of her offspring at weaning may be simply saying 'No, you cannot nurse'. In human terms, the act of ignoring is often an intentional means of snubbing, rather than an attempt to cause the recipient to believe they were not heard. The human use of the term 'deception' places a premium on the intent of the agent and it would seem wise to make a similar distinction as we look for roots of this behaviour in non-human primates.

It is interesting and perhaps significant that nearly all of Whiten's and Byrne's examples of deception are, at one level or another, deceptions about immediately imminent social actions. That is, they entail situations in which an animal's behaviour does not render a veridical account of what he/she is about to do.

The one type of behaviour which cannot be classified as a deception about impending action is that of 'pretending' to see or hear something that is not there. This sort of behaviour is found in a variety of different categories in Whiten and Byrne's classification system; however, they all have in common the fact that an animal is conveying inaccurate information by acting as though an event has happened when in fact it has not. Such an animal is not conveying false intent, but rather false factual information about the environment. It is noteworthy that in all instances of this sort, false information is conveyed in only one way, by the agent acting as though he sees or hears something that is of general concern when in fact there is nothing there.

It is probably significant that in Whiten and Byrne's schema there are no examples of one animal misleading another by giving the false location of a genuine object. There are no examples of one animal pretending to have done something that was not done, or pretending to know something that it does not know. All of these forms of deception are common among human beings and their absence among non-human primates suggests that language plays an important role in man's ability to deceive his fellow man.

Does deception occur in symbol using primates?

What of non-human primates who have learned to use symbols? Have they also learned to become more artful in the ways of deceit? Do they engage in a greater variety of deceitful acts than their less educated peers?

One thing that can be said of such apes is that the individuals whom they deceive are generally human beings. Because of the close association between humans and apes in these projects, it is possible to obtain reports from human beings who feel themselves to have been intentionally deceived. This takes at least half of the inference out of any given example of deceit. The human being can verify whether or not they were in fact deceived though they cannot verify the ape's intent to deceive. (In spite of the fact that some language using apes have achieved remarkable skills, none of them is so advanced that they can be queried with regard to whether they 'intended to deceive' and be expected to produce a reliable report.)

Over the past 15 years there have been many opportunities to observe deception in apes at the Language Research Center. During this time five common chimpanzees (*Pan troglodytes*) and four pygmy chimpanzees have been housed and reared in the research programme. (For more complete descriptions of this work see Savage-Rumbaugh 1986; Savage-Rumbaugh *et al.* 1986). The best way to present a picture of how symbol-using apes deceive is to give some examples. Many of these examples are taken from the behaviour of Kanzi, a pygmy chimpanzee (*Pan paniscus*). Other examples are taken from Kanzi's mother Matata, who was wild-caught as a juvenile and who learned a few symbols. Still others come from Sherman and Austin, common chimpanzees (*Pan troglodytes*). None of the chimpanzees frequently uses symbols to deceive or lie, though this has happened on occasion. More frequently, deception takes simpler forms. In cases where symbols are used, the symbols themselves are geometric designs located on a board which is connected to a speech digitizer. When a symbol is touched, a spoken word is produced. Each symbol is equivalent to a single English word and the chimpanzees 'talk' by touching different symbols. Sherman and Austin were taught to do this (Savage-Rumbaugh 1986), but Kanzi has learned without being taught. His linguistic knowledge is acquired through observing how people use the keyboard to communicate with him and with each other.

In describing instances of deception, it is often necessary to impute motivational states to the perpetrators, an act that is frequently criticized as 'anthropomorphic'. Anthropomorphisms (the attributing of human characteristics to animals) are frequently discarded out of hand by many researchers who view any attribution of feeling or intent to animals as the antithesis of science. However, when describing the chimpanzee, an animal that is an *anthropoid*, and one which shares 99 per cent of our DNA, we are faced with the fact that the majority of behaviours that are found in man are also found in the chimpanzee. That is, many supposedly 'human characteristics' also happen to be 'chimpanzee characteristics'. If one avoids describing such behaviours for fear of committing the dreaded error of 'anthropomorphism' it becomes impossible to adequately convey what chimpanzees are doing. Furthermore it denies in them, *a priori*, behaviours which most assuredly exist.

The facial anatomy of the chimpanzee is very similar to that of the human being with the exception of the large protruding jaw and brow ridges. The muscles which move the different facial structures and the nerves which innervate these muscles are nearly identical to those in man. For this reason, once one is visually accustomed to the facial alteration caused by the large jaw and brow ridges, it becomes as easy to read the chimpanzees' facial expressions as it is to read human facial expressions. This is especially true in the pygmy chimpanzee, whose brow and jaw are much smaller. Expressions in this species are generally interpreted accurately within a few days even by people who have never seen these animals before.

The circumstances which promote different feelings in chimpanzees vary, just as they vary across different human cultures. Nonetheless, feelings of anger, happiness, sadness, fear, hesitancy, apprehension, interest, etc., are evoked by markedly similar circumstances in ourselves and in chimpanzees. Chimpanzees are less likely than human beings to hide their displays of emotion, and consequently, one can often tell straight away how they feel. The ability to accurately read such emotions develops rapidly when one is engaged in direct social encounters with chimpanzees. Chimpanzees can lie, by hiding their emotions; they find this is much easier to do with strangers, than with individuals who know them well.

Doing things without being observed

Deception is a strategy that is not used by very young apes, but it becomes increasingly sophisticated with age. Prior to 10 months of age no instances of deception have been noted. Around 1 year old, the first realization that others have to be watching you to know what it is that you are doing, appears. At this point, infants begin to attempt, when no one is watching, to do things that they have been told not to do. For example, Panzee at 1 year of age often tried to drink and eat from the plates and glasses of others, but she was given her own glass and plate, and asked not to use those that belonged to others. She then began to monitor very closely whether this behaviour was being observed or not. If not, she would take someone else's glass from the table, drink out of it quickly and return it to the spot where she had found it, all the while watching the other party to make sure she was not seen. At 1 year of age, she could monitor only one person, no more. Thus, if a second person were in the room, she did not also check them to see if they were watching. By contrast, it did matter to older juvenile animals if three or four people were in the room; they could and would monitor everyone present before doing something that they did not want to be seen doing (cf. Chapter 5).

Austin has, across years, perfected a means of deceiving Sherman that he uses only at night. Sherman has always been somewhat scared of the dark and hesitant to go outside at night. Austin, who is smaller than Sherman and subordinate, is not afraid of the dark and he has learned to use Sherman's fear to his own advantage by making strange noises—particularly if Sherman is bullying him indoors. Through the use of these noises, Austin can, at night, reverse the normal dominance order between them. He accomplishes this by going outside and making unusual noises which sound as though someone is scraping or pounding on the metal. After making such noises Austin runs back indoors and looks outside as though there is something out there to fear. Sherman then becomes fearful, runs over and hugs Austin, and stops the bullying. It is extremely difficult to observe Austin making these unusual noises since he stops immediately if he sees anyone watching him. However, he was observed once making a very soft tapping noise on a pipe that was hollow. He then quickly scurried inside and looked back in that direction as though he had heard a very suspicious noise. Other noises have been very unusual, but it has been impossible to observe how Austin makes the sounds. He must produce them, however, as no other individual is present and the noises occur only when Austin is outside.

Similarly, both Austin and Sherman have, across the years, perfected a number of different ways of getting out of the cage—but they never allow themselves to be observed when actually engaged in these attempts. For example, they discovered that by repeatedly slamming a heavy duty plastic cube against the lexan wall (lexan is an extremely strong clear plastic building material) they could fatigue the material to the point that it would break. Heavy pounding sounds were heard from the area, but every time anyone would enter, Austin and Sherman were found sitting quietly on these cubes, looking about as though they did not know what was making all the noise. When we left, the noise would resume. They apparently attempted to conceal these activities because they did not want to be seen when the lexan actually gave way and provided them the opportunity to escape.

Matata's most frequently used deceptive tactic was that of attempting to gain social advantage by making it look as though someone else had grievously wronged her when no such thing had taken place. A common strategy was to send me out of the room on an errand, then while I (S.S-R unless otherwise stated) was gone she would grab hold of something that was in someone else's hands and scream as though she were being attacked. When I rushed back in, she would look at me with a pleading expression on her face and make threatening sounds at the other party (see Type 1 examples, Table 15.1, Chapter 15). She acted as though they had taken something from her or hurt her, and solicited my support in attacking them. Had they not been able to explain that they did nothing to her in my absence, I would have tended to side with Matata and support her as she always managed to appear to have been grievously wronged.

Deception in Kanzi

Kanzi engages in a wider variety of deceptive tactics than Austin, Sherman, or Matata. Perhaps this is a reflection of his greater communicative skills, his overall intelligence, or simply his personality. Many of his deceptions are difficult to verify at this point, such as when he promises to be 'good' in response to our queries, by saying 'good' at the keyboard and assuming an angelic expression. When he does not keep his promise, it is not yet possible to determine whether he really does not understand the implications of such a promise, or whether he is intentionally attempting to deceive. In the descriptions which follow, we attempt to concentrate on things that are less difficult to interpret. Nonetheless, it should be noted, that Kanzi, like many young children, does use symbols to make promises that don't get fully carried out.

Hiding self

Kanzi enjoys hiding himself. This behaviour has occasionally been observed also in Matata, Sherman, and Austin, but it is much more frequent and elaborate in Kanzi. He will hide in bushes, behind trees, under blankets, in cabinets, under tables, and anywhere else he can find. He often asks one caretaker to hide with him, while another is left to find them both. If the searcher gets too close, Kanzi may quietly move to a second hiding spot.

The first time that such hiding behaviour occurred we were genuinely fearful that something had happened to Kanzi, not realizing that he was capable of intentionally hiding and keeping quiet for a very long time. On this morning one of the staff came in and found Kanzi asleep in bed. After checking on him, she went to get some things ready for the day while waiting for him to wake up. When she went back into the bedroom he was gone. She looked under the bed and everywhere in the room. Thinking that perhaps he had somehow walked past her into another room without her seeing him, she quickly began to search all of the other rooms of the laboratory, but could not find Kanzi. Concluding that he must somehow have gone outside she rushed outdoors and began looking everywhere, even on the roof, but still could not find him. By this time she had been searching for at least 20 minutes and became afraid that something had happened to Kanzi. She went back into his bedroom once more and noticed that one of the blankets that Kanzi used to build his night nest moved just a little bit. She pushed on them, and immediately Kanzi appeared from under this pile of blankets with a big grin on his face. He had been hiding the entire time by lying flat on his stomach and pulling the blankets over him, so that only a pile of blankets appeared to be on the bed.

Hiding objects

Another frequent type of deception is object hiding. For example, on one occasion Kanzi wanted to go outdoors and play early in the morning before enough people had arrived to be able to supervise him. I was watching Kanzi and two babies and did not want to try and watch all three chimpanzees as they played about outside, since, like very young children, they can rapidly put things in their mouths that might be dangerous, wander in front of vehicles, etc. Kanzi repeatedly asked to go outside, but was told that we could not go right then. I informed him that in the meantime he could sit in the outdoor enclosure and watch what was happening outdoors until it was time to go out. The door to the outside is surrounded by a wire enclosure to keep chimps from running outside unsupervised. A small square tool, generally attached to a large knob, must be used to open this enclosure. Kanzi can easily and quickly insert this tool in the proper place on the enclosure door and open it, so we try not to leave it lying about.

Kanzi walked over and sat down in the wire enclosure at my suggestion. I noticed a strange look on his face, as if he were planning something. As soon as he saw me looking at him this expression vanished. Thinking that perhaps someone had left the opening tool on the floor I walked over to look because I did not want Kanzi running out alone. I searched the entire enclosure while Kanzi watched me very patiently and with a most innocent expression on his face. I asked him if he had seen the opening tool and he did not respond. I then asked him to help me look for it, and he began to act as if he was diligently helping me search for it. I did not find it. Finally, after satisfying myself that the opening tool could not possibly be there I walked out and turned my back on Kanzi for a moment to attend to the younger chimpanzees. Within 30 seconds after I had taken my eyes off him, Kanzi produced the tool from nowhere and let himself out. I never discovered where he had hidden this implement, but he had surely hidden it well for I had thoroughly searched him and the area just moments before he opened the door.

Mushrooms are another object that Kanzi repeatedly hides. He is not allowed to eat wild mushrooms for fear that he will accidentally ingest a poisonous one. However, he loves them and, consequently, has devised many ways to hide them. Often he will secretively conceal them between his thumb and forefinger as he walks, never breaking stride so that it is not obvious that he has picked a mushroom. He will then keep the mushroom hidden in his hand until the person who is with him looks elsewhere for a moment. Just at that moment, Kanzi will quickly pop the mushroom into his mouth. There is no way to tell either that he has picked it or that he has eaten it unless one should look back and catch him chewing it before he has swallowed it. The only tell-tale sign is the distinct mushroom odour which lingers upon his breath.

Once Kanzi has secretively picked and managed to hold onto a mushroom without being seen, he will create opportunities to consume it, if they do not arise spontaneously, by hiding behind or under objects, being careful to do so in a very casual manner. For example, one day Kanzi was walking about outside looking for acorns and wild strawberries which he is allowed to eat. I was watching him to see that he did not pick up any mushrooms and put them in his mouth when I noticed that he kept spending a bit of extra time behind trees that came between us as we walked. I waited till he did this once more and then sprang behind the tree myself to find him eating a mushroom that he had hidden in his foot.

On occasion, Kanzi seems to take great delight in flaunting his ability to pick and eat mushrooms without being detected. He has, for example, picked a mushroom without being seen and then right while someone is looking at him he will hold it up to show it off. Before anyone can grab it, however, he will quickly pop it in his mouth, apparently taking great satisfaction in his ability to have displayed the forbidden treat before eating it. Generally, this sort of behaviour is reserved for mushrooms that he really does not want to eat anyway since he knows that we will make him open his mouth and spit out the mushroom if possible.

At times, Kanzi will go to even greater lengths to display his ability to eat mushrooms. One day he was playing in the woods beneath a treehouse when he suddenly seemed very quiet. I asked him to come back up to the treehouse as I could not see what he was doing underneath. He did not come. I repeated my request several times and waited about 10 minutes for him to come. When he did not I finally climbed down to see what he was doing. As I got far enough down to look under the treehouse I spied Kanzi sitting there with a 6-inch mushroom in hand. As soon as he saw me, he held it up to show off and then took a big bite. Kanzi must have been waiting for me with the mushroom in his hand the whole time, since he had been in that same spot for 10 minutes and could not have failed to notice such a large mushroom.

Kanzi not only hides objects that are of interest to him, but also objects that are used by others in ways that he does not like. For example, one effective way of stopping Kanzi from picking a wild mushroom is to toss a rock in the direction of the mushroom. This generally causes him to dodge and keep away from the mushroom long enough for someone to reach the spot and destroy it. Since we can never be sure when Kanzi might spy a mushroom and run ahead to grab it, we may carry a rock to be prepared. Kanzi has noticed this and will often deceive us by using the keyboard to ask for the rock. He acts as though he wants to incorporate the rock into his nesting activity, but when no one is looking, he hides it. Often these rocks are never found, sometimes they are found quite well hidden. For example, one small rock was found hidden in an old mop bucket under several scrub brushes that had been carefully placed over the rock to conceal it.

Kanzi also hides objects apparently for the fun of it, when he does not appear to be attempting to deceive anyone. He frequently hides balls and peanuts in the woods, and remembers the locations days later. That he remembers the locations, as opposed to simply happening upon the object later is evidenced by the fact that he will use the keyboard to announce that he is going to get a 'ball' or some 'peanuts' and then lead directly to a place where only he knows that these items are hidden. Just as we begin to think that he is confused and doesn't know what he is talking about or where he is going he will reach into a pile of leaves or pine needles and produce a ball or some peanuts that was hidden days earlier when he was with another person.

Non-existent objects

The hiding of objects takes on an especially interesting form when it becomes a game that is played with pretend objects in place of real ones. Kanzi frequently pretends to hide invisible objects in piles of blankets or vegetation. Later he will take them out and pretend to eat them, at times even acting as though he has gotten a bad bite. He then spits out the invisible object and comments 'bad'. All the while there is really nothing there. Kanzi also engages the participation of others in these 'invisible object' games by giving them the pretend object and then watching to see what they do with it.

Sherman and Austin engage in similar pretend eating games. They have also pretended that a fearsome animal was housed in an empty cage by making Waa barks and attacking the empty cage. This behaviour was not demonstrated for them, but occurred after they watched a tape of King Kong who was housed in a similar cage.

Deceptive distraction

Deceptive distraction is a favourite tactic of Kanzi's during keep-away games. When he is having difficulty obtaining the target object (often a ball), he will feign disinterest in the game by acting as though his attention has wandered to something else. This often results in the other party putting the ball down while approaching to see what Kanzi is doing. Immediately, Kanzi will grab the ball and run away with it, suddenly intensely interested in the game once again because he now has the ball. If the tactic of appearing disinterested does not work, Kanzi will then do something forbidden, such as take a handful of vaseline out of the bottle, spill the soap on the floor, etc. As soon as the other party drops the ball to attend to the problem which Kanzi has caused, Kanzi immediately grabs the ball and runs away with it.

Tactics of deception are used not only when others have the ball, but also when Kanzi has the ball. He deceives us by acting as though he has dropped the ball and does not notice, then just as someone reaches for it, he will, with great speed and dexterity, shove them aside and receive the ball. Such games of keep-away can become very complicated if both parties utilize tactics of deceit. When I have a great deal of trouble getting the ball from Kanzi, I may act as though I am no longer interested in the game and really want to do something else. Simultaneously, in order to bait me, Kanzi will pretend to drop the ball about 4 feet away from me and act as though he too is no longer trying to keep the ball away. However, he will be careful to stay within reaching distance of the ball should I try to grab it. We are then in a sort of stalemate, me attempting to look disinterested until Kanzi moves far enough away for me to grab the ball, and Kanzi trying to look disinterested, though he is really watching my every move. Such stalemates may last 3–10 minutes during which both of us continue to feign disinterest in the ball, then suddenly we will both dive for the ball. Before that time, every glance and slight body movement is weighed by both parties, and both attempt by every move and glance to deceive the other in a game where deceit is the ground rule.

Deception at the keyboard

Although it is infrequent that Kanzi uses the keyboard to deceive, it does occur—though it is always difficult to ascertain for certain that he has indeed used symbols to intentionally 'lie'. Nonetheless, the human recipients of such deception believe that Kanzi's demeanour in such situations reveals that he has purposefully misrepresented his intentions.

One way in which Kanzi has repeatedly used lexigrams in a deceptive manner is to induce others to travel with him to locations that they did not intend to go, or to places that he just been told he could not go. Kanzi's presence can prove disruptive in some portions of the laboratory when other studies are in progress with other subjects. Consequently, Kanzi is allowed in these areas only during certain times. He does not understand this constraint and requests to go at all times; consequently, it is frequently necessary to deny these requests. As he has grown older, he has begun to use deceptive requests when he realizes that it is not possible to go where he wishes.

For example, on several occasions Kanzi has asked to visit Sherman and Austin. When this request is denied, he has responded by asking to go get melon instead. Melon is a food, like many others, that can be found at a specific location in the woods around the laboratory. Kanzi has learned the locations of all of these foods and from time to time asks to travel to various locations to retrieve different foods (Savage-Rumbaugh et al. 1986). The significance of a melon request, after a request to visit Sherman and Austin, is that one must go right past the Sherman and Austin building in order to reach the melon site. When it has been agreed that Kanzi may travel to melon, we often find that just as he passes the Sherman and Austin building, he breaks away from the group and rushes over to play with Sherman and Austin. When he is brought back and travel to melon continues, Kanzi accompanies very reluctantly and upon arrival he refuses to eat any melon.

Of course, such a trick will not work each time, and Kanzi has also used other foods (tomato, carrots, and potatoes) which require passing near the Sherman and Austin building. Each time, his real intent becomes clear as we reach the Sherman and Austin building and, if forced to continue on to the food, Kanzi shows no interest in eating it. Additionally, this is not the only location where Kanzi uses such deception. Apples are typically found just beyond the childside (a portion of the lab where research with children is carried out), and Kanzi often asks for apples, but once he passes the childside, he rushes over to look in the windows at the children and shows no further interest in apples. Indeed, all five of the 'off limits' locations have received similar treatment at one time or another.

False alarm

There is a type of deceit which has been practised toward Kanzi by some of the people who work with him. It is the use of alarm calls to alter the social situation. At times, when Kanzi is playing with people, he becomes too rough, and though he does not actually attack, rough play, if not tempered, can increase to a point where Kanzi is, in effect, hurting and bullying people. Some of the people have learned to avoid this situation by giving alarm calls and looking toward the trees, down the road, etc., as though they have seen something to be worried about. Kanzi stops his bullying in such circumstances and comes over to put his arm around the person and stare in the same direction to see what has caused them to worry.

When Kanzi reached 6 years of age, he began to use this tactic himself. Occasionally, when a large group of people were playing with Kanzi, he would begin to feel a bit intimidated physically. In a large group, there are many individuals to chase Kanzi and to co-ordinate a keep-away game. The odds of 5 to 1 make it difficult for Kanzi to win any game. When Kanzi begins to feel intimidated by the rough playful behaviour of a large group, he may look out toward a wooded area and make several alarm calls. These calls do not have the loud, clear sharp sound that they have when he is really alarmed. Instead, they sound forced or strained. Moreover, he does not show pilo-erection as he normally does when he is alarmed. However, he is able to make his facial visage appear quite concerned and he will continue to stare in a particular direction until people stop what they are doing and look to see what Kanzi is calling about. Often he will then ask to leave where he is and go look in the particular direction of the vocalization. Nothing is ever found, but the rough play has ceased, and the group has become unified and engaged in some other activity. Whether or not Kanzi would have used this tactic had it not been employed around him is impossible to guess. In any case, he has come to appreciate its value.

Conclusion

Monitoring others, and doing something while they are not looking seems to be the most frequent and simplest form of deception in the chimpanzee and it is a very common event, occurring many times each day. It requires the realization that other individuals must be able to see one's actions to object to them or hold one accountable for them. If others cannot see, then it is possible to do things that they might otherwise object to. Such behaviour necessitates the emergence of the concept that what others see is not necessarily what you see. It also requires an understanding that what the other party sees is dependent upon their line of regard, which differs from one's own. This entails the mental imaging of the line of visual regard of another individual (see Chapter 16).

For example, when Kanzi hides behind a tree to eat a mushroom, he must hide on the side that is directly opposed to the caretaker's line of regard. Hiding on any other side is ineffective. To know which side of many different trees to hide on with the position of the caretaker constantly changing, Kanzi must be able to compute the caretaker's line of regard and know that it is different from his own. This becomes far more complex when he is with two or three caretakers, yet he does so quite easily.

Hiding oneself, one's actions, or objects is a more complex form of deception, but also relatively common. This requires not only computation of the line of regard, but also the imaging of what others will do at different points in time. Thus, when Kanzi hides under a pile of blankets and presses himself flat against the mattress he is computing line of regard, and what the pile of blankets must look like to the caretaker. As he stays there quietly while the area is searched, even as the caretaker goes out of the room, he evidences understanding that the caretaker is continuing to look for him, even though he cannot hear or see her.

Purposefully and clearly misleading others by behaving in such a manner as to make them think X when Y is actually the case is much rarer, but definitely does occur. When Austin makes an unusual noise outside, he is very careful that neither Sherman nor anyone else sees him do this, since if they did they would not be startled by the noise. Only unexplained noises are worrisome, not ones whose origin is known to be Austin himself. Thus, Austin shows not only that he understands Sherman's line of regard, but also that he realizes that Sherman must imagine or search for a cause. When he rushes back in with pilo erection and looks outside as though he too is scared, his behaviour clearly reveals that he understands the importance of portraying the noise as something that was caused by other than himself. He not only realizes that Sherman's line of regard is different and that Sherman's assessment of what caused the noise will be different, but also that he can influence Sherman's perception of what caused the noise by how he behaves (see Chapters 12–14).

Using symbols to mislead others entails all of the above concepts, but also requires the realization that the opinions of others can be formed not only by what one does, but also by what one says. When Kanzi says that he wants to go to get melon or that he is going to be good, his caretakers take him at his word and behave accordingly. The ability to anticipate that this will happen requires the capacity to misrepresent one's goals, selecting instead goals that are known to be acceptable to others, but that can be used indirectly to accomplish one's own ends.

Are symbol-using primates capable of more artful forms of deception than non-symbol-using primates? There is not yet a clear answer to this question, as forms of deception are just beginning to be seriously studied in a variety of non-human primates. Symbol-using primates have not yet pretended to know something that they in fact do not know, nor have they used symbols to give false information about things other than their own intentions. However, the hiding of objects (including the use of symbols to request that others give objects of value that are then hidden) has not been reported in other non-human primates. Perhaps the emphasis on the use and transport of objects which occurs when chimpanzees are raised in a human cultural system fosters the development of schema that are object orientated. This would permit the development of more elaborate tactics of deception with regard to objects.

The manipulation and use of imaginary objects also appears to be unique to symbol using primates, and is, perhaps, a predecessor of more complex forms of deception. Certainly, the ability to pretend that something is there when it is not is the basis of much human deception.

Future research in the area of deception with apes will hopefully develop ways to better objectify and quantify these complex skills of deception. We need to be able to predict when deception will occur, the kinds of conditions which precipitate deception, and the type of deception that will be selected. As we better understand these things in apes we will gain clearer insights into this elusive, but important phenomenon.

Acknowledgements

Special thanks are extended to Elizabeth Rubert, Penny Nelson, and Karen Brakke for their help in the preparation of this paper. This research was supported by National Institutes of Health Grants NICHD-06016 and RR-00165 to the Yerkes Regional Primate Research Center, Emory University.

18

The ontongeny of tactical deception in humans
PETER J. LaFRENIÈRE

Developmental studies, interesting in their own right, can help us to understand at a more fundamental level the ways in which simple deceptive capacities can change into sophisticated ones, so they have a double interest for students of human evolution. Here LaFrenière reviews progress in research on the development of deceptive competence, including his own recent experiments and naturalistic observations.

A central issue to the study of communication, from both evolutionary and developmental perspectives, concerns the psychological concept of intentionality. Signals that communicate one animal's intentions may, on occasion, be used by another to gain a strategic advantage, especially in situations of competition or conflict. It follows that false signals may be strategically advantageous to their sender on just such occasions, particularly if they are relatively rare and likely to be interpreted as true (Dawkins and Krebs 1978). Such considerations lead to a more complex view of the social aspects of individual adaptation, in which the possibility of deception between familiar or unfamiliar individuals must be taken into account. Such a position also suggests a number of critical evolutionary and ontogenetic issues, including the development of the appearance-reality distinction and the emergence of a theory of mind in which the perceived intent of social partners becomes a basis for responding to them (see Chapters 12–14, 16, and 17).

While deception appears to be a widespread solution to the problem of competition in nature, in this chapter I shall distinguish the broad construct of deception, which may be taken to include Batesian mimics and other phylogenetic adaptations (Wickler 1986) from 'tactical deception' which implies a short-term behavioral adaptation (Chapter 15). For our purposes tactical deception will be confined to any interaction in which sender deliberately conceals or falsifies information in order to manipulate receiver's behaviour. This definition rules out self-deception because of the problematical nature of its observation and instead defines deception within a social context. It also rules out learned associations that are not intentional or deliberate actions, but merely conditioned reflexes. A further criterion is the instrumental nature of the action. The qualifier, 'tactical', suggests purposeful behaviour that is designed to generate a specific effect. Finally, the definition entails two basic forms of deceit, one in which information is withheld or concealed, and one in which false information is provided (Ekman and Friesen 1969).

The importance of deception in human affairs has long been recognized by political philosophers, military analysts, playwrights, novelists, and other observers of human behaviour. For example, deception has been considered central to intergroup conflict from the biblical accounts of the seige of Ai and the Greek legend of the Trojan horse, to the modern examples of Pearl Harbor in 1941, Normandy in 1944, and Czechoslovakia in 1968. (Handel 1982; Whaley 1969). This fact is so much in evidence that one is inclined to agree with Sun Tzu, that 'all warfare is based on deception' or with Churchill, who is reported to have said that 'in time of war, the truth is so precious it must be attended by a bodyguard of lies'. While warfare is certainly one of the most dramatic illustrations of the human capacity for deceit, the intricacies of political intrigue from Machiavelli's classic work in the early renaissance to the unfolding of Watergate in our own times, provide further examples of the extent of duplicity in public life.

In everyday life, social psychologists such as Mead (1934) and Goffman (1959) view all social interaction as involving an element of deception, in the sense that participants are engaged in a dramatic performance to control the impressions of themselves that are presented to others. Perhaps the most extreme view of human deception in everyday life was articulated by a contemporary sociobiologist who views human society as a 'network of lies and deception, persisting only because systems of conventions about permissible kinds of lying have arisen' (Alexander 1977). This one-sided view of human sociality, ignores the vital function of reliable communication in human intercourse. Without then overstating the case, the relevance of this chapter to the theme of this book is that the possibility of deceptive, well calculated communications and the necessity of detecting such machinations and manipulations, provided a major impetus for the evolution of primate and human intelligence (see Chapters 1 and 2).

This social function of intellect may be considered a primary context for its ontogeny as well. At birth, the human infant is well equipped with reflexes designed through natural selection to provide communication with an adult, and to elicit and assure adequate caregiving. In time, such behaviours are brought under the voluntary control of the infant and different elements of the infant's expanding behavioural repertoire may be combined and co-ordinated to achieve a goal (Piaget 1936). With the onset of intentional, goal-orientated behaviour and the voluntary control of expressive behaviour, tactical deception becomes possible. In its mature form, tactical deception calls upon the individual's capacity to co-ordinate cognition, affect, and behaviour in order to successfully pull off a dramatic performance designed to mislead a perceptive audience regarding something of possibly great importance to the individual. Such an act implies a series of cognitive achievements, beginning with decentration and social perspective taking and leading to an awareness that others are social beings with their own capacity to decentre and comprehend one's own perspective.

This cognitive capacity for second-order representations of intention (Chapter 5 and 14) is a significant achievement from both an evolutionary and developmental perspective. Because of the logical sequence in the development of prerequisite subskills required for tactical deception and without, in any way, requiring that 'ontogeny recapitulates phylogeny', developmental analyses may shed light upon evolutionary progress in the same domain. In this chapter, I have chosen to analyse the developmental sequence underlying the emergence of tactical deception in children. Such an analysis should lead to a developmental or evolutionary model and generate testable hypotheses, that may in turn provide empirical data for revising or elaborating the theoretical model. I shall begin by reviewing the literature on cognitive, affective, and behavioural factors underlying deception ability. Finally, naturalistic and experimental data on age related changes in the form and function of tactical deception in young children will be reported.

Early research on children's deception

The first large scale study of deception in children was published in 1928 by Hartshorne and May. Previous studies of deceit were philosophical, anecodotal or based on adults' accounts of children's lying (see Vasek 1986, for a review). Hartshorne and May used a battery of classroom tests to indirectly assess the frequency and extent of cheating among 11 000 children between the ages of 8 and 16. While methodologically flawed, their data revealed that cheating (defined from the adult point of view) was highly prevalent in schoolchildren. They also concluded that deceit is not a unitary character trait, but rather a function of the specific situation. However, deception was thought to 'run in families to about the same extent as eye colour, intelligence and height'. Between the lines Hartshorne and May were presenting a vague case for the heritability of deceitfulness and were inclined to believe that if 'nurture were held constant', individual differences would remain. While large scale in design, Hartshorne and May's work did little to advance our understanding of developmental processes underlying the emergence of the ability or motivation for children to engage in deceptive behaviour. The first studies on deception from a developmental perspective were conducted by Krout (1932) and Piaget (1932). Krout's developmental analysis involved a progression from misapprehending the real, confusion between wishes and reality, and finally the intentional substitution of the fanciful for the real. However, Piaget was the first to formulate methods to study children's conceptions of lying and to take into account their understanding and appreciation of intentionality in making moral judgements.

Intentionality

Piaget's clinical method involved extensive interviews with children while presenting them with a series of hypothetical situations, varying in degree of intentionality. While relying on the linguistic ability of his subjects and using verbal lies as his definition of deceit, Piaget's work does show that children's conceptions of lying follow a developmental sequence from equating lies with naughty words, to any untrue statement, and finally, by the age of 10, to the conventional definition of a lie as an intentionally false statement. More generally, Piaget demonstrated that the child's ability to engage in or cope with lying rests upon his or her communicative competence, perspective-taking skills, and appreciation of intentionality. This placed the study of deception within the broader domain of social cognition.

Because of the widespread influence of Piaget's work, the study of intentionality was, until very recently, restricted to its role in the formation of moral judgements. A more recent line of research involves the study of deception as a sure indicator of intentional communication (De Vries 1970; Shultz and Cloghesy 1981; Wimmer and Perner 1983; Woodruff and Premack 1979). Each of these studies attempts to draw the further distinction between awareness of another's intention and a recursive awareness of intention, that is, an awareness of the other as possibly aware of one's own intentions.

In their ingenious study of chimpanzee communication, founded upon earlier work by Menzel (see Chapter 12), Woodruff and Premack (1979) found evidence of intentional communication in 3-year-old subjects. In their experimental paradigm, communication concerning the location of a hidden incentive was studied in chimpanzee-human dyads. This study demonstrated that chimpanzees were able to adjust their behaviour depending upon the co-operative or competitive nature of their human partner, producing and comprehending behavioural cues that conveyed accurate information with co-operative partners, while learning to withhold information or mislead a competitive partner, or to discount or controvert that partner's own midleading cues. These laboratory observations offer incontrovertible evidence of tactical deception in chimpanzees. Woodruff and Premack consider that it is 'rarely possible' to infer intentional deceit from field observations alone. It should also be noted that the capacity for deception in these 3-year-old chimpanzees appeared only in some subjects, and then only after a period of many months and numerous trials of testing. To deceive their competitive partners, the animals first learned to suppress behaviours that provided accurate information, such as orientating responses to the hidden food. Woodruff and Premack note that, in the four chimpanzees tested, comprehension and production appeared to develop independently, with neither ability showing developmental priority.

Turning to recent studies of intentionality and deception in human children, several parallels may be noted (with one important exception) to the chimpanzee data. De Vries (1970), using a binary choice social guessing game outlined a five-stage model of the development of deception. This paradigm, appropriate for very young children, called for one child to hide a penny in his/her hand while another child guessed its location. Children had the opportunity to play each of the two roles over a series of trials. In the first stage, there is no recognition of individual perspective nor, consequently, the need for secrecy. At the second stage, the child becomes aware of the other's perspective regarding behavioural roles, but without an awareness of different motivations. There is still no appreciation of the competitive nature of the situation and the partner's goals may be confused with the child's own goals. At the third stage, the child recognizes the opposition implicit in the goal structure of the situation, but shows no capacity to infer intentionality, instead using a regular and easily predictable hiding or guessing strategy. Finally, during the fourth and fifth stages the child acquires an irregular hiding strategy and, subsequently, an irregular guessing strategy. According to De Vries,

The fact that the competitive and deceptive hiding role develops before the competitive and deceptive guessing role suggests that the child is able to take account of the other's perspective before he is able to take account of the-other's-taking-account-of the child's perspective (1970, p. 769).

In this paradigm, production appears to precede comprehension in accordance with Piaget's general model of cognitive development.

This behavioural distinction between production and comprehension rests upon the cognitive distinction between awareness of the other's intention and a recursive level of awareness of intentionality. In a series of recent studies, Shultz and his colleagues have found that children of 3–5 years of age can distinguish intended actions from unintended behaviour such as mistakes, reflexes, and passive movements (Shultz *et al.* 1980). Subsequently, the recursive awareness of intention, essential to tactical deception, was investigated in children from 3 to 9 years (Shultz and Cloghesy 1981). In their paradigm, a card game was designed such that one player, after noting the colour of the top card, pointed to either a red or black card as an indication to the other player, whose task it was to guess the colour of the top card. Like the Hide-a-Penny task of De Vries (1970) the task was competitive and called for the use of an irregular deceptive strategy. The investigators found that only children aged 5 years and older showed a recursive awareness of intention. The 3-year-olds showed great interest and enthusiasm for the game, appeared to understand the rules and the object of the game, but did not at all apply the appropriate recursive strategy. Deceptive hiding of the card to be guessed was consistently shown by one-third of these very young children confirming the relatively early acquisition of first-level awareness of the opponent's visual perception. This ability to successfully hide an object from another is acquired as early as 2–3 years (Flavell *et al.* 1978), but merely hiding an object does not imply a recursive awareness of intention. Rather, this ability was discerned by spontaneous changes in the child's guessing strategy. Such changes reveal that receiver recognized that sender had guessed his intended strategy and that sender was about to change strategies to fool receiver. In the children's own language, 'Before she tricks me, I'm going to trick her' (Shultz and Cloghesy 1981).

Wimmer and Perner (1983) also investigated the understanding of deception in children in the 3–9-year age range. They told children stories in which the protagonist was led to believe that an object was hidden in one place while they knew that it was actually hidden elsewhere. When asked where the protagonist would search, 3-year-olds ignored the protagonist's false beliefs and predicted he would search the correct location, while 4–6-year-olds predicted he would search the incorrect location. Finally, a recent study by Harris and his colleagues also confirms that children's ability to make distinctions that imply an understanding of deception emerges in the 4–6 age range (Harris *et al.* 1986). Their results show that young children understand that false beliefs may be induced not just about an observable fact, such as the location of a hidden object, but also about something more private, such as another person's state of mind. While 6-year-olds consistently demonstrate their grasp of the difference between real and apparent emotion, even 4-year-olds showed a limited grasp of this distinction. Taken together, these studies offer convergent data showing that the essential cognitive skills for tactical deception emerge within the period of 4–6 years.

It is interesting to speculate about the developmental processes underlying this achievement, as well as its timing. Piaget stressed the role of peer interaction, as opposed to adult-child interaction, in facilitating cognitive growth. In his view, the sociocognitive conflict that arises when confronting peers with opposing points of view, provides an impetus for understanding the other's perspective. Peers are also likely to compete for status and resources. On this basis, social interaction with peers may provide a particularly salient environmental stimulus for greater decentration, perspective-taking, and a recursive representation of intentionality. At an age when children begin to interact much more extensively with the wider social community outside the family, such abilities may begin to serve an adaptive function in such settings.

Behavioural inhibition and affective control

A second domain of abilities critical to successful deception involves the capacity to inhibit behavioural responses or affective expression. Both Darwin (1872/1965) and Freud (1925/1959) were convinced that deception may be detected by close observation of non-verbal behaviour. Freud commented that

he who has eyes to see and lips to hear may convince himself that no mortal can keep a secret. If his lips are silent, he chatters with his fingertips: betrayal oozes out of him at every pore (Freud, 1925/1959).

It is clear that there are considerable differences between adults and children in terms of their ability to deceive or detect someone else's deception, and a number of experimental studies have been conducted to examine these difference. Ekman and Friesen (1969) were the first to formulate a broad theoretical statement on deceptive behaviour in adults. They introduced the concepts of deception clues (which inform receiver that deception is in progress, but do not reveal the concealed information) and leakage (the betrayal of the withheld information). According to Ekman and Friesen, in initiating a deceptive act, sender has two choices: inhibition or simulation, though most often the result is some blend of the two. Sender may choose to modulate or adjust the intensity of an emotion to show either more or less than what is actually felt. The extreme form of this is neutralization or complete inhibition. While cutting off communication entirely is the surest way to forestall leakage, it is usually a strong clue that deception is being attempted. It is more strategic for sender to maintain the communication flow, pretending that nothing has been

concealed, while selectively omitting the expression of certain emotional reactions or states. This more fluid form of deception, involving simulation and masking, is more effective because it:
(1) fills gaps left by selective omission which if left unfilled are obvious clues to deception;
(2) erects a barrier against the breakthrough of the inhibited emotion (this was clearly anticipated by Darwin (1872/1965) in his discussion of antithesis as a general principle of expression);
(3) the substitution of the false message may be instrumental to the goal of the deception.

Ekman and Friesen identify the face as the focal point in affective expression: it has the greatest sending capacity. The legs/feet are usually the poorest non-verbal senders, while the hands are intermediate. Feedback from receiver closely parallels this order. Feedback from receiver that informs sender what receiver has perceived and evaluated is *external*. *Internal* feedback is defined as the sender's conscious awareness of what he's doing and his ability to recall, repeat or specifically enact a planned sequence of behaviour.

Most people look most often at the best sender, the face, which is typically under the most conscious control of the sender. There are, however, limits to the extent to which one can attend to the face; too much looking can be uncomfortable and may suggest a power struggle, interrogation or intimacy. There are also taboos against too much looking at the hands. While many gestures are used for purposes of illustration, other hand acts, which may be nervous or unconscious and which involve contact with the body, clothing, or hair, are not usually reacted to by receiver. Even less commentary is reserved for sender's legs or feet. Thus, in terms of those behaviours with high information value, those that may elicit scrutiny and evaluation, and those that are controllable, facial movements are first, then the hands, and lastly the legs/feet.

What is relevant about these concepts for our purposes is that Ekman and Friesen derive their hypotheses about non-verbal sources of leakage from them. Sender will surely not bother inhibiting expression in areas of the body that are largely ignored by receiver, nor can sender inhibit behaviours about which he has learned to ignore internal feedback or from which he receives minimal feedback. In short, non-verbal leakage is hypothesized to reverse the pattern described for differences in capacity and feedback. The face, which is equipped to betray the most, can also conceal the most and thus becomes the most confusing source of information during deception. Other parts of the body, though they have less capacity, also receive less feedback, are less controllable, and are hypothesized to be a primary source of leakage.

While this model does not address developmental issues, a number of researchers have employed it to examine age differences in the encoding of deceptive communication. In one study (Feldman *et al.* 1979), first graders, seventh graders, and college students were instructed to display a pleased, positive expression while drinking sour fruit juice. Subsequently, naive raters more readily detected deception in the youngest age group. Such results would be expected since young children are less experienced in using both internal and external feedback to control their facial expression than older children or adults. Saarni's work (1979, 1984) also shows a steady progression in children's ability to manage emotional expression. In her paradigm, first, third, and fifth grade children were presented with a disappointing gift when expecting a desirable one. This study provides data on spontaneous attempts to control expressive behaviour when confronted with a situation that calls for a conventional display of sham positive affect. Her results indicate that the youngest children (especially boys) were more likely to look displeased on receiving the disappointing gift (a drab baby toy), while the older children (especially girls) were more likely to look pleased. As discussed by Saarni (1984), this paradigm does not allow one to distinguish between awareness of the display rule and motivation to carry out the rule. As a result, these studies may significantly underestimate young children's ability to control their affective expression.

In the following two sections, results from new experimental and naturalistic studies of deception are presented. The experimental study was designed to explore developmental changes in children's capacity for deception, asking primarily, 'what *can* children do?' The naturalistic observation, on the other hand, addresses the question 'what *do* children do?' In addition, these observations in the everyday context, as opposed to a specifically designed experimental context, provide an invaluable source of hypotheses concerning the possible function of deception in individual adaptation.

An experimental study of children's deception

Because of the practical and ethical problems of observing or eliciting deception in children, researchers often turn to a competitive game context. Our interest in children's ability to control affective expression and inhibit behavioural responses that could be interpreted as non-verbal leakage, led us to choose a paradigm in which the child must conceal the location of a hidden object from an adult interrogator. In addition, we employed three trials in order to improve our estimate of the child's capacity to deceive, as well as to provide a measure of regular versus irregular placement strategy. Our hypotheses all concerned age changes in performance on the task. Specifically we hypothesized that older children (5–6 years) would be

(1) more successful in deceiving the adult;
(2) more able to control affective expression and inhibit non-verbal leaks;
(3) more likely to use complex strategies involving simulation and unpredictable patterns of hiding the object across trials, than younger children (3–4 years).

Forty pre-school children (20 girls, 20 boys) from 40 to 79 months of age were video-taped in a laboratory task requiring the deception of an adult confederate in a mildly arousing game context. After hiding a toy bear in one of three hiding places, the child was instructed to 'try to fool' an adult who questioned the child in order to find the bear's location. The child was free to choose between three 'hiding places' located in line on a game board approximately 50 cm apart. After the bear was hidden, the adult confederate entered the room and questioned the child about each location, looking the child in the eye, and asking 'is the bear hidden in the house (tower, truck)'? After each series of questions, E made his best guess as to where the toy was hidden. The game was repeated three times, each using the above format. Tapes were analysed for number of successfully deceptive trials, placement strategy, types of deception strategy (inhibition versus simulation), verbal and non-verbal leakage, and clues to deception.

Results revealed that age was significantly correlated with success ($r = 0.42$, $P < 0.01$). With one exception, children younger than 48 months were unsuccessful in all attempts. These children were unable to conceal information and some were delighted to reveal the hiding place to the adult (see Fig. 18.1).

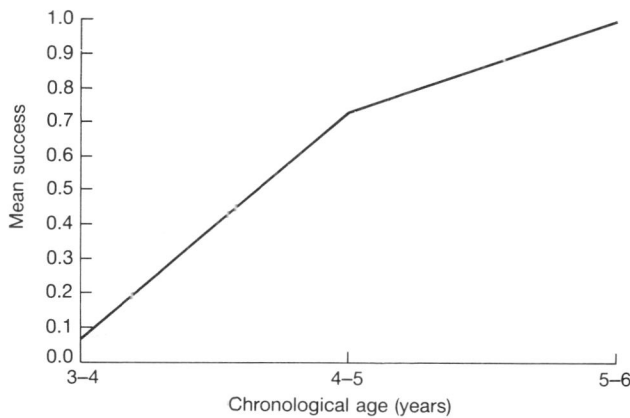

Fig. 18.1 Mean number of successful deceptions as a function of age.

Children of all ages (90 per cent) used a consecutive placement strategy in which the toy was hidden in a different location for all three trials. Thus, on the third trial, children were unsuccessful at all ages since the experimenter could infer the correct location on the basis of the previous two trials. Maximum success (2) was achieved by only 25 per cent of the older children, though 69 per cent were successful on at least one trial. How was this success achieved? As shown in Fig. 18.2, older children most often used the strategy of neutralization or inhibition (56 per cent), but were not often successful because of detectable non-verbal leakage (e.g. side glances at the correct location). Only five children used simulation as a modal strategy and they achieved the highest success rate at the task. In order to further establish age effects for the dependent variables of success and strategy, 2 × 2 contingency tables were created by dichotomizing age into younger (less than 48 months) versus older groups and dichotomizing the dependent variables into successful versus unsuccessful or strategic versus non-strategic. Chi-square analyses revealed significant effects for age versus success ($\chi^2 = 10.37$, df = 1, $P < 0.01$) and age versus strategy ($\chi^2 = 13.14$, df 1, $P < 0.001$).

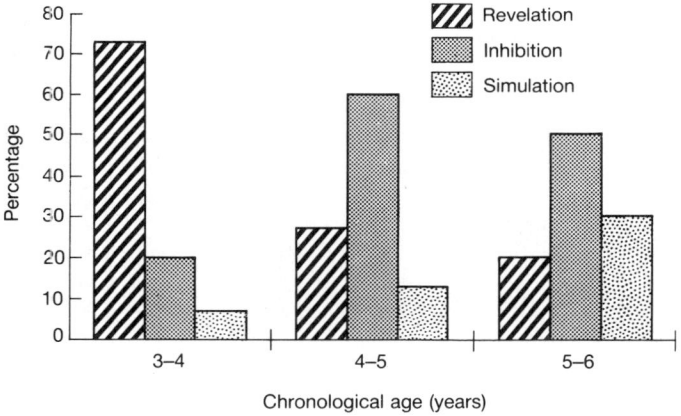

Fig. 18.2 Modal strategy as a function of age.

Results from this experiment thus demonstrate that children younger than 4 years are not yet competent in deceiving an adult in this game playing context. As with Premack's young chimpanzees, these young children had considerable difficulty in suppressing a behavioural response such as orientating towards a hidden object while being questioned concerning its location. This capacity for behavioural inhibition, critical for success in this deception task, develops relatively slowly in young organisms (White 1965). Young children also have less fine muscular control of the type that is necessary to manipulate their expressive behaviour (Charlesworth and Kreutzer 1973). These difficulties, together with the cognitive limitations discussed earlier, account for children's limited skill in formulating and enacting a tactical deception. In addition, their lack of success in a competitive game context may be partially attributed to motivational factors. Some young children were unable to sharply distinguish their goals in the game from those of the adult and were often pleased when the adult guessed correctly. However, older children appreciated the competitive nature of the game, were pleased at their own success and often disappointed when the adult guessed correctly. Their strategies for misleading the adult included feigning ignorance or, more rarely, intentionally misleading the adult. While these older children clearly showed some elements of tactical deception, they were not highly successful owing to frequent non-verbal leakage, and a predictable placement strategy. However, their motivation, cognition, and expressive control were demonstrably different than their younger counterparts.

Since decentration and a second-order representation of intentionality are clearly prerequisites for successful deception, future work should include a number of cognitive measures, as well as groups of slightly older children, in order to complete our understanding of just what is developing as the child learns how to conceal or simulate information and affect in face-to-face social interaction when it appears to be advantageous. The oldest children in this study averaged only one success for every three attempts, with considerable individual variation. Few children chose to provide false information to the adult. It is not clear whether this reflects a general incapacity for simulation or, perhaps more likely, an inhibition against outright lying to an adult authority figure. As Ekman (1985) notes, anxiety about the consequences of a failed attempt to deceive may make successful deception all the more difficult. In everyday life the stakes may be much different than in an experimental game situation. In this experiment, adult interrogators were instructed to behave in a restrained serious manner and to look the child directly in the eye while posing the questions in an effort to assure that the context was mildly arousing, thus approaching one dimension of a real life situation. Nevertheless, the child's motivation to deceive was probably not very great. As stated earlier, such limitations may be inevitable for experimental studies of deception in children for ethical reasons. This assures an important role for naturalistic observation, despite the inconveniences and difficulties inherent in such work.

Naturalistic observation of deception in children

A class of behaviours as complex and diverse as contained within our definition of tactical deception, requires further categorical treatment than simply dichotomizing it into two basic forms. What is needed is a more detailed taxonomy that will, in turn, permit finer discrimination between different types of deception. Such a taxonomy may be generated inductively by collecting many different examples of behaviour that fit the definition and sorting them into meaningful groupings or deductively on the basis of theoretical or logical analysis.

Viewing deceptive behaviour as somewhat natural, in the sense that it appears early in the repertoire of human behaviour without specific tutoring and may have a genetic basis, implies that it may be useful in promoting individual adaptation. Such a statement is best regarded as an hypothesis rather than a philosophical assumption or proven fact. As an hypothesis it allows us to move quickly towards what is of central importance to human ethology, namely the study of adaptation. If deception, as a particular form of social intelligence, has acted as an evolutionary ratchet, as Humprey and others suggest (Chapters 1– 4), then a functional solution to the taxonomic problem of studying deception in children would appear to be worthwhile. Such a solution allows, even demands, for observers to study behaviour in relation to the environment, specifically a reactive, social environment.

Naturalistic studies of problem-solving behaviour in young children like their counterparts in primatology (de Waal 1986; Chapters 14–16) have failed to uncover problems of the complex gadgetry type typical of laboratory studies of cognition. In his ethological approach to the study of cognitive development, Charlesworth (1983) observed spontaneously occurring problem-solving episodes of children in the home or pre-school. A problem was inferred on the basis of an observable interruption of ongoing behaviour. The observer's task included the identification of the stimulus conditions responsible for the interruption. As with non-human primates, the majority of young children's problems involve social, as opposed to informational or physical, needs and demands (Charlesworth 1983). Based on these naturalistic observations of children encountering problems, one can make a strong argument that social interaction rather than object manipulation provides the basic stuff from which higher intelligence develops.

Informal naturalistic observation of deception in children from 2 to 6 years-old in the home and pre-school environments, interacting with parents, teachers, and peers, has generated the functional classification scheme outlined in Table 18.1.

These examples are all spontaneously generated by the children in situations of varying degree of control by the investigator. As with anecdotal data from the animal literature, it is not possible to draw firm conclusions regarding the intention of the communicator: nevertheless, each is an example of a false communication embedded in a social context,

Table 18.1. *A functional taxonomy of deceptive behaviour in children*

Type	Example	Function
Playful	Humorous, to provoke laughter, maintain attention	Child offers toy, then pulls it away as mother reaches for it, laughing (19 month, Chevalier-Skolnikoff 1986)
Defensive	Escape blame or punishment	Child spills milk. Father asks 'who did this?' Child blames little brother (who was not present) before her parents (who were) (2 years). (Dr White, personal communication and film)
Aggressive	Inflict harm without being held responsible	Child bites hand, shows teacher, indicating that another child, present, did this to her (2½ years) (Dr C. Piché, personal communication)
Competitive	Achieve desirable outcome/goal	Child cheats peer at counting game, claims innocence, wins game (4 years)
Protective	Protect the feelings or well being of others	Child smiles to adult after receiving a disappointing gift (8 years) (Saarni 1979, 1984)

and each suggests a possible function. While this taxonomy is by no means exhaustive, it is illustrative of the types of situation in which children are likely to attempt a deception. Certainly, one of the most common early situations involves the child's attempt to escape blame or possible punishment for actions he or she believes to be wrong. One may speculate that different socialization practices regarding discipline may underly individual variation in the frequency of lying among children. Defensive deception may be quite prominent in some children, as they attempt to adapt to a situation where accidents or misdeeds are frequently or severely punished. Tactical deception of the aggressive type, with intention to harm another, appears to be much less common in young children. The example in Table

18.1 was reported by a researcher investigating the development of children at risk for psychopathology. The behaviour was noted in a child who was herself a victim of child abuse (C. Piché, personal communication).

Finally, this functional taxonomy does not imply positive or negative connotations concerning deceptive behaviour. One may imagine with equal ease, situations where such behaviour would be positively or negatively evaluated by societal convention, and situations where it would be clearly functional of dysfunctional for the individual. Rather, this initial taxonomy is offered as a means of cataloguing various descriptions of deceptive behaviour, as an heuristic device to stimulate further research efforts.

Acknowledgements

I am grateful for the competent assistance provided by several students during the period of data collection, notably Diane Dubeau, Jean-Marc Ménard, and André Pronovost. I also wish to thank Denyse Rompré for her sparkling efficiency in the preparation of the manuscript, as well as the delightful Quebecois children and day care staff who generously contributed their time. The project was supported by grant 410–85–0651 from the Social Sciences and Humanities Research Council of Canada. Finally, I wish to thank William Charlesworth for the many stimulating hours of discussion that inspired this initial empirical effort.

Social or non-social origins of intelligence?

19

Social and non-social knowledge in vervet monkeys*
DOROTHY L. CHENEY and
ROBERT M. SEYFARTH

The concern of this section of the book is to ask whether there exist good reasons to prefer the Machiavellian Intellect hypothesis to other —non-social—ideas to explain the evolution of human intelligence. This chapter shows that the famous work on Amboseli's vervets makes a direct test of this: are vervets more intellectually advanced in social or non-social domains?

We think of ourselves as relatively intelligent creatures for at least two reasons. First, humans not only learn rapidly when taught, they also acquire information without active instruction, by observing objects and events in the world around them. Secondly, human intelligence is not domain-specific, and knowledge acquired in one domain can readily be applied to another. Thus a principle derived from a social problem can easily be applied to a logical similar problem involving objects.

Comparing human intelligence with the intelligence of non-human primates is difficult, because primate intelligence has thus far been measured almost exclusively by performance on learning tests. Comparatively little is known about the knowledge that monkeys and apes acquire naturally, in the absence of human intervention. More importantly, animal intelligence is generally tested only in one domain, using biologically arbitrary objects as stimuli. Most of the problems confronting non-human primates under natural conditions, however, are ones that derive from competitive and co-operative interactions with conspecifics. There is reason to believe that primates may reveal greater intelligence when dealing with each other rather than with irrelevant objects. In this paper, we examine what free-ranging vervet monkeys

* Reprinted from *Philosophical Transactions of the Royal Society of London*, **B308**, 187–201 (1985).

have learned, without human intervention, about their environment. We do so by means of observations and experiments that attempt to compare primate performance in social and non-social domains.

Primates tested in the laboratory, with objects, often face problems that are logically similar to the social problems confronting primates in the wild. Despite this similarity, however, the performance of primates in these two settings often seems to differ strikingly. To cite just one example, McGonigle and Chalmers (1977) and Gillan (1981) demonstrated transitive inference in captive squirrel monkeys and chimpanzees, respectively, but were able to do so only after considerable training with paired stimuli. In contrast, field observations suggest that, even from a very young age, monkeys are readily able to deduce a dominance hierarchy among conspecifics from their observation of dyadic interactions (Cheney 1978; Seyfarth 1981; Datta 1983 *a–c*; Gouzoules *et al.* 1984). Observations and experiments have also suggested that primates regularly classify individuals on the basis of kinship or close association (for example, Bachmann and Kummer 1980; Cheney and Seyfarth 1982*a*; Judge 1982; Smuts 1985). Moreover, numerous examples from field studies (admittedly anecdotal) suggest that primates can predict the consequences of their own actions on others, and understand enough about the behaviour and motives of others to be capable of deceit, and other subtle forms of manipulation (for example, Goodall *et al.* 1979; Kummer 1982; Cheney and Seyfarth 1985*b*; Chapter 10). Such observations are both intriguing and frustrating, because they suggest the existence in the wild of striking mental abilities that, with some notable exceptions (Woodruff and Premack 1979; Premack and Premack 1982; Chapter 13), have not been documented or duplicated in the laboratory.

Because of the qualitative differences in field and laboratory stimuli, one possible explanation for the animals' differing performance suggests that primate intelligence is relatively domain-specific. This hypothesis argues that group life has exerted strong selective pressure on the ability of primates to form complex associations, make transitive inferences, and predict the behaviour of fellow group members. Thus abilities that seem to emerge only with human training in captivity may readily occur in primates under natural conditions, but mainly in the social domain (Chance and Jolly 1970; Kummer 1971, 1982; see also Rozin 1976; Chapter 23). Similarly, when captive chimpanzees solve technological problems that require foresight and an understanding of the consequences of past decisions (Dohl 1968), they may be demonstrating abilities for which they have been preadapted as a result of the need to make equally strategic decisions about each other (Chapter 10).

This domain-specific hypothesis posits that natural selection may have acted to favour complex abilities in the social domain that are, for some reason, less easily extended or generalized to other spheres. The hypothesis does not, however, specify exactly how elaborate a monkey's social knowledge is, or what processes underlie it, nor does it claim that social knowledge can never be extended to other spheres. It argues simply that certain problems are solved more easily in the social domain as compared with other areas.

An alternative view argues that the contrast in ability between social and non-social behaviour derives not from any fundamental difference in ability between the two domains, but from the animals' lack of motivation to perform under laboratory conditions. Research on captive primates has been plagued by motivational problems, and it has often been difficult to distinguish between a lack of ability and a lack of incentive to perform the task at hand (for example, Terrace *et al.* 1979).

We have begun an investigation of these hypotheses by presenting free-ranging vervet monkeys with logically similar problems involving 'social' and 'non-social' stimuli. Although our field experiments are less precisely controlled than laboratory tests, they have at least two advantages. First, problems of motivation and human training are circumvented. Secondly free-ranging primates daily encounter similar social and non-social problems, thus permitting a direct test of performance in the two domains. Thirdly our subjects regularly deal with objects in the external world that may be either relevant or irrelevant to their survival. It is, therefore, possible to compare social knowledge both with non-social knowledge of biologically relevant objects and with non-social knowledge of objects that are apparently unrelated to the animals' survival.

At the outset, two points should be emphasized. First, throughout the paper we draw a distinction between the performance of primates in the 'social' and 'non-social' domains. While we believe that this distinction is a real and heuristically important one, we recognize that the boundary between these spheres of activity is ill-defined. Secondly, in evaluating observations and experiments we make no claims about the mechanisms underlying performance. Our experiments define knowledge operationally; they measure only the responses that particular stimuli evoke, and not the processes (mental or otherwise) that underlie such responses. Many of the results we describe could, for example, result either from relatively simple associative learning or from more complex cognitive processes. Our aim is not to argue for one of these alternatives. Instead, we use experiments to determine which of two stimuli is more salient, and to suggest that animals form some sorts of associations more readily than others.

1. Study site and subjects

Experiments were conducted on three free-ranging groups of vervet monkeys in Amboseli National Park, Kenya. Vervet monkeys live in stable social groups consisting of a number of adult males, adult females, and their juvenile and infant offspring (Cheney et al. 1985). As in most Old World monkeys, female vervets remain in their natal group throughout their lives, maintaining close bonds with maternal kin. Males, in contrast, emigrate to neighbouring groups at sexual maturity, often in the company of brothers or natal group peers (Cheney and Seyfarth 1983). Within each group, males and females can be ranked in linear dominance hierarchies that predict the outcome of competitive interactions over food, water, and social partners. Offspring acquire dominance ranks immediately below those of their mothers, such that all members of a family typically share adjacent ranks (Cheney 1983).

2. Relevant aspects of other species' behaviour

Vervets in Amboseli are preyed upon by leopards, a number of small carnivores, two species of eagle, baboons, and pythons (Cheney and Seyfarth 1981). Predation is a major cause of mortality (Cheney et al. 1981, 1985). The monkeys give acoustically distinct alarm calls to different predators, and experiments have shown that each of these alarms evokes qualitatively different escape reponses (Seyfarth et al. 1980 a, b). Calls given to leopards, for example, cause monkeys to run into trees, while calls given to eagles cause monkeys to look up in the air. The monkeys' alarm calls therefore function to designate different types of danger in the external world.

Vervets are not the only species to give alarm calls to predators, however, and it would seem advantageous for the monkeys to distinguish among alarm calls given by other species. We investigated the vervets' knowledge of other species' alarm calls through playback experiments with three different calls of the superb starling (*Spreo superbus*). Starlings give two acoustically distinct alarm calls to predators, neither of which bears any acoustic resemblance to the vervets' own alarms. One starling alarm—a harsh, noisy call—is given to various terrestrial predators (including vervets), all of which prey on starlings or their eggs, but only some of which prey on vervets. The second alarm—a clear, rising tone—is given to many species of hawks and eagles, two of which prey on vervets.

In conducting playback experiments of starling alarm calls, we followed the same protocol previously used in tests of the vervet's own alarm calls (Seyfarth et al. 1980 a, b). First, we hid a loudspeaker near a group of one to five vervets ($\bar{x}$ = 17.4 m; SD, 5.3). The monkeys were then filmed for 10 s, to establish the probability that they would show a given response in the absence of any call. We then played one of the starlings' calls, and continued to film the monkeys' responses for another 10 s. Three starling calls were used: their ground predator alarm call, their aerial predator alarm call, and, as a control, their song. Individual monkeys generally appeared only once in all trials, and successive experiments on the members of a given social group were always separated by at least 48 h. Results are presented in Fig. 19.1.

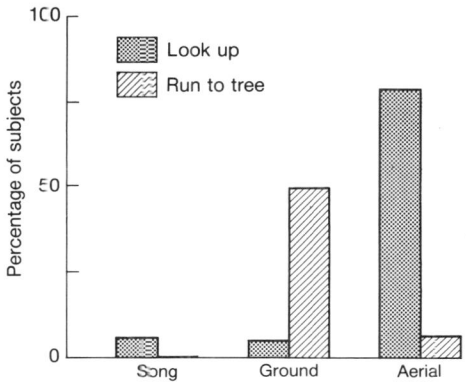

Fig. 19.1 Responses of vervet monkeys to playback of three different starling vocalizations. Number of subjects for playback of song, terrestrial predator alarm, and avian predator alarm were 17, 18, and 15, respectively. Ground predator alarms evoked significantly more running to trees than either song (χ^2 = 11.4, $P < 0.01$) or avian predator alarms (χ^2 = 7.3, $P < 0.04$); avian predator alarms evoked significantly more looking up than either song (χ^2 = 18.2, $P < 0.01$) or terrestrial predator alarms (χ^2 = 19.0, $P < 0.01$).

Playback of the starling's ground predator alarm caused a significant number of monkeys to run into trees, while playback of the aerial predator alarm caused a significant number of vervets to look up. In contrast, the starling's song elicited little response. In this test, therefore, where the behaviour of another species was relevant to the vervets' survival, the monkeys' knowledge of another species' calls was similar to their knowledge of their own calls.

3. Apparently irrelevant aspects of other species' behaviour

The alarm calls of other species represent one end of a continuum of biologically relevant or irrelevant stimuli in the external world. It is perhaps not surprising that vervets discriminate between such alarm calls, since these calls are so obviously important to their survival. Can similar knowledge be demonstrated, however, for aspects of another species' behaviour that are apparently unrelated to the monkeys' survival? This seems an important question, because one striking feature of human intelligence is our inclination to accumulate information about the world that is not directly relevant to our survival. Can the same be said of vervet monkeys? Are vervets as good naturalists as they are primatologists?

To address this question one must first identify two comparable features of the monkeys' environment, one social and biologically relevant, the other non-social, and apparently irrelevant to the monkeys' survival. Then experiments must be designed to compare the monkeys' knowledge in these two domains. As a social, biologically relevant test, we asked the monkeys how much they knew about the ranging behaviour of other vervets. As a non-social, apparently irrelevant test we asked the monkeys how much they knew about the ranging behaviour of other species that neither compete nor interact with vervets in any obvious way.

Vervet monkeys aggressively defend their group's range against incursions by other groups. Females and juveniles are active participants in intergroup encounters, and give a distinctive vocalization when they spot the members of another group (Cheney 1981; Cheney and Seyfarth 1982*a*). In testing the vervets' knowledge of other groups' membership and ranges, subjects in one group were played the intergroup call of an animal from a neighbouring group, either from the true range of the vocalizer's group or from the range of another neighbouring group. In these paired trials, subjects responded with significantly more vigilance to calls played from the 'inappropriate' range than to calls played from the 'appropriate' range (Cheney and Seyfarth 1982*a*).

Subsequent experiments followed the same design, but used as stimuli the calls of other species. Vervets were played the calls of two species that are habitually found in or near water, the hippopotamus and the black-winged stilt (*Himantopus himantopus*). The hippopotamus's call is a territorial call, while the black-winged stilt's is a low-intensity alarm given to a wide variety of potentially disturbing species. These two species were chosen because neither competes or interacts with vervets, and both are, therefore, of little biological importance to the monkeys. Nevertheless, each is a species that is so restricted to wet areas during the day that any indication of its presence in another habitat might be regarded, at least by humans, as anomalous. Black-winged stilts are never found away from water, and although hippos do emerge from water to feed on dry land, they do so only at night (Olivier and Laurie 1974).

Hippo and stilt calls were played to vervets either from the edge of a swamp or from a dry woodland area that contained no permanent water.

Social and non-social knowledge in vervet monkeys 261

All subjects were members of groups whose ranges bordered both types of area, and all had regularly heard the calls of both hippos and black-winged stilts when foraging near the swamp. Subjects were played hippo or stilt calls in paired trials, from either the swamp ('appropriate') or dry woodland ('inappropriate') habitat. Hippo calls were played at a mean distance of 91.9 m (SD, 18.3) from the subjects, while stilt calls were played at a mean distance of 40.4 m (SD, 11.2). As in previous experiments, these distances reflected the different calls' relative amplitudes. Order of presentation was systematically varied, and no individual appeared as a subject in more than one pair of trials. Because the calls were relatively long in duration, subjects were filmed for a total of 25 s following the onset of each call. Results are presented in Fig. 19.2.

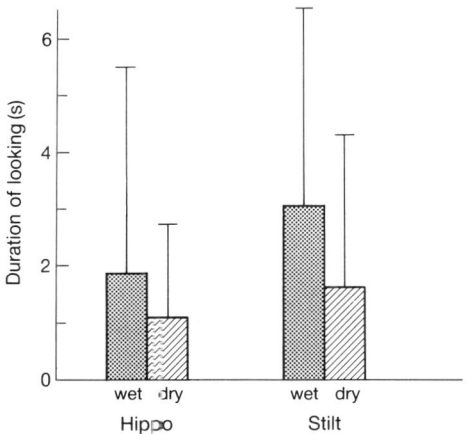

Fig. 19.2 Duration of looking towards speaker (in seconds, measured at 18 frames per second) after playback of hippopotamus and black-winged stilt vocalizations from wet and dry habitats. Values shown are means and standard deviations. For hippopotamus calls $n = 10$ subjects; for stilt calls $n = 18$ subjects. Duration of responses to calls played from different habitats did not differ significantly $(P > 0.10.)$

Subjects responded to the playbacks either by looking in the direction of the loudspeaker or by apparently ignoring the call. In the case of hippo vocalizations, subjects generally showed little response to playback, regardless of the habitat from which the calls were played. In the case of stilt vocalizations, subjects typically responded to playback by looking in the direction of the loudspeaker, but with no significant difference in the duration of response in the dry or the wet habitat. There was some indication that the vervets recognized that the stilt's call was an alarm call:

five out of 18 subjects looked up and three subjects ran towards trees or stood bipedally when they heard the calls. Again, however, the monkeys did not respond more strongly in one habitat than another. Vervets responded to both hippo and stilt calls as if they did not recognize that calls played from a dry habitat were anomalous.

These negative results, of course, cannot distinguish between the failure to recognize an anomaly and the failure to respond to one. It is entirely possible, for example, that vervets recognize that hippos belong near water, but that hippo calls played from a dry area simply fail to evoke any measurable response. Negative results are of interest, however, when contrasted with similar experiments that do evoke responses. Although vervets fail to respond to hippo or stilt calls coming from an inappropriate area, under comparable conditions they respond strongly to the calls of another vervet. The different performances are particularly striking given that the trials with conspecific calls asked subjects to assess the appropriate location of different individuals, whereas the hippo and stilt calls required only a gross understanding of the appropriate location of different species.

4. Associations between other species

Previous playback experiments have demonstrated that vervets can associate the screams of particular juveniles with those juveniles' mothers (Cheney and Seyfarth 1982b). Vervets, therefore, seem capable of forming associations between other group members, based on observations of their social interactions. To test whether vervets can form similar associations outside the social domain, we tested their understanding of the relationships that exist among other species.

Vervet monkeys regularly come into contact with Maasai tribesmen, who bring their cattle into the park to graze. Although the Maasai do not prey on vervets, they occasionally throw sticks or rocks at the monkeys, with the result that their approach causes increased vigilance and flight. Cows themselves pose no danger to the monkeys. Nevertheless, since cows never enter the park without Maasai, a cow alone potentially signals the approach of danger. To test whether monkeys have learned to associate cows with Maasai, we played the lowing vocalizations of either cows or wildebeest (*Connochaetes taurinus*, a common ungulate) to vervets in paired trials.

Calls were played to subjects from a mean distance of 72.3 m (SD, 18.0). Each subject heard each type of call only once, with order of presentation varied. Because the calls were of relatively long duration, subjects were filmed for 25 s after the onset of each call type.

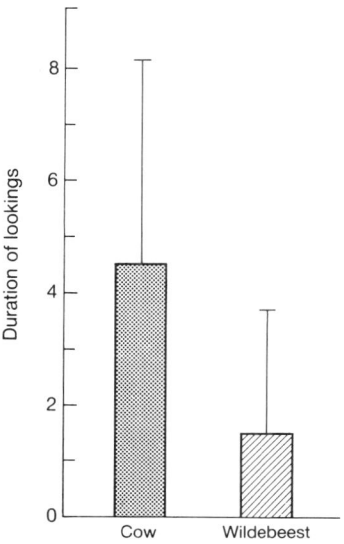

Fig. 19.3 Duration of looking towards speaker after playback of wildebeest and cow vocalizations. Legend as in Fig. 19.2. Duration of responses was significantly longer after playback of cow vocalizations (two-tailed Wilcoxon test, $n = 19$, two ties, $T = 20.5$, $P < 0.05$).

As Fig. 19.3 indicates, playback of cow vocalizations caused subjects to look towards the speaker for significantly longer durations than did playback of wildbeests' calls. This increased vigilance suggests that vervets associated cows with danger, and that they responded to the apparent approach of cows as they would to the approach of Maasai themselves.

5. Secondary cues of danger

When leopards make a kill, they frequently drag their prey into trees, where they can feed without harrassment from other predators. This behaviour is peculiar to leopards and local humans recognize that the sight of a fresh carcass in a tree denotes the proximity of a leopard. Each of the three vervet groups in our study has often seen leopards with carcasses in trees, and each time they have responded with prolonged alarm calls. We examined whether vervets knew enough about the behaviour of leopards to understand that a carcass in a tree in the *absence* of a leopard represented the same potential danger as did a leopard itself.

In conducting the experiment, we first procured a limp, stuffed carcass of a Thompson's gazelle, a species that is frequently preyed upon by leopards. This carcass was then placed in a tree, before dawn, approximately 50–75 m from the monkeys' sleeping trees. The carcass was positioned in such a way as to mimic its placement by a leopard (indeed, our attempt fooled at least one tour bus driver into thinking that a leopard was in the area). At first light, we observed the behaviour of the monkeys for a period of 2 h, noting at 5-min intervals the direction of gaze of as many group members as could be seen. In total, we presented the carcass to one group of baboons and four groups of vervets. One of these vervet groups had seen a leopard in a tree with a carcass only 4 days earlier, and had uttered prolonged alarm calls even when the leopard temporarily left the tree.

Despite all of the groups' experience with leopards and carcasses, neither baboons nor vervets alarm-called at the sight of the carcass alone. Moreover, there was no increased vigilance in the direction of the carcass over that which might have been expected by chance. In all cases the monkeys behaved as if they did not recognize that a carcass in a tree denoted the proximity of a leopard.

As a further test of monkeys' knowledge of secondary cues of danger, we tested the vervets' recognition of python tracks. Pythons are a frequent predator on vervet monkeys (Cheney *et al.* 1981), and when vervets encounter a python, they give alarm calls to it and closely monitor its movements through the area (Seyfarth *et al.* 1980 *a, b*). Pythons lay distinct, wide, straight tracks which cannot be mistaken for those of any other species, and which are easily recognized by local humans. It is possible to determine the freshness of a python track by noting both the clarity of its outline and whether or not other species have walked across it. Indeed, on many occasions when we have encountered a fresh track we have subsequently been able to find the python in a nearby bush. Vervets in the three study groups have often watched and alarm-called at a python as it laid down a track and then disappeared into nearby bushes. Do vervets therefore recognize that a fresh python track represents potential danger?

To investigate this issue we relied upon both observation and experiments. Over a 5-month period, we noted eight separate occasions when a python laid down a track in the dust and then disappeared into a nearby bush when there were no monkeys in the area. We then waited until the monkeys approached the area, and recorded their behaviour. In no case did any individual show vigilance or change its behaviour when it approached and crossed the track. Indeed, on two occasions at least one individual subsequently entered the bush where the track led, encountered the snake, and alarm-called at it. Five subsequent replications of these conditions were made by laying down an artificial python track in an area which the monkeys were approaching. Again, the animals showed no increased vigilance towards the track, and behaved as if they did not recognize that the track signalled danger.

In the preceding experiments, vervets performed well when the secondary cues of danger were auditory stimuli like alarm calls, but performed poorly when the secondary cues were visual stimuli like carcasses or tracks. There are at least two explanations for these results. First, auditory cues may be more salient than visual ones. Auditory signals have a more rapid onset time, and it has been shown that rats are more likely to associate sudden events with other sudden events, and gradual events with other gradual events (Testa 1974). Thus, it may be easier for vervets to associate (and respond to) secondary cues of imminent danger when these cues are in the auditory modality. This explanation is limited, however, because it fails to explain why, in the first instance, natural selection has favoured different abilities in the visual and auditory domains.

Alternatively, it may be argued that the vervets' use of communication has evolved mainly to solve social problems, and that this has both shaped and limited their use of signals outside the social domain. Consider, for example, differences in the way primates use auditory and visual signals during social interactions. Vervets use vocal signals both in the presence and absence of visual contact. If animals are foraging in dense bush, a vocalization can tell them that another group is approaching, or that a snake has been seen nearby, without any supporting visual information (Cheney and Seyfarth 1982a).

In contrast, although vervets make extensive use of visual signals when communicating with each other, such signals are limited to occasions when animals are in sight of one another. Vervets do not, for example, make use of each other' tracks when foraging or monitoring incursions by neighbouring groups, nor do they visually mark aspects of their physical environment to denote their rank or group membership. As a result, their lack of attentiveness towards the visual cues of predators may be related to their limited use of visual signals as secondary cues in their social interactions. Conversely, the monkeys' regular use of auditory signals to designate objects and events may facilitate their use of auditory signals as representational cues when dealing with other species.

6. Co-operation and reciprocity

Co-operative alliances among humans are characterized by the exchange of goods or services between individuals. Significantly, such exchange is not limited to any particular domain. Exchange may involve actions (for example, reciprocal support in an aggressive coalition), individuals (the exchange of spouses between two villages), or material goods (the donation of money or food to cement an agreement). In contrast, while non-human primates frequently reciprocate past affinitive acts with future co-operation, the exchange of objects is rare (Chance 1961; Chance and Jolly 1970; Kummer 1971; Reynolds 1981).

A variety of studies has demonstrated that, in interactions involving both kin and non-kin, monkeys and apes may exchange grooming, alliances, and tolerance at food sites (for example, Packer 1977; Seyfarth 1977; Chapais and Schulman 1980; de Waal 1977, 1982). Primates seem both to remember past interactions and to adjust their co-operative acts depending on who has previously behaved affinitively towards them (Seyfarth and Cheney 1984). While monkeys and apes often reciprocate previous affinitive acts, however, such altruism rarely involves the use or exchange of objects. Primate tool use, which has received considerable attention because of its relevance to human evolution (for example, Beck 1974), is striking in part because it is relatively rare. By comparison, observers of primates are continually struck by their extraordinary ability to use other individuals as 'social tools' to achieve a particular result (for example, Kummer 1968; Chance and Jolly 1970; see Chapters 9, 15, and 16). Similarly, although parties of baboons and chimpanzees often hunt and kill prey, there is little evidence that such hunts are truly co-operative, or that meat is genuinely shared (Kummer 1968; Altmann and Altmann 1970; Wrangham 1975; Busse 1978; Strum 1981; Teleki 1981).

Reciprocity among monkeys and apes therefore appears to occur more commonly in the form of social interactions, such as grooming and alliances, than in the exchange of material goods (Chance 1961; Reynolds 1981). Before we conclude, however, that non-human primates differ from humans in restricting their co-operative acts mainly to the social domain, a number of caveats should be mentioned.

First, the relative rarity of food sharing among non-human primates may result at least partly from the fact that, with the exception of meat, the food of non-human primates is simply not worth sharing. Monkeys and apes feed mainly on leaves and fruit that are distributed in such a way that they are not easily monopolized by one individual. There may, therefore, be little benefit in acquiring food directly from another. Individuals may derive greater benefit through tolerance at a particular feeding site or fruiting tree, and indeed, grooming, copulation, and other affinitive behaviours do occasionally increase the frequency with which subordinate individuals are able to feed near dominant animals (Weisbard and Goy 1976). Secondly, while non-human primates seldom exchange material goods for future beneficial acts, such patterns of exchange do occur in other species. For example, in the courtship displays of many birds and insects, the male offers food to his mate (reviewed in Wittenberger 1981).

Social and non-social knowledge in vervet monkeys 267

Finally, while non-human primates do not provision each other with food, a number of species of carnivores bring food to a central den or gathering point, where young and other individuals are fed (reviewed in Wittenberger 1981).

Co-operative behaviour in some animal species, therefore, is occasionally characterized by the exchange of material goods. We do not know, however, whether such patterns of exchange are at all modifiable. While humans can readily substitute a behavioural altruistic act for a material one, such flexibility in the 'currency' of reciprocal acts has seldom been convincingly documented in other animals. More research is clearly needed before co-operation and reciprocity in non-human species are fully understood. For the moment, however, we may hypothesize that, as in other aspects of their behaviour, reciprocity in monkeys and apes appears to occur more often in the social than in the non-social domain.

7. Discussion

When interacting with each other, vervet monkeys are apparently able to form complex associations between individuals. Within a local population, vervets can both recognize individuals and associate them with particular groups. Within their own groups, the monkeys appear to understand dominance and matrilineal kinship relationships, and also remember who has behaved affinitively towards them in the past (and see Chapter 6).

Vervets seem less able, however, to form similar associations about non-social aspects of their environment, even when to do so would confer an obvious selective advantage. Although the monkeys do recognize and respond to the different predator alarm calls given by birds, they appear to ignore the visual or behavioural cues associated with some predators. They do not seem to recognize the relationship between a python and its track, nor do they understand that a carcass in a tree indicates a leopard's proximity, even though they have had ample opportunities to learn such associations.

Similarly, although vervets and other primates exhibit many forms of co-operation and reciprocity in their social interactions, comparable behaviour using non-social currency (for example, food sharing) is relatively rare. Monkeys readily behave altruistically and form alliances to achieve social goals, but they seldom co-operate to find or exploit new types of food resources.

Finally, vervet monkeys are poor naturalists. They seem disinclined to collect information about their environment when that information is not directly relevant to their own survival. Vervets do not seem to know that hippos stay in water during the daytime, or that particular shorebirds do not occur in dry woodlands. These data are perhaps not surprising, but they do point out a potential difference between monkeys and human beings, who are naturally curious about much of their environment, and who engage in many activities that have little practical value to survival.

We believe that these results can help us to understand the intelligence of non-human primates, and to specify more precisely how the minds of monkeys and apes differ from our own. We also recognize that any interpretation is likely to be controversial. In the following section we therefore state our own hypothesis in its strongest form, then consider alternative explanations.

The primacy of social knowledge

It is now widely agreed that species-specific predispositions affect animal learning (Seligman and Hager 1972; Hinde and Stevenson-Hinde 1973; Johnston 1981). As a result of evolution in different habitats, the behaviour of different species depends not only on the logical structure of the problems they face, but also on the particular stimuli involved. We suggest that, among primates, evolution has acted with particular force in the social domain. As a result, while monkeys are able to form and make use of complex associations in their social interactions, the same sorts of associations are formed less readily when dealing with other species. Within the social group, the behaviour of monkeys suggests an understanding of causality, transitive inference, and the notion of reciprocity. Despite frequent opportunity and often strong selective pressure, however, comparable behaviour does not readily emerge in dealings with other animal species or with inanimate objects.

The special sensitivity of non-human primates to social events is not surprising. Human infants, after all, show special sensitivity to social as opposed to non-social visual stimuli (Sherrod 1981), to speech sounds as opposed to other auditory stimuli (Eimas *et al.* 1971), and to human interactions as opposed to other relationships in their environment (see below). In a similar manner, non-human primates appear to exhibit their most subtle discriminations when dealing with conspecific faces, sounds, and social relationships.

Because primate intelligence has evolved mainly to solve social problems, monkeys often show surprising gaps in their knowledge of the non-social world. For example, in social interactions visual cues are not used to represent objects or individuals in the absence of face-to-face encounters. Auditory cues, in contrast, often function during social behaviour to designate objects in the absence of visual information. These differences are reflected in the non-social domain, where monkeys are 'prepared' (Seligman 1970) to recognize auditory cues that are secondary indicators of predators, but appear 'unprepared' to associate visual cues with danger. Differences in the monkeys' use of visual and auditory information when dealing with other species may therefore result from the different way they communicate in these modalities when dealing with their own species.

Alternative arguments

(a) Differences between apes and monkeys Some of our generalizations about domain-specific performance may be less applicable to apes than to monkeys (Premack 1976), since apes do appear to make occasional use of visual symbols in their social interactions. Free-ranging chimpanzees, for example, make sleeping nests each night, and, when the members of one group make incursions into the range of another, they have been observed to make aggressive displays upon encountering their neighbours' empty nests (Goodall *et al.* 1979). The captive chimpanzee Vicki was able to sort pictures of animate and inanimate objects into distinct categories without previous training (Hayes and Nissen 1971). Whether or not a monkey would be capable of similar classification is now known, because the relevant experiments have not yet been conducted.

(b) The importance of ecological factors The food exploited by non-human primates, particularly ripe fruit, is both spatially and temporally dispersed. Field data on many species, especially orangutans and chimpanzees, indicate that primates frequently range over large areas, and that they remember the locations and phenological patterns of both water and a variety of plant foods (Clutton-Brock 1977; Rodman 1977; Wrangham 1977; Sigg 1980; Sigg and Stolba 1981). As a result, it has been argued that ecological pressures have played a major role in the evolution of primate intelligence (for example, Clutton-Brock and Harvey 1980; Chapter 21).

This hypothesis emphasizes that the distinction we have drawn between social and non-social knowledge is not a simple one. Primate memory has no doubt evolved as a result of the need to remember both the location of spatially dispersed food resources and previous social encounters. The point is not to oppose one unifactorial ecological argument against an equally unifactorial social one, but to gain a better understanding of precisely how ecological and social factors have combined to give non-human primates an intelligence that appears simultaneously to be superior to that of other mammals and inferior to our own.

Although ecological factors are undoubtedly important, primates do not appear to manipulate objects in their environment to solve ecological problems with as much sophistication as they manipulate each other to solve social problems (see above). The challenge of exploiting widely dispersed and ephemeral food items may thus have led to increased intelligence not simply because food collection itself becomes more difficult, but also because ecological complexity sets the stage for increasingly complex social competition.

Social knowledge and non-social knowledge in human infants

In the past, many students of human development believed that infants' knowledge of the social and non-social world developed at similar rates (for example, Piaget 1963). Recent studies question this view, and suggest that an understanding of certain concepts may appear at an earlier age when the stimuli involved are animate. For example, Hood and Bloom (1978) examined the development of children's expressions of causality, using 2 and 3-year-olds as subjects. Previous work had indicated that causal understanding develops slowly, and is often not apparent until age seven or eight (Paiget 1963). Hood and Bloom, in contrast, found that children readily discussed the intentions and motivations of people in causal terms. They did not, however, talk about causal events involving objects (see also references in Gelman and Spelke 1981; Hoffman 1981). In a study of naming behaviour in children 17–22 months old, MacNamara (1982, p. 30) concluded:

> ... by the time the child comes to learn language, he has already learned that objects in certain categories are important as individuals, those in other categories are merely exemplars of the category. Person is the preeminent category of the first sort. (See also Chapter 18.)

These results, together with those presented above, suggest that there is, in both human and non-human primates, an evolutionary predisposition which makes it easier for organisms to understand relationships among conspecifics than to understand similar relationships among things. Compared with humans, non-human primates exhibit this predisposition in an extreme form: they show sophisticated cognitive skills when dealing with each other, but exhibit such skills less readily in their interactions with objects. Among humans the predisposition is more subtle, but nevertheless may appear in the earliest years of childhood, when infants exhibit remarkable social skills while at the same time remaining ignorant of much of the world around them. For a few brief years, children reveal the results of selection acting on the primate brain: selection that has made them particularly sensitive to the emotions, behaviour, and social relationships of their conspecifics.

20

Tools and the evolution of human intelligence
THOMAS WYNN

The traditional hypothesis for the evolution of human intelligence, to which Machiavellian Intellect is an alternative, is that the selection pressure for intelligence was to construct more sophisticated tools and technology, which in turn fed back to heighten the same selection pressure. To evaluate the current status of this pervasive idea, we turned to an anthropologist who had shown how Piaget's system of developmental stages can be used to ask 'How intelligent was the maker of this stone tool?' This enables hims to test the fit of the Tool Making hypothesis with data, giving an answer which we had not expected.

Introduction

Modern humans are more intelligent than any of our primate cousins. In none of the tests that we have devised to get at this elusive quality, whether set laboratory tasks or the more free-wheeling opportunities posed in field problems, has a non-human approached the human ability to solve problems (Passingham 1982). Human brains are relatively far larger than chimpanzee brains; and in absolute terms, certain parts of the human brain, the neocortex for example, dwarf the comparable anatomies of our nearest relatives (Holloway 1983; Tobias 1981; Jerison 1973). A correlation between relative intelligence and relative brain size is not, of course, surprising. When we search for an evolutionary reason for this vast difference in brain power we are immediately struck by another difference between humans and apes: technology. Of all of the differences between human and ape behaviour the most visible is in the domain of technology. Humans rely on tools; without them we would perish. Recently, we have developed our technology to a level of complexity, size, and power that is so far beyond the technology of apes that it has led some scholars to discount the comparative value of ape technology altogether (Gowlett

1984). Moreover, we know from archaeology that human technology has developed more or less concurrently with the human brain. It seems, then, a reasonable hypothesis that technology was the factor that selected for the evolution of the human brain and intelligence.

This hypothesis has been very influential in the study of human evolution. In the aptly titled book *Man the Tool-maker*, Kenneth Oakley (1959) argued for the central role of tool-making in hominid evolution.

When the immediate forerunners of man acquired the ability to walk upright habitually, their hands became free to make and manipulate tools—activities which were in the first place dependent on adequate powers of mental and bodily co-ordination, but which in turn perhaps increased those powers (Oakley 1959, p. 2).

Washburn argued in a similar way.

It follows that the structure of modern man must be the result of the change in the terms of natural selection that came with the tool-making way of life (Washburn 1960, p. 62).

Neither Oakley nor Washburn considered tools alone to be sufficient but they did consider them to be the fulcrum for a complex of behaviours, including language and hunting, that levered intelligence onto a higher plane.

Twenty-five years later this bias toward tools is still influential, though in a slightly different guise. Parker and Gibson (1979), for example, suggest that a particular kind of tool-use selected for greater intelligence and linguistic ability. Certain kinds of intelligence

... facilitated expanded tool use in extractive foraging. Protolanguage was selected as an adaptation for food sharing, necessitated by the long apprenticeship for extractive foraging with tools (Parker and Gibson 1979, p. 381).

The terms and concepts of ecology and ethology now elaborate the earlier and simpler notion of 'tools maketh man'. It is unfashionable to focus on tools themselves, but very fashionable to focus on the natural systems in which tools play a central role. Many archaeologists prefer to emphasize the selective force of early 'culture', a suite of learned activities that included tool-making, tool-use, aimed throwing, butchery of large animals, transport of meat, delayed consumption, and perhaps, sharing. Tobias (1983) contends that this 'holistic' culture selected for an intelligence that was much more powerful than any known for apes. All of these ecological or holistic approaches emphasize a range of behaviours, but for most advocates this range, including language, revolves around one indispensable behaviour—tool-use. Changes in other behaviours, including social behaviour, are seen as a consequence of tool-use in foraging. Only recently have some anthropologists begun to suspect that social behaviour may have played a central role (see, for example, Holloway 1983).

The evidence of fossil brains

One way to investigate the possible connection between technology and intelligence is to ask whether or not there is a correlation between the evolution of tool behaviour and the evolution of the brain. If one uses this approach one is, in a sense, using the brain as an independent measure of intelligence. It is axiomatic, of course, that the behaviour of an animal is tied somehow to its brain. If we can document that the hominid brain evolved hand-in-hand with hominid technology, that is, if dramatic developments in one correlate with dramatic developments in the other, then we can make a case for technological selection for the brain, at least. On the other hand, if brain developments occurred in the absence of any technological developments, or vice versa, then the case would be seriously weakened.

There are three ways to look at the evolution of brain anatomy: changes in size, changes in overall shape, and changes in the position of various landmarks. The relevant data consist of the endocasts of fossil hominid crania. Such endocasts are not ideal subjects of study. They are not actual casts of the brain; rather they are casts of the impression that the brain made on the internal surface of the cranium. They preserve very little detail. At best they capture only the major features of the external shape of the brain. Moreover, a complete endocast is rare. Most extant endocasts reproduce only portions of the brain and often these are distorted because the cranium itself was distorted during fossilization. However, the major problem in studying endocasts is the size of the sample. The most complete study to date used a sample of only 41 endocasts *for all of hominid evolution* (Holloway 1981a). This sample was not controlled for sex or age of the individuals, making comparisons even more difficult. Nevertheless, as with all evolutionary samples, we must make do with what we have.

The most reliable evidence supplied by endocasts is the size of the brain, but even this is not a simple matter of measurement. Brain size is related to body size—large animals have larger brains than small animals. As a consequence we must have a measure of relative brain size. Complicating the matter further is the fact that brain size does not increase at the same rate as body size. In any given taxon, the larger species will have a relatively smaller brain. One cannot then compare brain size by comparing the percentage of body weight contributed by the brain. However, for a given taxon, brain weight and body weight are related in a regular fashion. A regression line can be fitted to the log values of brain and body weight for a particular taxon, say primates, and one can predict what the brain size should be for an animal of a given body size. One can then ask whether or not the brain is larger than predicted and also how much larger than predicted, a measure that Jerison termed the EQ or encephalization quotient (Jerison 1973; Hofman 1983 *a*, *b*). Using such a measure we find that modern humans have a brain that is three times as large as predicted using other primate data (Passingham 1982).

All currently known fossil hominids fall between the EQ values of modern apes and humans. Some fall closer to humans, some closer to apes. The earliest hominid hypothesized to be on the human line is an early form of the genus *Australopithecus* which lived around three million years ago. Measures of cranial capacity and estimates of body size give an EQ for *Australopithecus* that is clearly higher than that of the chimpanzee (3.42 *v.* 2.63 using Passingham's figures). The earliest known member of the genus *Homo*, which dates to 1.8 million years ago, has a higher EQ still (4.45), as does *Homo erectus*, the fossil species immediately antecedent to *Homo sapiens* (the Peking *Homo erectus*, dating to about 500 000 Ma had an EQ of 6.01). Modern humans have the highest EQ of the series (6.93). The specific values of the EQ can only be used for comparative purposes. Such a comparison yields two conclusions relative to my argument. First, even early in hominid evolution selection had yielded larger brains than any known for apes. Secondly, selection for large brains was especially dramatic in the *Homo* line by 1.8 million years ago.

The EQ is simply a measure of relative brain size. It is a number that lumps together all of the evolutionary changes in the brain and, as a consequence, tells us very little about specific behaviours. One can also ask what has changed in the shape of the hominid brain. Differences in brain shape between taxa reflect, largely, differences in niche. Animals that rely on the sense of smell have relatively large olfactory lobes and animals that rely on vision have relatively large occipital lobes. On endocasts such differences would appear as differences in proportion for the relevant areas of the brain. One result of such niche differences in brain shape is that distantly related animals can share some proportional characteristics. The large flying reptiles of the Mesozoic, for example, have some features of brain shape in common with modern birds (Jerison 1973). It is possible to examine hominid endocasts and assess changes that occurred in the relative proportions of parts of the brain. One can compare the results to what we know of modern apes and modern humans, and based on knowledge of the functioning of modern brains, attempt to explain the differences in evolutionary terms, including the evolution of intelligence.

In general, the evidence for changes in brain shape corroborate the evidence for changes in size. All fossil hominid endocasts differ from those of apes, but some differ more than others. Perhaps the best known difference in brain shape between humans and apes lies in the realm of cerebral asymmetry. While many primates demonstrate some differences in the anatomy of the right and left hemispheres (Falk 1980; Passingham 1982), humans possess a distinct kind of asymmetry in which an area of the left occipital is enlarged and an area of the right frontal is enlarged (Holloway and De la Coste-Lareymondie 1982). This is a pattern of asymmetry not found in the living apes. However, it has been identified on the endocasts of the earliest *Homo* fossils, dating to 1.8 million and, though the pattern is much more equivocal, may also be true of 3 million-year-old australopithecine endocasts (*ibid.*). The evolutionary significance of this pattern is far from clear, but is likely to be related to important aspects of hominid niche. Studies of brain trauma and 'split-brains' in modern humans suggest that there is right hemisphere dominance in matters requiring manipulation and location in space and left hemisphere dominance in processing of symbols and also physical dexterity (LeDoux 1982; Passingham 1982). This is a controversial field of study that has produced many hypotheses but most focus on these two behaviours—symbolic ability, including language, and spatial ability, including tools (see, for example Hewes 1973; Falk 1980; LeDoux 1982; Frost 1980). Based on such information, Holloway (Holloway and De la Coste-Lareymondie 1982) has suggested that the asymmetry identified in endocasts resulted from selection for tool-making to a standard pattern and social communicative skills. Whatever the reasons for the *Homo* brain shape, it appears to have been established by 1.8 million years ago.

One final bit of brain anatomy corroborates the evidence of size and shape. Certain brain landmarks, most notably the lunate sulcus, are not in a pongid position on the early hominid endocasts. Identifying such landmarks on endocasts is so subjective that Holloway terms it the 'new phrenology' and questions its usefulness. It is not clear, for example, that the lunate sulcus is in a human position either. Nevertheless, what little evidence there is does place three-million-year-old australopithecines out of the range of living apes (Holloway 1981*b*).

What, then, do the endocasts tell us about the evolution of hominid intelligence? Even if early hominids relied heavily on spatial behaviour and symbol processing, as suggested by brain shape, it does *not* follow that these behaviours selected for general intelligence. Brain shape always reflects niche (Jerison 1973; see above). Problem solving ability is not localized in any quadrant or lobe of the brain but seems to be distributed throughout the neocortex, the external layer of the brain (Passingham 1982). As a consequence it may well be that size, as crude a measure as it is, may in fact be the best indication of overall intelligence (Parker and Gibson 1979; Jerison 1973). Size, in terms of EQ, does correlate with number of neurons and cortical volume, two aspects of brain anatomy that are probably relevant to general intelligence (Gould 1975). Size cannot of course tell us what has selected for greater intelligence; for that we must look elsewhere. However, it can give us some idea of when important developments occurred.

In sum, fossil endocasts of the brain give us a crude look at the evolution of the anatomy of intelligence. Both size and shape of endocasts indicate that even early hominids, over 3 Ma had brains outside of the range of modern apes. By 1.8 million the brains of early *Homo* had a human-like shape and a size that was closer to modern humans than to apes. By 500 000 years ago the size of *Homo erectus* brains was very close to modern.

The evolution of hominid tool behaviour

We can now select certain points in the evolution of the hominid brain and see if there is any reason to believe that tool behaviour, as opposed to some other factor, was the selective agent responsible.

3.5 Ma

The earliest well documented fossil hominid has been assigned to the species *Australopithecus afarensis* (Johanson and White 1979). Alternative thinking argues that these are East African forms of *Australopithecus africanus* (Tobias 1983). In either case, this was a small, bidepal, savannah-dwelling creature whose relative brain size, as we have seen, was larger than that of apes. Moreover, the shape of its brain also sets it apart from the apes. While its phylogenetic relationship to the genus *Homo* has not been certainly established, its anatomy is that of a possible ancestor and is considered as such by many palaeoanthropologists. We know little about the behaviour of this early hominid. It has not been found in association with processed animal bones, as later forms have, and we have nothing that could be called an archaeological site. There are also no certain stone tools from this period. The earliest stone tools date back to about 2.5 Ma and these are not associated with fossil hominid remains (Harris 1983).

Given the evidence, we cannot conclude that *Australopithecus afarensis* relied heavily on tool behaviour. Tool behaviour is, of course, well documented for our nearest relatives, the chimpanzees, who use it in both agonistic behaviour (branch throwing, stone rolling) and feeding (termiting, ant dipping, nutting) (McGrew *et al.* 1979; Boesch and Boesch 1984). Some of this behaviour is quite sophisticated (see below). One is tempted to argue, based partly on brains, partly on later hominid behaviour, that *Australopithecus afarensis must* have relied more on tools than does the modern chimpanzee, but that these tools were made of perishable material. This is, however, no more than speculation; worse, it leads to a circular argument about brains. It is essential to have independent evidence of greater tool reliance if one wants to argue for tool behaviour as a leading selective agent. Otherwise, it is only one among many possible 'invisible' behaviours, including social behaviour.

In sum, we have no reason to conclude that tool behaviour was responsible for the changes in brains seen in *Australopithecus afarensis* and, by extension, no reason to conclude that tool behaviour selected for intelligence early in hominid evolution. This does not, of course, eliminate the possibility that it played a role later on.

1.8 Ma

By 1.8 Ma we have fossils that appear to be members of our own genus, *Homo* (Tobias 1983). While there are differences in teeth between this early *Homo* and *Australopithecus*, the more striking anatomical difference is in relative brain size. The EQ of these hominids was far outside the range of apes and, as we have seen, the shape of the brain had assumed a pattern more like that of modern humans. We also have, for the first time, an extensive archaeological record of the behaviour of these hominids. This record includes abundant stone tools and also the refuse of hominid activity. Is there anything in this record that suggests that tool behaviour was the factor selecting for larger brains? We must look separately at two behaviours—tool manufacture and tool use.

The stone tools themselves do not argue for an intelligence greater than that known for apes. Most stone tools result from the fracture of stone, producing sharp edges that can be further modified into particular shapes. By examining the patterns produced in manufacture, we can make some assessment of the thoughts of the maker. Holloway (1983), for example, argues that tools manufactured according to a standard pattern imply an intelligence greater than any known for apes. However, these 1.8 million-year-old tools were not manufactured according to a standard pattern. Indeed, the hominids' only interest appears to have been the creation of sharp edges on stones or flakes of appropriate size to the task at hand (Wynn 1978; Isaac 1981). There are really no discernible types and certainly no standard patterns (Isaac 1976). There are, in fact, no more arbitrary shapes among these tools than among chimpanzee tools. What is different is the medium. Gowlett (1984) suggests that these early stone tools required longer chains of action than anything known for apes and that this is evidence for greater intelligence. While this is intuitively reasonable, a similar argument would lead us to conclude that we are more intelligent than our great-grandfathers; after all, modern automobiles require longer chains of procedure than bicycles. I see no reason to accept simple length of action chains as a measure of intelligence. The tools are not, however, entirely mute.

One can, for example, analyse stone tools using a formal theory of intelligence. Such an approach has the advantage of using well established criteria for assessment. In my work I have used the developmental theory and method of Jean Piaget to assess the spatial competence of prehistoric hominids (Wynn 1979, 1981, 1985). One of the facets of Piagetian theory is a description of a sequence of stages through which children pass in their intellectual development. The stages constitute a typology of increasingly powerful forms of reasoning, a typology that has been applied in comparative studies as well as ontogeny (for example, Parker and Gibson 1979). Piaget developed his scheme by observing how children solved particular tasks. Among the tasks he posed were ones requiring the child to use spatial relationships: copying figures, describing the route to school, reconstructing a scene from another perspective, and so on (Piaget and Inhelder 1967). During his early years, a child constructs more and more complex geometries to aid in solving such spatial problems. For example, when copying figures the very young child will only scribble in an unordered fashion. Later, his first true copies will attend only to topological qualities of the figures—whether the figure is enclosed (squares and circles are copied in the same way) or open (crosses, for example), whether elements are connected to one another, and so on. Later still the child adds a distinction between curved and rectilinear, perspective, parallels, and measurement. Just as drawing and arranging require spatial notions of some sort, so does stone knapping. In particular, the knapper employs spatial concepts to arrange trimming blows so that they produce a useful result, a sinuous edge or a steep edge or a projection and so on. When one looks at 1.8-million-year-old stone tools one is struck by their geometric simplicity. Indeed, the hominids used only simple topological notions to trim the artifacts. Some trimming blows are 'near' others, some are 'connected' to one another, the tool's edge sometimes acted as a 'boundary' 'separating' groups of trimming blows, and so on. Nowhere do we have evidence of more sophisticated spatial notions such as symmetry, or perspective or measurement. Children master such simple topological relationships relatively early, at the beginning of Piaget's second major developmental stage, that of preoperations.

I do not mean to argue that these hominids were like human 3-year-olds, but the analysis does underline the simple intellectual requirements of these stone tools. Moreover, modern apes also test at the level of early preoperations (Parker and Gibson 1979). It does not appear, then, that stone tools themselves could have selected for a more powerful intelligence.

Of course, it may have been the uses to which the tools were put that selected for intelligence. Such an argument is more in keeping with current archaeological fashion. The archaeological record provides an interesting picture of early *Homo* and his use of tools. There is good evidence for two behaviours that are rare in chimpanzees: carrying and eating meat from large mammals. Several East African sites present stone tools in association with broken up animal bone. Both the stone tools and the animal parts had been carried to the sites, some from as far away as 3 k. There is evidence on some of the bones that they had been butchered by stone tools and also cracked for marrow (Bunn 1981). However, many of the bones had also been gnawed by carnivores and few of the body parts carried to the sites were very large. Potts (1984) argues that these sites were not home bases—geographic focus points for social groups practising sharing—but were relatively safe spots visited by hominids for the consumption of body parts scavenged on the savanna. Early *Homo* visited these sites intermittently over several years, but so did carnivores. Modern chimpanzees practice a comparable kind of behaviour. One group of West African chimpanzees have been observed returning again and again to the same stations, where they break open nuts. They even transport stones to the site to use as hammers (Boesch and Boesch 1984). Once again the parallel is striking: chimpanzees transport nuts and tools, early *Homo* transported meat and tools. The behaviours are not, of course, identical. It does appear that this kind of behaviour was far more important in the behaviour of early *Homo* than it is for modern chimpanzees. However there appears to be very little difference in the intellectual requirements for the two sets of behaviour, and that is the point at issue here.

It is possible that my assessment of early *Homo* tool making and tool use underestimates their intelligence. Archaeologists have a rather severe logical difficulty. We can only assess the minimum abilities necessary for a particular bit of archaeological evidence. Many anthropologists believe that a more 'holistic' view of early *Homo*—savannah living, tool use, meat eating, carrying, and so on presents a picture of complex culture beyond the abilities of apes. However, the *minimum* competence required for these is within the abilities of ape intelligence as we know them. If early *Homo* was more intelligent, and I believe that the endocasts argue that he was, then it must be for reasons not apparent in the archaeological record.

1.5 Ma—300 000 years

Homo erectus appears in the fossil record of East Africa about 1.5 Ma (Rightmire 1986). From endocasts we know that *Homo erectus* had a larger cranial capacity than the earlier *Homo* fossils and a pattern of left occipital asymmetry that is essentially modern (Holloway 1983). *Homo erectus* also has a suite of skull features that included pronounced brow ridges and a very low, broad skull vault. This anatomy is quite distinctive and makes *Homo erectus* one of the most easily recognized hominid taxa.

It is not until the appearance of *Homo erectus* that we have evidence of a technology that is well outside of what is known for apes. A number of developments distinguish tools of this period from the simple tools of early *Homo*. One is the invention of a technique for manufacturing very large flakes (Jones 1981). Hominids then modified these flakes into a number of relatively discrete categories of tool. The manufacture of such discrete tool 'types' implies that some previously existing idea of the overall shape of the tool existed in the heads of the stone knappers. These shapes were relatively few in number and were standardized; archaeologists find the same shapes repeated again and again in site after site. Now, there were only a few such standardized types, two or three in fact, and the remainder of the tools resemble the earlier stone tool assemblages. However, the existence of even a few standardized types means that the technology was not entirely *ad hoc* and was therefore clearly beyond anything we know for apes. If we look at the spatial concepts necessary for these tools, we again see a clear development. In addition to the topological notions used earlier, these hominids employed a rudimentary concept of symmetry and a primitive notion of measurement (Wynn in press). We usually think of symmetry as the repetition of a congruent shape, usually by rotation or mirroring across a midline. However, such a symmetry is a euclidean notion that is fairly sophisticated. A simpler notion of symmetry involves only the reversal of a configuration without maintaining congruency. Only certain simple qualities of the shape are maintained during reversal— recurves, breaks in the outline, and so on. The shape is similar but not a precise copy. The key spatial notion is that of reversal, which is a topological notion that is not tied to quantity of space, only quality. Nevertheless, it is a more sophisticated notion than 'nearness' or 'connectedness' and is achieved later than these in ontogeny. The second notion we see at 1.5 Ma is a primitive notion of measurement in the guise of the constant diameters on stone balls and tools termed discoids. The knappers used the diameter as a reference amount of space to which the tool was compared during knapping. Such an interval is not an abstract type of measurement like a metre, but it is, nevertheless, a quantity of space and quantity of space is crucial in some forms of geometry, especially euclidean geometry with its measured angles and distances. However, even this simple notion of quantity is outside the realm of topology. Both symmetry and spatial interval are characteristic of later preoperational intelligence in Piaget's scheme. This level of intelligence appears to be beyond what is known for apes (Parker and Gibson 1979). We may conclude, then, that both the technology and the spatial competence of these hominids were significantly different from that of earlier *Homo*. Unfortunately, we know very little about the context of use of these early standardized tools because we lack appropriate kinds of sites.

Tools and the evolution of human intelligence 281

Nevertheless, the correlation between anatomical developments (*Homo erectus*) and technological developments is provocative and we must consider the possibility that tools played a role at this point in the evolution of hominid intelligence.

However, if it did play a role, it was a fleeting one. Over the next 1 Ma we have evidence that intelligence evolved, but that technology changed very little. Here the endocasts do not present a clear picture. Rightmire (1981, 1986) argues that there was no demonstrable increase in cranial capacity from the first appearance of *Homo erectus* 1.5 Ma to the late *Homo erectus* of 400 000 years or so ago. Wolpoff (1986), on the other hand, believes the evidence does, in fact, argue for a substantial increase in cranial volume. The situation has not been resolved. What we do know from the endocasts is that the EQ of some late *Homo erectus* fossils was very close to that for *Homo sapiens* (Passingham 1982). If the endocasts are inconclusive on the matter of intelligence, the geometry of the stone tools is fairly clear. The spatial concepts necessary for 300 000-year-old stone tools are much more sophisticated than those necessary for 1.5 million-year-old tools. The fine handaxes of the later period required such spatial understandings as perspective and congruency (Wynn in press). Perspective is the ability to construct alternative points of view. Perhaps more importantly, it also requires the ability to coordinate several points of view simultaneously. We use such abilities every day when, for example, we approach a familiar scene from a new direction and yet can still recognize it. How do we do this? We take the evidence we see and construct, in our imagination, the familiar point of view, in the process rearranging order of elements and the masking of elements. This is not a simple cognitive task. The hominids used this ability to create fine cross sections on their tools, even though the cross sections could not be directly observed. Congruency requires concepts of spatial amount that can be used as an independent means of comparing shapes to one another. The fine handaxes often have congruent symmetry where the mirrored shape is a virtual duplicate. Such requires a notion of spatial quantity that is more abstract than that of diameter. These concepts are typical of concrete operational intelligence, Piaget's penultimate stage (Wynn 1979). The use of concrete operations places the makers at an essentially modern level of intelligence, a level clearly beyond that evident at 1.5 million.

Fine geometry is only one characteristic of tools, however, and one that probably had little relevance to the mechanical task at hand. Other aspects of technology changed little over the million plus span. Hominids made and used more or less the same range of tools at 300 000 years as they used at 1.5 million. Indeed, there are few things that one can point to as inventions in the entire period. This implies that the hominids used the same range of tasks at 300 000 years as they did at 1.5 Ma. It was a remarkably conservative time for technology. The context of these tasks may well have changed but this is hard to document given our ignorance of the early phase. We do know that during this period hominids moved out of the tropics, developed the use of fire, probably began to hunt large animals, and used specialized sites. These are important developments. Indeed, by 300 000 years ago the context of tool use appears fairly modern, but the tools themselves had not changed much, and so it takes a bit of faith to lay the responsibility for these developments solely at the feet of technology.

After 300 000 years

The transition from later *Homo erectus* to early *Homo sapiens* was a gradual one. Several 'transitional' fossils, sharing characteristics of both, are known from Europe and Africa and date to about 300 000 years ago (Jelinek 1980). Most of the evolutionary changes occurred in facial and cranial anatomy. There was no great jump in brain size. Indeed, the EQ of *Homo sapiens* is little greater than that of late *Homo erectus*, many of which fall within the range of modern humans. One group of archaic *Homo sapiens*, the Neanderthals, even had an average cranial capacity larger than that of fully modern humans (*Homo sapiens sapiens*), though this difference is perhaps misleading given the probable greater 'lean body mass' of Neanderthals (Holloway 1983). Compared to earlier brain evolution, the developments after 300 000 years were minor.

From the perspective of spatial concepts and Piagetian theory, the record after 300 000 years shows no developments. Indeed, tool geometry at 35 000 and at 15 000 years argue for the same suite of spatial concepts and, by extension, the same level of intelligence—that of concrete operations (Wynn 1986). I must add, though, that it may well be impossible to recognize formal operations, Piaget's controversial final stage, in a lithic technology. Nevertheless, the evidence we have argues against any marked increase in intelligence after 300 000 years. This is not as surprising as its sounds. If we measure from early *Homo* 2 Ma, the last 300 000 years is only the last 15 per cent of human evolution.

It is in this final 15 per cent of human evolution that all of the major developments in human technology have occurred. I include here both numbers of inventions and complexity of invention, in the sense of numbers of parts and their relationship to one another. A list of a few includes compound tools designed for several functions, trapping devices like fish weirs, alloyed metals, printing presses, and personal computers. Our technological mediation of the environment has also undergone profound changes—from simple extractive foraging to irrigation agriculture and chemical fuels. I think my point is fairly obvious. Most of our technology developed after the evolution of modern brains and modern intelligence. Therefore, it seems an unlikely selective factor in and of itself.

Conclusion

Given the evidence of brain evolution and the archaeological evidence of technological evolution, I think it fair to eliminate from consideration the simple scenario in which ability to make better and better tools selected for human intelligence. At almost no point in hominid evolution was there even a provocative correlation. The earliest known hominids, *Australopithecus afarenis*, had a brain larger than an ape's of equivalent size, but as far as we know, no greater reliance on tools. Early *Homo* at 2 Ma had a much more 'encephalized' brain, but the tools and even the context of use were not beyond the capacity of modern apes. *Homo erectus* did possess technology that was outside the range of ape behaviour, but by this time, 1.5 Ma, much of the encephalization of the *Homo* line had already occurred. In sum, most of the evolution of the human brain, the presumed anatomy of intelligence, had occurred prior to any evidence for technological sophistication and, as a consequence, it appears unlikely that technology itself played a central role in the evolution of this impressive human ability.

Discussion

Some may counter, especially the archaeologists who read this, that I have unfairly focused on tools and not on the cultural context of tool use, especially foraging patterns, information exchange, and so on. Increasing reliance on tools must have had ramifications in other realms of behaviour, they would argue, and it was precisely such ramifications that selected for a more powerful intelligence. I am not arguing that tools were not important in human evolution. They clearly played a role in changing subsistence patterns, geographic expansion, and so on. However, the question here is intelligence, not ecological success, and we cannot assume that the two are tied directly to one another through the agency of tools. We actually have only a limited knowledge of this context of use, at least until relatively late in prehistory, and what we do know is often inconclusive. Recall, that it was not until the time of *Homo erectus* that we have evidence for both tools and a context of use that was clearly beyond the range of what we know of modern apes. I have focused on the tools themselves because tools are abundant and, presumably, were an important aspect of behaviour. The tools do not argue for a central role for technology in the evolution of intelligence. Once we step away from the tools themselves and attempt to see context we are in the realm of archaeological invisibles. Some are subject to inference—butchery patterns for example—but others are highly resistant to archaeologists' efforts. However, we have no reason to conclude that the more easily inferred behaviours must also have been the more important. Were butchery patterns more important to human evolutionary success than social sophistication? As soon as we move away from the tools themselves we must consider all aspects of 'context' as equally likely to select for intelligence, including such behaviours as language and social organization, both of which are just as likely as foraging to be pivots for human evolutionary success.

In trying to identify what in human behaviour selected for intelligence we might be better off looking at what modern humans do that taxes their reasoning power. Once again tool behaviour fails to loom large. Recent studies of modern tool behaviour indicate that the everyday, mundane use and manufacture of tools is a relatively simple kind of behaviour. For example, one learns tool behaviour in a sequential manner as a series of actions chained together. (Daugherty and Keller 1982; Gatewood 1985). Practice and apprenticeship are essential in order to construct and memorize, by rote, the appropriate sequences. While one may later conceive of the behaviour linguistically as a hierarchy—routines and sub-routines—it is not learned in this manner (Gatewood 1985). I am referring here to mundane tool behaviour, not computer design, but almost all tool behaviour is mundane and repetitive, and always has been. The cognitive prerequisites for such behaviour do not match the prerequisites for other everyday behaviours such as language and complex kinship systems. Given this discrepancy in organizational underpinnings it is hard to see how technology could have selected for the kind of reasoning humans use every day.

Many anthropologists feel that complexity in human culture lies not in tools and subsistence but in politics and religion. 'The intentionality of human action implies that, where human beings are concerned, we can never predict what will happen next' (Leach 1973, p.764). The machinations of human political systems are legion; one rarely sees the term Byzantine applied to technology. Complex socio-political systems are not a recent development. The distributional patterns of stone age artistic styles 20 000 years ago suggest complex differences between and interrelationships among groups (Gamble 1982; White 1985). It seems just as reasonable to suppose that this kind of behaviour selected for intelligence as it is to suppose that tools selected for intelligence.

21

Foraging behaviour and the evolution of primate intelligence
KATHARINE MILTON

Recently, an alternative hypothesis for the evolution of primate intelligence has suggested that the intellectual difficulties (but potential foraging returns) of remembering locations and availability of widely dispersed, ephemeral, but high-quality plant foods in tropical forests were crucial. Milton not only put this forward, but backed it with good field data (Section 1*). We asked her, in the light of the most recent evidence, how well this hypothesis stood up and, in Section 2, she responds powerfully to the challenge.

Section 1: Distribution patterns of tropical plant foods as an evolutionary stimulus to primate mental development

Framework for the hypothesis

1. Body size, diet, and home-range size Larger primates require absolutely more food than smaller primates. Thus, in general, they require a larger supplying area (Milton and May 1976). Young leaves are more patchily distributed in space and time in tropical forests than mature leaves; ripe fruit is more patchily distributed than young leaves (see below; see also Milton 1977a, 1980). Primates eating primarily young leaves should have larger supplying areas than those specializing more on mature leaves, while fruit specialists should have larger supplying areas than leaf specialists (with body size and all other considerations being equal) (Milton and May 1976; Clutton-Brock and Harvey 1977a, b).

2. Spatial distribution of potential plant foods Tropical forests are generally characterized by a high diversity of tree species. Most species have very low densities. An analysis of data collected on all trees 60 cm

* Section 1 extracted from *American Anthropologist* **83**, 535–543 (1981).

and over in circumference breast height in a set of quadrats (6 one-hectare quadrats) representing 60 000 m^2 of lowland tropical forest in central Panama showed that 65 per cent of all species encountered occurred less than once per hectare. Only one of the 135 different tree species identified in this study had a relative density in the total sample greater than 5 per cent. When tested for pattern, data showed that most species tended toward a significantly clumped rather than random or uniform distribution (see Milton 1977a, 1980). Thus, potential food sources for plant-eating primates in such forests can be described as *patchy in space*.

3. Temporal distribution of potential plant foods Phenological data were collected on leaf, flower, and fruit production in this same lowland tropical forest for a 5-year period. Each week the phenological status of 394 different trees representing 145 species was recorded. An analysis of these data showed that there were pronounced peaks and valleys in the production of seasonal items (new leaves, flowers, fruits) and that individual trees of most species generally showed some degree of intraspecific synchrony in phenology (Milton 1977a, 1980; Leigh and Smythe 1978). To quantify how long particular categories of seasonal dietary items might be available from particular species, I drew 12 species at random from the 145 species represented in the phenological sample, and noted the months in which new leaves, fruits, and flowers, respectively, were observed on these species for one annual cycle. Young leaves were available on particular species for a mean of 6.8 months of an annual cycle, green and ripe fruits for 3.7 months, and flower buds and flowers for 2.7 months. Ripe fruits, however, were available on particular species for only 1.1 months. Further, these items were available on individual trees for even shorter periods of time: new leaves for a mean of 5.3 months, green and ripe fruits for 2.1 months, and flower buds and flowers for 1.8 months. Ripe fruits were available for individual trees for only 0.8 month. Examination of the nutrient content of some of these dietary items showed that in individual cases some of these items apparently were optimally edible for no more than 72 hours per annum (e.g. flush leaves for *Ceiba pentandra*) (Milton 1980). Thus, potential foods for primates in such forests can be described as patchy in time and generally *ephemeral* in terms of optimal nutritional quality.

These features of plant foods in tropical forests should make it difficult for large plant-eating primates to specialize on only one or a few food species. An enormous supplying area would be required for such a specialist and dietary items would have to be available on a year-round basis. Thus, in general, larger primates are under considerable pressure to diversify their diets and show considerable dietary flexibility.

4. Predictability in space and time Though plant foods, particularly preferred seasonal foods such as young leaves, flowers, and fruits, are very patchily distributed in space and time in tropical forests, they share an important feature that could work to the advantage of a primary consumer. This is the degree of predictability associated with their spatial and temporal distribution (Milton 1977*b*, *c*). Once the location of a particular food tree is known, it becomes a dependable seasonal resource in terms of its location for the lifetime of a primate. Further, though each tree species has a particular phenological pattern, most such patterns show some degree of predictability as well (Augspurger 1978). This element of predictability with respect to the spatial and temporal distribution of potential foods helps to counteract the patchiness component associated with such a diet. It may have served as a critical stimulus in the development of cerebral complexity in higher primates for it places them under somewhat different selective pressures than secondary consumers with respect to certain important features of their foraging behaviour.

5. Food search efficiency Secondary consumers, carnivores, typically are faced with mobile and evasive prey items which must first be located and then pursued and captured. Models dealing with foraging strategies of secondary consumers generally emphasize the importance of *both* the search and the pursuit components of foraging (MacArthur and Pianka 1966; Schoener 1971). Primary consumers, however, do not have to devote any notable expenditure of time or energy to the pursuit of prey since their foods—leaves, fruit, and flowers—are sessile and pursuit costs are small and similar for all items (Westoby 1974). We might, therefore, predict that primary consumers are under strong selective pressure with respect to increased food search efficiency since this should be a feature of major importance to overall foraging success. Increasing food search efficiency would lower foraging costs related to both time and energy expended in foraging. It would also reduce the risk of exposure to predators. In this respect, the predictability in space and time of particular items of diet could work to the advantage of primary consumers. Rather than wasting time and energy seeking out patchily distributed foods in a random fashion, such animals could move directly to particular dietary items when and where they were available.

How might selection function to improve food search efficiency when dealing with patchily distributed plant foods? It seems maladaptive in general, given the dynamic nature of tropical forests, to try and code a great variety of dietary information genetically. Rather, what appears to be required is a great deal of behavioural flexibility—flexibility to respond to continually changing forest conditions. Increasing mental complexity with a strong emphasis on learning and retention is one direction selection could take. This would accomplish the dual purpose of improving an animal's ability to recognize and remember the locations and timing of a variety of preferred foods as well as provide the required behavioural flexibility. In time, 'wise' primates that could seek out the most nutritious foods with the least expenditure of time and energy should outcompete 'less wise' primates relying primarily on a strategy of chemosensory cues or random search.

For maximum efficiency, some means other than genetic coding should also be developed to transmit such essential information to offspring since such foods are not only patchy within home-range areas but also vary from one home range to another (Richard 1977; Milton 1977a, 1980). Membership in a relatively cohesive social unit, that utilized essentially the same supplying area over successive generations, would greatly enhance efficient food search by serving to transmit information (either through imitation, learning, or a combination of the two) on types and distribution patterns of preferred foods to new generations of kin.

If the implications presented above are valid, when we examine the behaviour of forest-living primates that are primary consumers, we should find that they fall along a continuum. All else being equal, those primate species dependent on the most hyper-dispersed and patchy foods should show greater evidence of mental development than primates eating more uniform dietary resources. To test this hypothesis, I examined the diet and foraging behaviour of two primate species, both primary consumers, living sympatrically in the lowland tropical forest on Barro Colorado Island in central Panama and then, using data from other sources, I related my results to estimates of their relative cranial capacities as well as the cranial capacities of other primate species. The two species selected for detailed examination were howler monkeys (*Alouatta palliata*) and spider monkeys (*Ateles geoffroyi*).

Howler and spider monkeys

Howler and spider monkeys are large Neotropical primates, averaging approximately the same adult body weight (7–9 kg). They are found living sympatrically over much of their extensive geographical range. Both species are highly arboreal, eat an exclusively plant-based diet, and have relatively unspecialized digestive tracts when compared with certain other plant-eating primates, particularly the leaf-eating colobines and indriids (C.M. Hladik 1967; Milton *et al*. 1980). Both species live in relatively closed mixed social groups; strange conspecifics are generally repelled (Milton 1977a, 1980; Klein and Klein 1977). My own long-term data indicate that a high degree of inter-relatedness exists between troop members of both species.

Troops of both species live in specific home-range areas that overlap with home ranges of conspecific troops (Milton 1977a, 1980; Klein and Klein 1977). My data indicate that these home ranges are used over various

Foraging behaviour and the evolution of primate intelligence

generations by members of the same social network. Therefore, animals living in a particular section of the forest may be the descendants of animals living there some generations ago. Thus, in a great many important respects, these two primate species share common characteristics.

Howler monkeys and spider monkeys differ considerably, however, in dietary focus. Though both species eat primarily leaves and fruit, howlers are more folivorous while spider monkeys are more frugivorous. On Barro Colorado howlers spent an annual average of 48 per cent of feeding time eating leaves, 42 per cent eating fruit, and 10 per cent eating flowers and flower buds. At certain times of the year, however, when fruit was in short supply, howlers switched to a diet consisting almost entirely of leaves (Milton 1977a, 1980). At such times, they spent as much as 90 per cent more of daily feeding time eating only leaves. Similar high percentages of leaf-eating over prolonged periods of time have been reported for howler monkeys living in other habitats (Glander 1975, 1978).

In contrast to howler monkeys, spider monkeys on Barro Colorado spent an annual average of 72 per cent of daily feeding time eating fruit, 22 per cent eating leaves, and 6 per cent eating flowers and flower buds. When fruit was scarce in the forest and howlers turned to a heavily leaf-based diet, spider monkeys still were able to spend a daily average of 60 per cent of feeding time eating fruit by carefully seeking out what appeared to be all available fruit sources in the habitat. Hladik and Hladik (1969) have estimated that 80 per cent wet weight of the annual diet of spider monkeys on Barro Colorado comes from fruit pulp; at no time do these monkeys become strongly dependent on leaves. Similar feeding data have been obtained for spider monkeys living in other locales (Klein and Klein 1977).

As noted above, fruit is a more patchily distributed food resource in tropical forests than leaves—even young leaves. This difference is reflected in the considerable difference between the two species in the size of their supplying area. On Barro Colorado, where troop size for howlers averages some 19 animals, average home-range size is 31 hectares. In contrast, one troop of 15 spider monkeys uses a home-range area of around 800 ha. Supplying areas of similar size have been reported for similar-sized troops of *Ateles belzebuth* in evergreen forest in Colombia (Klein and Klein 1977).

Thus, spider monkeys are faced with a far more complex problem than howlers with respect to locating their food sources since, in effect, they are dealing with a supplying area over 25 times as large. How might selection have operated on the foraging behaviour of these two primary consumers so as to improve their respective foraging efficiencies?

Howler monkeys: The social group as the information unit

Data show that over an annual cycle, howlers eat foods from more than 150 different plant species, primarily large trees. An average of seven different plant species are eaten each day and daily turnover of species averages 51 per cent. Chemical analyses confirm that howlers generally choose foods of a high nutritional quality. Given the wide variety of potential foods in tropical forests, how do howler monkeys know what to eat and where to find it?

Data from several lines of investigation indicate that food choices are determined in large part by learning. Howler troops living in different areas of the Barro Colorado forest had different dietary traditions that were not totally explicable in terms of the relative abundances of particular tree species or relative availability of foods in each area (Milton 1977a, 1980). Similar patterns have been reported for other primate species eating strongly plant-based diets (e.g. Japanese macaques, Kawamura 1959; chimpanzees, Goodall 1973; sifakas, Richard 1977). Furthermore, temporarily caged wild howlers consistently refused to eat nutritious, but unfamiliar plant foods such as bananas, apples, and lettuce even though they were obviously hungry; they did, however, readily accept all familiar wild foods offered them (Milton *et al.* 1980). Young howlers raised in captivity will eat a wide variety of different foods, including eggs, cottage cheese, and meat. This too suggests that learning plays a critical role in determining what a wild adult howler will perceive as food.

Young howlers do not have a protracted period of maternal dependence. By 6 months of age a young howler, though still nursing, locomotes independently and feeds on solid foods with the troop. A long period of strong maternal dependence has been correlated with the need for a long period of information input and learning (Bartholomew and Birdsell 1953; Schultz 1968). Though howlers eat a diverse diet, young animals apparently do not have to maintain long-term close bonds with their mothers to master the dietary traditions of their area. I hypothesize that this is at least partially because they are part of a tight-knit social system. Members of a howler troop tend to perform all activities as a unit—when one animal is feeding, the rest of the troop is generally feeding too, typically in the same tree. Thus all a young howler has to do is stay with the troop and do what other members do in order to learn what to eat and how to locate such foods efficiently. After spending several years in the same home range, a young howler should master, through imitation and learning, the complexities associated with a diverse and patchily distributed diet.

Given that howler monkeys gradually learn to recognize preferred foods, the question remains as to whether they take advantage of the element of predictability associated with their food resources in order to locate them with a minimal expenditure of time and energy. There is considerable evidence that they do. First, data show that howlers tend to concentrate the bulk of their foraging activities in parts of their supplying area where the densities of preferred food species are relatively high. This increases their probability of locating preferred foods. Howlers show patterns of goal-directed travel and further can apparently recognize particular tree species as individuals and remember either their locations or routes leading to them (Milton 1977*a*–*c*). To quantify whether, in fact, howlers were locating sources of preferred food more frequently than would be expected from a pattern of random search, travel routes and encounters with individuals of two important food species (*Ficus yoponensis* and *Ficus insipida*) were analysed for one study troop for an annual cycle. It was found that howlers were significantly more efficient at locating sources of these preferred foods than if they had been travelling at random (Milton 1977*b*, *c*, 1980, unpublished data).

As leaf-eaters, howlers have relatively small home ranges, amenable to a program of constant monitoring. On Barro Colorado, average home-range size is 31 ha. The howlers' success at food location appears to depend, in large part, on the adherence to very regular patterns of activity in which some 10 per cent of the day is spent moving an average of 460 m through their supplying area over what appear to be traditional arboreal pathways connecting important clusters of food trees. Howlers may, therefore, not have to remember the types and locations of preferred foods, at least for more than a short time. (Some observational evidence, however, indicates longer-term memory.) All howlers have to do is learn these pathways and keep moving over their home range in such a way as to maximize probabilities for encounters with preferred foods. For howler monkeys, membership in a cohesive social group, living in a clearly delineated home-range area, and performing most essential activities as a unit appears to provide the mechanism whereby essential information on dietary traditions and the locations of preferred foods are efficiently transmitted from one generation to another.

Spider monkeys: the individual as the information unit

Like howlers, spider monkeys take foods from a wide variety of plant species and use a number of different food species each day. They, too, seek out seasonal foods of relatively high nutritional quality (Milton 1981). However, in various important respects the foraging behaviour of spider monkeys differs considerably from that of howlers. As noted, spider monkeys are very strongly frugivorous. Apparently in response to selection pressures related to the patchy distribution patterns of their principal dietary resources (ripe fruit), spider monkeys, unlike howlers, do not forage as a cohesive social unit. Rather, in a manner analogous to chimpanzees, spider monkey troops show a fluid grouping structure and members are often found foraging in small subgroups. Lone individuals are also encountered.

Young spider monkeys, in contrast to howlers, have an extensive and protracted period of strong maternal dependence. They apparently require a far longer period of maturation and development to master essential foraging skills. At 24 months of age a young spider monkey may still be carried for long distances by the mother and is still nursing. In addition, female spider monkeys spend time with their young offspring carrying out behaviours that appear to teach immature animals how to travel alone along foraging routes. A female spider monkey will begin to move along a foraging route, but will then sit down. Eventually, her infant grows tired of waiting and moves forward away from her a short distance through the trees along the route. The mother will then get up and move along behind the infant, reinforcing its independent locomotion through the trees in the position of leader, not follower. This can be a time-consuming process. It requires what a human observer would describe as considerable patience on the part of the mother, who, being hungry, should prefer to move rapidly to food trees rather than sit for many minutes waiting for her offspring to take the travelling initiative. This type of behaviour appears to represent active teaching of the infant by the mother.

Since spider monkeys are generally found in small subgroups, a young spider monkey probably acquires most early dietary information from its mother. Many food species are apparently learned through imitation but there is evidence that spider monkeys also acquire new dietary information by direct experimentation. I have found that captive young spider monkeys accept new foods far more readily than young howlers; captive adult spider monkeys also readily accept any palatable novel food. There may be a selective basis both for the flexibility of spider monkeys and the conservatism of howlers. As fruit eaters, spider monkeys are not likely to encounter toxic chemicals in most foods since ripe fruits are generally low in such substances. Leaf-eating howlers may often encounter toxic chemicals since leaves are generally high in such substances (Milton 1979). Furthermore, since spider monkeys tend to forage alone or in small subgroups, they may often encounter rare, unknown, but palatable fruit sources. It may be advantageous for spider monkeys to show greater behavioural flexibility than howlers with respect to sampling new foods since the risks are low and the gain could be considerable.

Data show that spider monkeys strongly utilize the element of predictability to improve food search efficiency. Like howlers, they often travel over the same arboreal pathways through the forest. Furthermore, their travel invariably seems goal-directed; single animals or subgroups or the entire troop will move directly from one individual of a fruiting species to another, often travelling a considerable distance between fruiting trees without stopping to eat along the way. The following day one or two of the same trees may be revisited by various spider monkeys and from four to eight new individuals of the same fruiting species added to the day route by moving over a route that minimizes travel distances between such fruiting trees. These data show that spider monkeys know the locations of particular trees, recognize them as individuals, and realize that when one individual of a species is producing ripe fruit that other individuals of this species are probably producing it as well. It further suggests that spider monkeys are capable of formulating a travel route in advance that takes them to a number of different fruiting individuals over the course of a normal day's travel such that they do not double back and cover areas already visited.

Given the distribution patterns of their resources, it is apparently most efficient for spider monkeys to forage in small subgroups rather than in one large unit like howlers (see also Klein and Klein 1977; Wrangham 1977). Each adult spider monkey would appear to use its own particular subset of the total pool of resource information represented by the troop to locate preferred foods. By foraging in small subunits, each spider monkey should obtain a more nutritious diet than otherwise would be the case. Many of the tree species favoured by spider monkeys (e.g. *Spondias mombin*, *Dipteryx panamensis*), unlike the fruiting trees favoured by howlers (i.e. *Ficus* spp.), ripen only a portion of their fruit crop each day. By dispersing over a wide area, and visiting a large number of different individuals of the same fruiting species each day, each spider monkey may obtain a better quality diet than if all members of the troop used the same few trees. Furthermore, spider monkeys eat absolutely more fruit each day than howlers (Milton 1981). Their greater overall fruit requirement may also best be served by using many rather than one or a few fruiting trees each day.

Thus, spider monkeys would appear to be under somewhat different selective pressures than howlers with respect to certain important features of their foraging strategy. There appears to be far more pressure on them as individuals to learn to recognize and remember types and locations of a great many different food species and to be able to add to this number by showing considerable behavioural flexibility. Spider monkeys also show a more complex set of social behaviours than howlers, including a rich repertoire of facial gestures, elaborate bouts of allogrooming, and various distinctive vocalizations, some of which carry long distances and appear to convey information of a dietary nature to members of the troop in other parts of the home range.

Discussion

If, as hypothesized, a more patchily distributed plant-based diet is, in some manner, correlated with advanced mental abilities in primates, all else being equal, spider monkeys should show more evidence of mental development than howlers since they are dealing with a more complex foraging matrix. One way of approaching the question of mental development is to examine the absolute and relative brain size of each species. A study by Quirling (1950), which examined brain size in a number of primate species, showed that mean brain weight for 63 male and female *Ateles* was *107 g* ($\bar{x}$ body wt = 7.6 kg). In striking contrast, mean brain weight for 28 male and female *Alouatta* was *50.34 g* ($\bar{x}$ body wt = 6.2 kg). Bauchot and Stephan (1969) also showed that mean brain size for spider monkeys was almost double that of howlers. Jerison (1973), who compared both relative brain size [encephalization quotient (EQ) = the ratio of an animal's actual brain size to its 'expected' brain size based on body size (see Jerison for a full discussion of techniques used in computing encephalization indices)] and relative neural complexity for a large number of primate species, found that species of Ateles showed a relative brain size and degree of neural complexity approximately double that of *Alouatta* species.

Since howler monkeys eat considerable foliage, it might be assumed that the difference in relative brain size between the two species is an artifact effect produced by the presumably greater weight of the howler digestive tract. However, data compiled on the relative surface area of different sections of the digestive tract of howler and spider monkeys show that only in the area of the large intestine is there any notable difference in size between the two species (C.M. Hladik 1967; Milton 1981). Furthermore, field observations suggest that spider monkeys consume a greater volume wet weight of food per day since they spend a greater percentage of their daylight hours feeding than howlers and spend more time eating fruit which generally can be ingested approximately twice as rapidly as leaves per unit wet weight (Milton 1980). Therefore, though there is some difference in gut size between the two species it does not appear sufficient to account for the considerable difference in brain weight between them [see Leutenegger (1973) for a discussion of exponents of allometry for brain/body size in primates].

It is recognized that indices of relative brain size as well as measures of neural complexity do not necessarily reflect relative mental abilities. The problems inherent in using such indices as evidence of increased mental abilities have been the topic of many elegant and thoughtful discussions as, for example, those of Jerison (1961, 1963, 1973), Holloway (1973), and Gould (1975). Avoiding speculation as to the precise significance of larger brain size in spider monkeys, it does appear clear that selection has favoured an increase in brain size in this lineage far more than in howlers. The brain is an expensive organ to maintain and its considerable size in spider monkeys must in some manner by compensated for by benefits accrued. The most striking difference between spider monkeys and howler monkeys in terms of their basic ecology is their different dietary focus. Most other behavioural differences as well as many morphological differences seem secondarily related to this fundamental difference in diet.

Selective pressures related to increased efficiency in exploiting an extremely patchily distributed but high-energy fruit-based diet would appear to be the most likely factor initially involved in the cerebral expansion of the *Ateles* lineage. [For an interesting discussion of encephalization ratios of various divisions of the brain in a wide range of primates see Douglas and Marcellus (1975), who conclude 'man must have a platyrrhine ancestry . . . more like that of an American woolly or spider monkey than like that of either the chimpanzee or the gorilla' (p. 179).]

Turning from a consideration of two primate genera to a wider examination of many primate genera, the same basic trend can be discerned. The comparative charts of Jerison (1973) present data on primate body size, brain size, and indices of cranial size and neural complexity. When viewed with respect to dietary focus, these indices offer support for the hypothesis that primate groups exploiting a more complex foraging matrix show greater evidence of cerebral complexity. Furthermore, a recent paper by Clutton-Brock and Harvey (1980), which uses a method of encephalization specifically designed to avoid some of the problems pointed out by Jerison (1973) [i.e. formula of Clutton-Brock and Harvey (1980) comparative brain size, CBS, (for a given genus) = log (brain wt) − (elevation for family + slope for family × log (body weight))] shows that, in general, more strongly frugivorous primates show a trend toward greater cerebral expansion than more folivorous genera. For example, the primarily folivorous Indriidae show a lower comparative brain size than the more frugivorous or omnivorous Lemurinae. *Alouatta*, the most folivorous cebid, shows the lowest value of all cebids tested. The leaf-eating Colobinae generally show lower values than the more frugivorous or omnivorous Cercopithecinae such as *Cercocebus*. The gelada baboon, *Theropithecus*, an animal that exploits a relatively uniform grassland substrate, shows a lower value than those terrestrial primates eating more hyperdispersed and patchy foods such as *Macaca* or *Papio*. The Hylobatidae and Pongidae show very high values and most are highly frugivorous as well (see Clutton-Brock and Harvey 1980).

As noted, all extant primate species take at least some portion of their diet from plants. Since all larger-bodied anthropoids are strongly dependent on plant foods, all of them face the same basic problems with respect to food search efficiency. However, some dietary groups, particularly larger-bodied frugivores, tend to face somewhat more complex problems in terms of efficient food location than other dietary groups and these are the primates generally exhibiting maximal cerebral expansion. Interestingly enough, a recent study of the relative cranial volumes of members of the order Chiroptera showed that nectarivores and frugivores (taking energy-rich foods from the first trophic level) had relatively larger cranial volumes than sanguivores, insectivores, or carnivores (taking foods from the second trophic level) (Eisenberg and Wilson 1978). Some features associated with hyperdispersed and patchy food resources, particularly high-quality plant resources, therefore, appear to stimulate an increase in brain size across ordinal boundaries though the precise functional reason for such an increase has not as yet been determined.

Section 2: Testing the plant-distribution hypothesis against recent data on brain-size and dietary constraints

This volume examines features of the social environment of primates, with the premise that primate intelligence[1] and, more specifically, human intelligence, has evolved primarily because it facilitates social expertise allowing subtle manipulation of others within the social group and avoidance of manipulation by others. In this argument, it is suggested that our intellect and that of our primate relatives can best be understood as 'a special aptitude for grappling with a uniquely social way of life' in that 'the most challenging problems faced and solved by the evolving intelligence of our ancestors . . . were those produced by social companions rather than by the *mere* (emphasis mine) physical environment' (*ibid.*). Because relatively large brain size and a high degree of sociality are characteristic features of primates, it is natural to try and link these factors together in an evolutionary cause-and-effect relationship. However, setting aside discussion of humans for the moment, there is little evidence to suggest that social systems (or breeding systems) *per se* show a strong primary relationship to either primate brain size or brain size in other mammalian orders. Indeed, all primates living in social groups should face approximately similar pressures with respect to the development of social expertise if this facility indeed confers a selective advantage. Since no other primate species shows cerebral expansion similar to that of the human lineage, sociality, in itself, is not a sufficient explanation for the evolution of large brain size in humans.

[1] Problems inherent in defining 'intelligence' have been discussed by many authors and presenting these arguments will not be attempted. For the purposes of this paper, we can refer to Jerison's suggestion that biological intelligence can be considered as a dimension of the information processing capacities of an animal . . . it is a construction of the nervous system designed to explain the sensory and motor information processed by the brain . . . a model of possible reality (Jerison 1976). For detailed discussion of the evolution of the mammalian brain and intelligent behaviour, readers are referred to Jerison (1973) and Macphail (1982). These authors carefully consider both topics and reading their work is an essential prerequisite for examination of the meaning of the relatively large brain size in non-human primates and humans. Though I touch on some of their ideas in this paper, it is best to read the original sources since quoting or paraphrasing ideas briefly does not do them justice and may even, unintentionally, distort the meaning of their authors.

Primates have long been large-brained mammals. The fossil record indicates that primates 50 Ma were about twice as encephalized as their land-mammal counterparts (Jerison 1976). It is unlikely that early primates, particularly if nocturnal, were high social. They did, however, have to deal with the efficient location of plant and insect foods in a complex three-dimensional environment, a task requiring integration of information from several sensory systems on an unusually wide array of different objects in space and time.

As discussed in Section 1, a detailed examination of the relationship between comparative brain size (CBS)[2] and certain environmental parameters (i.e. diet, stratification, activity timing, home range size, and breeding system) showed that folivorous primate groups had lower CBS values than frugivores, a difference not attributed to any difference in relative gut size between them (Clutton-Brock and Harvey 1980). Stratification and activity timing did not reveal significant differences between groups. A positive correlation was found between CBS and home range size, but this latter feature is typically related to diet (Milton and May 1974). In terms of breeding system, monogamous groups showed lower CBS than multi-male or single male groups though no significant difference was found between these latter two groups. Though the difference is statistically significant, the authors point out that the sample of three monogamous genera is too small for the result to be regarded as reliable. As, all else being equal, monogamous primates should exploit a smaller home range area than group-living primates (Milton and May 1974), this difference could be related to a less complex foraging matrix.

A similar association between small CBS and a folivorous diet likewise occurs in families of small mammals (rodents, insectivores, and lagomorphs) (Mace *et al* 1981). At the family level, an association between CBS and social living was noted for some families, but overall, the association between sociality and relative brain size is weak in small mammals perhaps, as suggested by the authors, because both of these features are related to diet. Three independent examinations of relative brain size and diet in bats have shown that frugivorous bats exhibit a larger relative brain size than insectivorous bats (Pirlot and Stephan 1979, Stephan and Pirlot 1970 cited in Eisenberg and Wilson 1978; Armstrong 1985). No comments were made on the role of sociality with respect to this finding. Recent examination of relative brain size in marine mammals (Worthy and Hicks 1985), has shown that the Sirenia, which feed on nutritionally poor forage relative to other groups, have relatively small

[2] Comparative brain size (CBS) is a measure of brain size derived by Clutton-Brock and Harvey (1980), which attempts to control for the more obvious effects of phylogenetic affinity: CBS (for a given genus) = log (brain weight) − elevation for family + slope for family × log (body weight). This brain size measure was also used by Mace *et al*. (1981) in their examination of relative brain size in small mammals.

brains. The odontocete whales, in contrast, are similar to primates with respect to large relative brain size. Their high degree of encephalization is suggested to be due in large part to the development of a sophisticated click-based echolocation system which they use to pinpoint the location of prey items. Odontocetes appear to live in relatively stable social groups. Worthy and Hicks (1986) suggest that a highly developed auditory system may also have facilitated sociality by permitting the use of graded acoustic signals. A more complex social group may require more interpretive ability, not only in recognizing graded signals but also the more complex contexts in which they may be used (*ibid.*). These authors suggest that a good auditory system therefore may not only have facilitated sociality to some extent, but also led to further increases in encephalization because of sociality. This point will be returned to later in my discussion of factors related to increased encephalization in our lineage.

Data sets on primates, small mammals, bats and marine mammals therefore suggest that diet (and complexities associated with its procurement) show an association with relative brain size. Little support is found for the view that social systems or breeding systems have a similar effect on brain size. Further, the social organization and breeding system of a mammal may show a strong association with diet (e.g. Milton 1985). In stressing the possible role of diet as an influence on brain size, I do not mean to imply that social systems and sociality do not also make a contribution to encephalization, for I believe that they do. However, it is important to realize that strong evidence for sociality as a primary influence on brain size has yet to be demonstrated.

Dietary items of high quality, particularly plant foods in a tropical forest, are likely to be rare relative to lower quality foods as well as highly attractive to a large number of animals. Therefore, it is not surprising to find that primate species committed to the exploitation of such foods as dietary staples might require foraging behaviour of unusual complexity both to locate such foods efficiently and to dominate access to them. Furthermore, due to the troop fragmentation often required to locate sufficient amounts of these high quality foods (e.g. spider monkeys, chimpanzees), we might also expect to find strong development of behaviours to facilitate recognition of other troop members, group bonding mechanisms, and perhaps a more sophisticated communication system, effective over long distances. In turn, selection for the potential for such highly developed social behaviour could likewise contribute to increased encephalization in these groups. However, *a priori*, one might assume that any primate, regardless of its dietary focus, might benefit from enhanced mental capabilities if these indeed confer selective benefits (e.g. enhanced mating success, better diet, more efficient escape from predators). Since most anthropoids live in social groups, at times large social groups, it is difficult to see why being a folivore *per se*, would select for less intelligence than being a frugivore if sociality lay at the heart of a highly developed brain. The pattern of relatively small brain size in folivorous lineages therefore suggests that some factor(s) may constrain mental development in these lineages. This brings us to the relationship of energetics to brain size or, as phrased by Armstrong (1985), the question not of what the brain can do for the body, but perhaps more fundamental, the question of what the body can do for the brain.

The cost of the brain

Brain tissue is metabolically expensive, requiring large supplies of oxygen and glucose. Furthermore, the brain's demand for these products is constant and unrelenting, regardless of the mental or physical state of the organism[3]. Most of the resting metabolism of an organism is derived from metabolic activity of the principal internal organs. The sizes of these organs—with the exception of the brain—are nearly linear functions of body weight and their metabolic rates also correspond with body size. The size and metabolism of the mammalian brain, in contrast, displays a wide interspecific variation that leads to a much more complex relationship to overall energetic consumption [above information from Hofman (1983,a, b) who cited various sources as well].

The mammalian brain consumes a much higher proportion of O_2 than brain to body weight ratios predict (Armstrong 1985). The brain uses O_2 for the aerobic oxidation of glucose and, unlike other organs, apparently has little choice for substrate (though ketone bodies can be oxidized to fuel the brain's activities during long periods of starvation). Because the brain stores almost no energetic reserves, severe hypoglycaemia leads to stupor with rapid and irreversible damage to the nervous system (Hofman 1983a, b; Armstrong 1985).

With increasing body size, the metabolic rate per unit of brain tissue declines. In *Macaca mulata*, for example (body wt 4.3 kg), the brain metabolic rate (mg O_2 g^{-1} min^{-1}) is 0.048 while for *Homo sapiens* (body wt 68.7 kg), this figure is 0.034 (Hofman 1983a, b). This phenomenon is believed to be due to a decrease in neuron density with increasing brain size (*ibid.*). It is possible to estimate the ratio of body to brain O_2 consumption. For most mammals, the metabolic ratio of these two factors (M_{brain}/M_{body}) is less than 10 per cent. For humans, however, this figure jumps to 20 per cent (*ibid.*). In humans, therefore, an impressively large proportion of total body metabolism is utilized by the brain. In human infants and children up to 4 years of age, this figure can reach 50 per cent (*ibid.*). Examination of the O_2 requirements of different portions of the brain reveals that cortex metabolism is about 43 per cent higher than the

[3] One exception may be hibernating mammals (Hofman 1983a, b).

weight-specific metabolic rate of the entire brain. This difference is associated with a difference in tissue structure, with grey matter possessing a much higher metabolic rate than white matter. The ratio of cortex-to-brain metabolic rate is independent of body size and increases with the evolutionary level of brain development (Hofman 1983a, b).

Given the constant high cost of the mammalian brain, it follows that if selective pressures favour an increase in brain size, a concomitant increase in energy to the brain must also be favoured. This could be accomplished in a variety of ways which are not mutually exclusive. An increase in body size, for example, would provide the brain with a larger cardiovascular delivery system; a higher basal metabolic rate would provide an organism with a faster turnover rate of O_2 (Armstrong 1985). Furthermore, animals differ in the amount of the energy and O_2 reserves that they send to the brain and may also differ in some as yet unstudied aspects of their physiology as, for example, the carrying capacity of haemoglobin (Armstrong 1985).

One of the strongest functions that has been statistically well established in physiology is the basal metabolic rate (BMR) of mammals in which the oxygen consumption or heat production of an animal per unit time is proportional to the ¾ power of body weight [Hofman (1983a, b citing Brody (1945) and Kleiber (1961)]. Recent work suggests that brain size may also scale to body size to approximately the ¾ power (0.73) rather than the ⅔ power as was formerly indicated by a smaller sample size (Martin 1981; Hofman 1983a, b; Armstrong 1985). The relationship between brain size and basal metabolism is less clear, but recently several authors have suggested that brain size scales isometrically with basal metabolic rate (*ibid.*). This finding indicates that enlargements of brain size may be limited by the availability of energy sources as determined by body weight and O_2 turnover (*ibid.*) Since diet apparently can influence metabolic rate (McNab 1980; Armstrong 1985, citing several sources) and metabolic rate is isometric with brain mass, one might predict an effect of diet on achievable brain mass (e.g. Worthy and Hicks 1986).

It therefore appears that brain size in mammals may relate to systems delivering 0_2 and glucose (Hofman 1983a, b; Armstrong 1985). Primates, however, pose a problem in this interpretation because they retain a relatively large brain to body size even after correction for their BMR. As yet it is not known how the primate brain maintains a large size relative to the body's total energy supplies. One possibility is that it receives a proportionately larger share of the body's total energy reserves (above information from Armstrong 1985).

Diet and the human brain

Regardless of selective pressures favouring a larger brain, the body must be able to provide the substrate for brain maintenance and, in humans, the cost is high. In addition, it is not only the metabolic demands of the body that must be met each day, but also the demands of active metabolism. The human lineage shows an increase in body size in comparison with australopithecines; further, early humans show a dramatic increase in brain size as well as refinement of the dentition. In combination, these features point to a change in diet.

It is unlikely that early humans fed entirely on meat since they lack metabolic specializations to diet characteristic of true carnivores (Morris and Rogers 1983). Furthermore, by analogy with the behaviour and diet of extant apes, it is unlikely that early humans fed on fibrous or low quality plant foods. In the hominoid line, commitment to a diet high in indigestible plant matter produces a relatively sedentary (though large-brained) ape, not noted for a high degree of sociality (i.e. orangutan, gorilla) (Milton 1986, 1987). In this respect, it is interesting to note that modern humans differ from all other extant hominoids in both the proportions and relative size of the gut (*ibid.*). In apes, the digestive tract is dominated by the volume of the colon (*ibid.*). Apes apparently require large colon size to cope with the masses of relatively indigestible matter taken in each day in their almost exclusively plant-based diet. Humans, in contrast, have a digestive tract dominated by the small intestine. The small intestine is the major site of nutrient breakdown and absorption. Its large size in humans suggests that their diet is both nutrient-dense and rapidly digested relative to the diets of extant apes. The large size of the human small intestine appears characteristic of all human groups, regardless of diet, and indicates
(1) that all modern humans eat a relatively high quality diet;
(2) that this feature is under genetic control;
(3) that it is derived from the common ancestor of all extant humans.
It further suggests that this common ancestor also ate a high quality diet for if this were not so one would expect to see gut proportions more similar to those of apes and presumably characteristic of the common ancester of all hominoids. There can be variation in the size of different sections of the gut; some human groups, for example, have a notably larger colon size than others (Wenger 1943). However, in so far as I am aware, the dominance of the small intestine is a characteristic feature of our species regardless of some variation in the size of particular sections of the gut.

It has been suggested that the human gut may also be small relative to body size in comparison with most other extant primate species (R. D. Martin, personal communication). Here we have an apparent anomaly in that we have a species with an unusually large brain, an organ regarded as an energetically expensive one, and yet at the same time this species has a gut that appears small for its body size. However, as noted, humans do have a capacious small intestine. In addition, the human gut may have some specializations which facilitate nutrient uptake and compensate for its small size. In any case, both gut proportions and gut size support the view that the human species, compared with other hominoids, eats an unusually high quality diet. Dental evidence from early humans supports the idea that the human lineage has long enjoyed a diet of high quality. The size of the teeth and the thinner molar enamel suggest that early humans were either chewing a fairly amenable nutrient substrate or that something was done to food prior to placing it in the mouth such that the teeth (and digestive tract) were spared from performing much of the work typically performed by the teeth and guts of other primate species.

As noted, increased body size can facilitate brain enlargement, but absolutely more food will be required by the organism. One way to meet this demand is not necessarily to eat more but rather to eat better, that is, to eat higher quality foods than before. Such foods should be correspondingly more costly to procure since they are presumed to be less abundant and/or to require heavy preparation costs. A focus on such foods should result in strong selection pressure to improve food search efficiency (and/or food preparation skills), thereby lowering procurement costs. How might this be accomplished? It could be accomplished in a variety of ways, which are not mutually exclusive, and which appear to depend heavily on advanced mental capabilities for their successful enactment. Co-operative hunting, for example, could make formerly inaccessible high quality foods into dependable dietary resources; technological innovations could aid with food procurement as well as food preparation. As stressed by Lancaster (1975), an ability to make environmental references, even on a very primitive level, would have had great adaptive value to members of a group providing that it was ecologically important for them to communicate information about environmental features (i.e. words or sounds suggesting major activities such as hunting or gathering, nearness of objects in space and time and the like). A division of labour by the sexes and food sharing would permit a species to specialize on foods from two trophic levels simultaneously—a dietary innovation that, among mammals, seems particular to humans and with major implications for human evolution (Lancaster 1975; Isaac 1978; Milton 1981, 1987). Thus, the foraging behaviour of early humans was probably marked by increased complexity in communication skills, social skills, and technological skills over the australopithecine level and all of these behaviours are assumed to relate to greater mental complexity in our lineage. The precise factor or factors leading to selection for greater cerebral complexity and the successful performance of these behaviours remains obscure, but a climatic change and factors related to foraging efficiency and dietary energetics in the face of such change continue to be compelling candidates (Batholomew and Birdsell 1953; Jerison 1973).

The human lineage is described as occupying the adaptive zone of culture and the foundations for this occupation rest on the abilities of humans for symbolic imagery and language. There is little that can be said with any degree of assurance on the evolution of language (Macphail 1982). However, its impact on intellectual activity is of incalculable importance in that many activities such as reasoning and communication—in the sense in which humans communicate—may be dependent on language (Macphail 1982). Jerison (1973, 1976) has suggested that the original function of 'language' was probably not the facilitation of inter-individual communication. Rather, language may have stemmed from pressures to place vocal labels on features of the environment (see also Lancaster 1975). If there were selective pressures toward the development of language particularly for intraspecific communication, one would expect the evolutionary response to be prewired language systems with conventional sounds and symbols (*ibid.*). In contrast, the very flexibility and plasticity of language systems of the human brain argue for their evolution having been analogous to the other sensory integrative systems, which are now known to be unusually plastic and modifiable by early experience (Jerison 1973). As stated by Jerison:

we can think of language as being merely an expression of another neural contribution to mental imagery, analogous to the contribution of the encephalized sensory systems and their association systems. We need language more to tell stories than to direct actions (Jerison 1976, p. 101) . . . the fact that (language) also resulted in improved communication should not be confused with the requirements for communication in troops of monkeys or apes. The quality of communication in living troops of non-human primates is such that it could be simulated with very little information-processing material, and it appears to be no more effective in the control of behaviour than the vocal communication of other species, for example, the vocal communication of birds (Jerison 1973, p. 427).

Though language originally may have evolved to help order reality and construct an unusually complex perceptual world, it also facilitates inter-individual communication, thereby creating a more complex social environment. This, in turn, requires a more complex level of discrimination for interpreting the meaning of such communication signals within the social matrix. In this sense, the human social environment, similar to the process described above by Worthy and Hicks (1986) for odontocetes, may have stimulated encephalization in early humans. It is presumably not accidental that many animal lineages which possess relatively large brains for their body mass are strongly social and possess an unusually elaborate communication system (e.g. honey bees, parrots, odontocetes, higher primates). Once human language began to develop and be used, with all that this implies, human encephalization presumably proceeded rapidly due to selective benefits accruing to those individuals best able to communicate and utilize this new medium efficiently. However, in the case of humans, language not only facilitated interindividual communication, but was carried to such an extreme that ultimately a given language structures reality for its speakers. However, I think it a mistake to see selection pressures related to subtle social manipulation as a primary factor influencing large brain size in the human lineage. For example, it is argued that because primates can perform complex tasks in a laboratory setting that are not seen in the wild, their mental capabilities must have arisen in response to pressures from the social rather than physical environment. In fact, birds and apparently most vertebrates are capable of performing tasks in a laboratory setting not seen in the wild, such that the suggestion is now advanced that all living vertebrates may possess comparable mechanisms of intelligence (Macphail 1982). To me, the present-day social environment of humans does not seem that much more complex than is the case for chimpanzees or savannah baboons and it is likely to have been far less complex at the stage when protohominids made the transition to the human grade of evolutionary development.

Early humans are believed to have lived as hunter-gatherers in small social groups, occupying an extensive geographical range. As discussed, the food acquisition system of human hunter-gatherers rests on an unusual system of foraging based on a division of labour by the sexes and food sharing. Selective pressures related to the development of such an unusual foraging system must have placed a strong premium on the ability of the individual to judge the net return in food obtained in exchange for the time and/or energy invested in food procurement. Cheaters (clever manipulators) in such a foraging system might survive over the short run, but would be unlikely to persist over the long run as eventually there would be no one left to carry out the work of hunting and gathering or no one wishing to hunt or gather with the selfish manipulator. Therefore, in my view, one of the strongest selective pressures on early humans would have been to develop the ability for co-operative behaviour, delayed gratification, and the sharing of highly desirable and essential goods. These social behaviours seem far more critical to immediate individual survivorship (and reproduction) in early humans than subtle manipulation of one's neighbour. In fact, what seems unusual about human evolution is not that humans display selfish behaviour—for this fact appears true of all social mammals. Rather, entrance into the human adaptive zone of culture, in effect appears to have forced each member of the social unit not only tacitly, but genuinely to co-operate with other members of the social unit in order to survive and reproduce. The fact that many members of such social units are likely to have been close kin could have facilitated the development of such behaviour (Trivers 1971, 1985).

Acknowledgements

I thank Allen Herre, Joseph Wright, and Mary Jane West-Eberhard for helpful discussion and comments with respect to this paper.

Exploiting the expertise of others

22

An experimental study of social knowledge: adaptation to the special manipulative skills of single individuals in a *Macaca fascicularis* group
EDUARD STAMMBACH

To what extent do the social initiatives of primates utilize knowledge of others' special abilities which can be exploited for their own benefit? Stammbach has pursued this question through the study of reactions to experimentally controlled changes in the specialist skills of specific low-ranking individuals in a social group of macaques. This work complements Menzel's more flexible approach (Chapter 12) to the same question.

1. Introduction

Knowledge plays a key role in the understanding of intelligence. Knowledge of other group members or of environmental conditions permits individuals to behave flexibly according to several variables in the social and nonsocial environment (see Chapter 5). Few studies have so far directly addressed monkeys' abilities to assess characteristics of others (see Bachmann and Kummer 1980; Cheney and Seyfarth 1980; Dasser, 1987 a, b; Chapters 6 and 7). These studies, however, focused on social characteristics of group peers.

For a systematic study of a primate's knowledge of others, I have chosen a procedure in which the skills of certain group members were controlled experimentally. In a first step, I ascertained that all members of a group of longtailed macaques (*Macaca fascicularis*) were able to attain some of their food by manipulating a popcorn dispensing apparatus, thus introducing the 'tradition' of lever pulling. Then, a single low ranking individual, who could not monopolize apparatus and food due to its dominance position,

was trained in a more complex manipulation task. This task consisted of manipulating three levers in a correct sequence, whereupon food became available for himself and other group members. This specialist thus was a producer of preferred food in the monkey group. The main questions of the study are:
(1) Is it possible for low ranking group members with privileged information to become specialized food producers?
(2) Will other group members become aware of the special skills of that individual and will they adapt their behaviour accordingly in order to benefit from his skills?
(3) Will there be any effects upon dominance or other social relationships in the group?

2. Methods

The study group consisted of 40 long-tailed macaques (*Macaca fascicularis*) and was naturally composed. Matrilineal kinship of the group is well known. The group was permanently housed together in a 50 m^2 indoor cage and a 1000 m^2 outdoor enclosure and was fed three times per day with cereals, seasonal fruits, and special monkey pellets. There was no special deprivation of food before trials were carried out. The experiments were carried out on subgroups of the colony. A subgroup was separated from the colony for each experimental trial. In each of three consecutive years (1983/4, 1984/5, 1985/6) one of three different subgroups was used. The subgroups consisted mainly of young males (subgroup 1), of young females (subgroup 2), and mainly of adult females (subgroup 3), respectively. Mothers had their youngest offspring with them.

The apparatus consisted of two parts.
1. The three levers that had to be manipulated in a correct sequence in order to release a food reward.
2. A food dispensing mechanism that provided preferred food items in three food bowls.

The three food bowls were installed beneath the levers, the middle one directly under the levers and the other two on the left and on the right in a distance of 50 cm. Thus, if a monkey manipulated the levers correctly, he himself and at least two other monkeys could eat from the food reward. Three levers were situated in a height that allowed a monkey to manipulate them in a sitting posture. A correct manipulation consisted of pulling first the left lever, then the central one afterwards and the right one last. Every 20 seconds, a monkey had 10 seconds of time available for a correct manipulation. During these 10 seconds, a beeper indicated that the levers were activated. For the whole period, when the apparatus was ready for manipulation, a yellow xenon flash light was blinking. For the basic training of operating a food dispensing apparatus as carried out with all animals of a subgroup, a simpler one-lever-apparatus was used.

In summer, trials were carried out in the outdoor enclosure, in winter in the indoor cage. A single trial lasted 1½ hours. In the first 45 minutes (feeding phase), the apparatus was switched on, which was indicated by the xenon flash, and successful lever manipulation was possible. In the second 45 minutes (social phase), the apparatus was switched off and social behaviour of the animals was observed. In winter, the social phase of the trials had to be reduced to 30 minutes. Each series of trials involving a particular specialist is referred to as a 'replicate'.

The training of a specialist took place within the subgroup: the subgroup was released into the experimental enclosure, but only the prospective specialist was now rewarded for each single pulling of one of the levers first. Since the food reward could be suppressed by switching off the food dispensing part of the apparatus, I could ensure that no other subgroup member than the specialist himself could perform a rewarded lever manipulation. In the next phase, not one single, but a triple pulling of levers in any sequence was rewarded. Then, the specialist was gradually guided to pull levers in a given sequence: in a first step, he had to pull the left lever once. Then he had to perform two pullings at one or both of the other levers. In the last step of training, he got the reward only if he pulled the correct sequence left → middle → right. The other subgroup members thus could observe how the specialist learned to manipulate the levers successfully.

In two of the subgroups (1 and 2) three replicates with three different specialists were established in sequence, in the third subgroup only two. Malen, Ukui, and Junus were the specialists of subgroup 1, Sanah, Djambi, and Saja the ones of subgroup 2, and Sakri and Mayun of subgroup 3. Specialists were chosen according to two criteria. They had to be as low ranking as possible and in spite of their low rank had to exhibit a satisfactory lever pulling activity. Before each replicate, during which only the lever pulling of the specialist was rewarded, a short series of trials took place, in which all the animals of the subgroup were allowed to manipulate one single lever successfully again. This phase served to 'dismiss' the specialist of the previous replicate who was not rewarded at all for the first three trials of these intermediate phases and to bring all subgroup members to lever pulling again before the next specialist was established.

3. Results

Figure 22.1 shows the successful lever activities of the first three specialists as examples. Activities are measured in 20-second intervals with successful

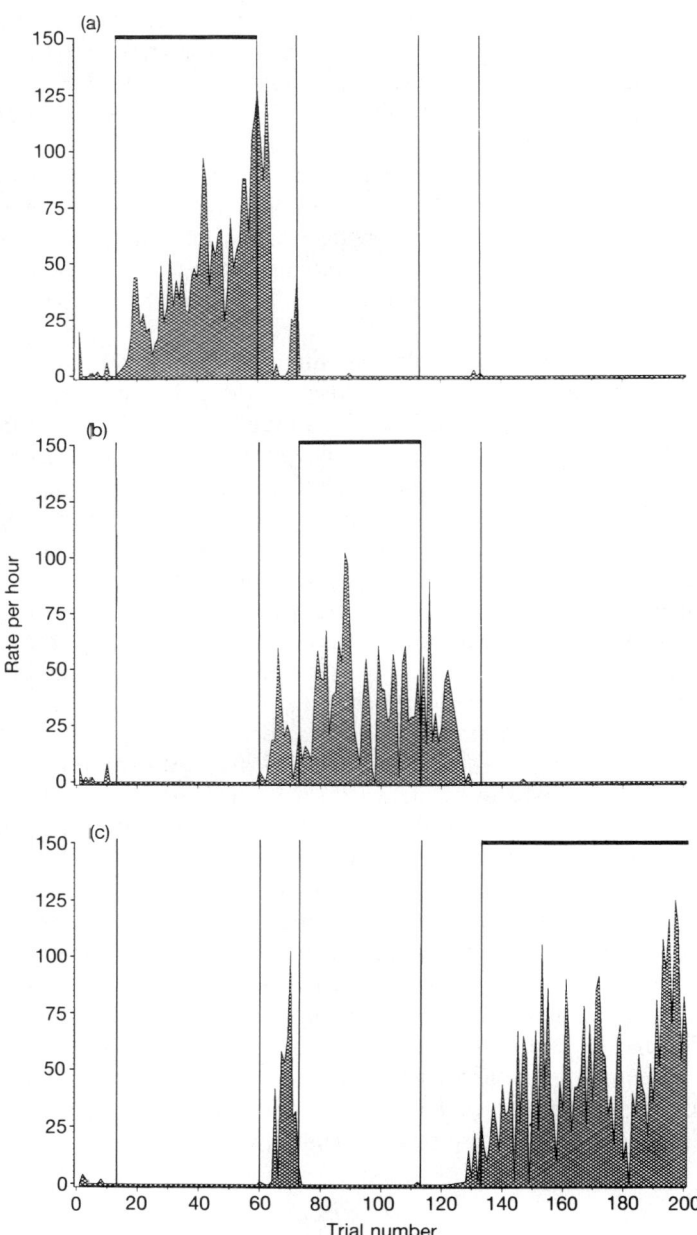

Fig. 22.1 Rates of lever pulling per trial for three of the eight specialists: (a) ML, (b) UK, and (c) JN. Vertical lines indicate the beginning or the end of a replicate. A black horizontal bar indicates the specialist's replicate.

lever manipulations per hour. The three male specialists, Malen (ML), Ukui (UK), and Junus (JN), and the two females, Saja and Mayun, pulled levers regularly and with quite a high frequency; a maximum of 180 lever manipulations per hour were possible. The three females Sanah, Djambi, and Sakri had lower frequencies.

Let us now deal with the way of measuring the amount of benefit that a group member gains from the specialist. Since I registered at the end of each 20-second interval in which part of the enclosure each animal was staying, I could determine whether two animals were present simultaneously near the food bowls. In each 20-second interval in which the specialist operated the lever successfully and in the following one, food was present in the food bowls. Thus, the frequency with which an animal was present together with the specialist in one or both of the two intervals following a successful lever manipulation is a rough measure of the benefit that an animal gains from the specialist. The *sum of joint presence* is defined as *the sum of intervals with joint presence within a replicate*. This figure varies between 0.0 and 30.1 per cent of the total intervals.

If a limited source of preferred food is available in a monkey group, the highest ranking group members initially try to monopolize the source. As soon as a specialist operated the levers correctly, high ranking group members approached and displaced or chased away the lower ranking specialist from the apparatus. Thus, the question arises, whether the high ranking animals were able to learn to approach cautiously and not to chase the specialist who is responsible for 'food-production'. I performed data analysis as follows. For each trial I determined how often in terms of 20-second intervals the specialist left the 1-m area situated around the apparatus. This number gives an estimated maximum number of intervals, in which the specialist could have left the apparatus, because it was displaced or chased by others.

To test for changes over time, Kendall τ correlations were computed between trial numbers and *displacings*, per total number of *leavings*; and for trial numbers and *chasings*. In 14 dyadic relationships out of 55 displacings were significantly reduced ($P \leq 0.05$) in the course of the replicates, while in 15 dyads out of 37 where the specialist is the lower ranking one, the higher ranking one significantly reduced his chases throughout the trials. These higher ranking animals thus could increase their benefit from the activities of the specialists.

These findings give a first hint that some of the higher ranking animals became somehow aware of the special role of the food producer and were able to hold back their chasing. This may be illustrated by a scene that I observed on December 13th in 1983 when Ukui was established as a specialist. Ukui manipulated the levers calmly and Ketut, the highest ranking subgroup member, sat near him and ate from the popcorn. As

Junus approached, Ketut chased him violently, passing behind Ukui's back by some 10–20 cm only. Ukui seemed not to pay attention to the situation; he appeared to know that this chase was not directed to him.

For the last five replicates and, thus, for the last five specialists two sets of apparatus were available, in the indoor and the outdoor enclosure. Alternately, only one of the two was switched on. Thus, both the specialist and the non-specialists had to move regularly between the two sets of apparatus if they wanted to continue to receive food rewards. What should we expect from an animal that, as we hypothesize, from trial to trial becomes more aware of the specialist's skills and role? An ignorant monkey should approach the apparatus only after a successful lever manipulation by the specialist when food is already available, thus responding to the visible food only. Informed monkeys should approach the apparatus as soon as the specialist himself starts to approach and thus follow or even pass the specialist. Three possibilities of movements of other group members in relation to the specialist's actual position have to be distinguished:

(1) another animal approaches after the specialist has already manipulated the levers successfully;
(2) other animals follow the specialist while he approaches the apparatus in order to pull levers; or
(3) they even pass him and arrive before him at the apparatus and wait for his lever pullings.

I defined two measures that reflect these manners of behaviour: if the specialist approached at least 1 m in the direction of the apparatus site, the movement of the other animals was observed. Other group members could either remain at their place or move in another direction than the specialist or they could perform one of two movements: they could follow the specialist (recorded if they moved in the same direction as the specialist for at least 1 m); or they could pass the specialist after having followed first and thus arrive first at the food site before the approaching specialist. Both types of behaviour were compared with the total number of approaches to the apparatus performed by the specialist. Whether an animal learned about the skills of the specialist or not is reflected by the changes of these proportions over the trials of a replicate: both proportions should increase in the course of the trial series. A Kendall τ correlation as described in the previous section was carried out to look for positive correlations among the above mentioned proportions and the trial numbers. In seven dyads we find significant correlations ($P \leq 0.05$) for following and in four dyads for passing of a total of 37 dyads that could be tested. An insufficient amount of data is available for the replicates of subgroup 1 as only one apparatus was installed. Of course, we can not conclude from a non-significant correlation that the respective dyad partner of the specialist was unable to assess the specialist's skills. He might have been prevented from following and passing by a higher ranking group member that had priority of access to the food bowls. For this reason, the significant correlations are found in dyads with high sums of joint presences.

An impressive scene illustrates that following and passing are, in fact, indicators of awareness of the specialist's capabilities. On November 13th 1984, when Mayun was the specialist, she was slowly approaching the apparatus after having hesitated for a couple of minutes. Djalan who was actually occupied with digging immediately ran over a distance of approximately 30 m, passed Mayun and sat near the apparatus, and waited for Mayun to operate the levers.

The next finding that I would like to present was quite surprising to me when I became aware of it and thus, data collection and experimental design were not specially adjusted to this question. When I performed the basic training of all the monkeys of the study group in pulling one single lever I found no indication that a monkey that first was confronted with the task in the indoor case had any difficulties in remembering the correct form of manipulation if he was tested in the outdoor enclosure next time and vice versa. However, later observations suggested that the site where a specialist was trained had some influence on whether he was respected as a specialist at any other site of the enclosure. Therefore, the question arose whether the animals utilize some generalized kind of knowledge about the specialists' skills or whether they just learn to behave in a convenient way at a specially defined site of the enclosure. In three of the eight specialists' series I had the opportunity to test this kind of knowledge: the three specialists were trained and established either in the outdoor enclosure (Ukui) or in the indoor cage (Junus, Mayun) and after they were fully accepted by the other group members as specialists, some of the following trials had to be carried out in the other enclosure for meteorological reasons. As shown in Fig. 22.2, all three specialists manipulated the levers almost exclusively at the original site. But after the change of enclosure, the exclusivity of lever pulling dropped in all three specialists: other animals tried to get a food reward for their own. These results can be confirmed for Junus and Mayun by a Wilcoxon rank sum test ($P \leq 0.05$), but not for Ukui. These findings suggest that the 'knowledge' about the specialist is not totally generalized, but connected with the site where the specialist performs his activities and is thus not an inference about general capabilities.

So far I have discussed changes in the behaviour of the specialists and their partners that happened while the apparatus was active. Other animals learned to stop displacing and chasing specialists away, and thus to increase the amount of reward. The same was true for following and passing the

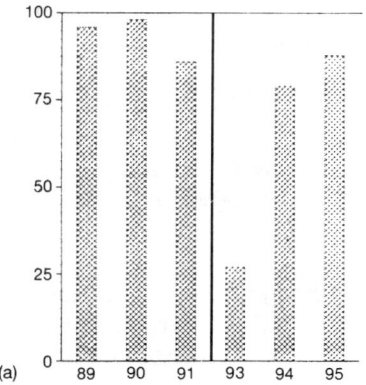
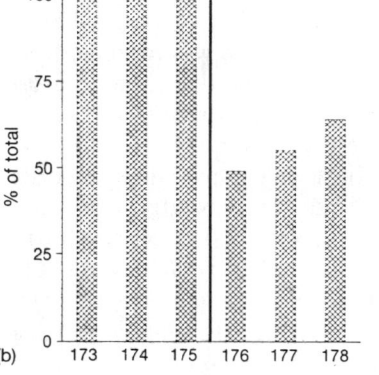
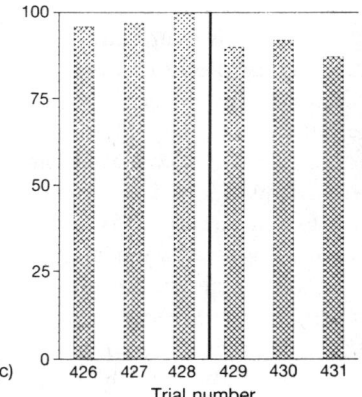

Fig. 22.2 Percentage of total lever pulling performed by the specialists in three trials before and three trials after the transfer of the experiments from the outdoor enclosure to the indoor cage or vice versa happened (a) UK, (b) JN, and (c) MY. Vertical lines indicate the moment of transfer.

specialist when he approached the apparatus. It is hard to doubt that a monkey would be able to succeed in these quite simple learning tasks. It would be less easy to explain if it could be shown that animals began to treat a specialist differently in the course of the ongoing trials even at times when no immediate increase of a food reward will be the result of the

action. No changes in dominance position of any specialist was observed in the course of their replicates. In order to test whether a different treatment occurred in some other way, we now focus on the social phase of each trial, when the apparatus was out of action and, thus, no immediate food reward could reinforce changes of behaviour. During this phase, data collection was concentrated on social behaviour. In primate ethology two main types of measure of social relationships are used, namely:
(1) proxemics, i.e. the spatial arrangement of the animals;
(2) interactions like grooming behaviour directed towards the social partner, aggressive behaviour, and helping behaviour such as giving support in aggressive episodes.

The spatial arrangement of the animals was registered as follows. During the social phase of each trial the nearest neighbour of each animal of the subgroup was recorded every 5 minutes. For grooming, the exact durations, and for aggressive interactions and alliances the rates per hour were available.

A sufficient degree of security about experimentally induced increases of neighbourship scores and grooming rates requires three conditions that have to be fulfilled in a group-member specialist dyad.
1. We must be sure that the group member effectively gained a satisfactory amount of benefit from the specialists. Thus, we can expect that animals that reached the highest scores of joint presence in terms of regularity and total frequency most likely might change their behaviour directed towards the specialist as a reaction to the obtained benefits.
2. The changes should coincide with the experimentally determined specialization of individuals. This was tested for each of the 55 specialist/non-specialist dyads by performing a Wilcoxon rank sum test where I compared the neighbourship scores and behaviour frequencies emerging from the trials during a specialist's replicate with the data emerging from the replicates when other individuals were established as specialists in the same subgroup.
3. Changes in social behaviour given to the specialist should correlate with the amount of benefit gained per trial owing to the specialist. The measure for the amount of benefit was referred as joint presence.

As a second statistical analysis I performed a Kendall τ correlation to detect possible correlations between the frequencies of joint presence, neighbourship scores and grooming rates per hour in each trial. At this point I would like to stress that the aim of this part of analysis is not to detect all possible changes in social behaviour that might be explained by the experimentally induced specialization, but to denote only these changes that with a large extent of security are due to experimental conditions.

Table 22.1 *Scores of Wilcoxon rank sum tests and Kendall τ correlation coefficients measuring the relation among status of specialists and neighbourship scores and grooming rates. For Wilcoxon tests the significance probabilities are given. A negative probability indicates a decrease of neighbourship scores or grooming rates during the specialist's phase. The dyads are arranged according to the sum of joint presence. Dyads revealing significant effects are contained in the table only*

Data Dyad			Neighbourships		Grooming given to specialist		Grooming received from specialist	
			Wilcoxon rank test P	Kendall τ coeff	Wilcoxon rank test P	Kendall τ coeff	Wilcoxon rank test P	Kendall τ coeff
Salim	Malen	+	−0.02	−0.18*	ns	ns	ns	ns
Ukui	Malen	+	0.02	ns	ns	ns	ns	ns
Malen	Ukui		0.02	0.17*	ns	ns	ns	ns
Titin	Ukui	+	−0.02	−0.16*	ns	ns	ns	nns
Ukui	Junus	+	0.01	0.20**	0.02	0.23**	ns	ns
Ketut	Junus	+	−0.02	ns	ns	ns	ns	ns
Titin	Junus	+	0.02	ns	ns	ns	ns	ns
Saja	Sanah		ns	ns	−0.04	−0.14*	−0.02	−0.16*
Tjat	Sanah	+	0.03	ns	ns	ns	ns	ns
Topi	Sanah	+	0.01	0.18*	0.01	0.31**	ns	0.26**
Sapi	Sanah		−0.01	ns	−0.01	ns	ns	ns
Djambi	Sanah		−0.02	ns	ns	ns	ns	ns
Jumi	Sanah		ns	ns	0.05	ns	0.02	ns
Saja	Djambi	+	0.01	0.31**	ns	ns	ns	ns
Tjat	Djambi	+	0.04	ns	ns	ns	ns	ns
Sanah	Djambi	+	−0.01	ns	ns	ns	ns	ns
Jumi	Djambi	+	−0.01	ns	ns	ns	ns	ns
Tjat	Saja	+	0.01	0.25**	0.04	0.17*	ns	ns
Rini	Saja	+	0.01	0.27**	0.01	0.27**	0.01	0.20**
Topi	Saja	+	0.01	0.19**	ns	0.17*	ns	ns
Sanah	Saja	+	ns	ns	0.01	0.19**	ns	ns
Sapi	Saja		ns	ns	ns	0.21**	ns	ns
Jumi	Saja		ns	ns	ns	ns	−0.03	ns
Djalan	Sakri	+	0.01	0.20*	0.01	0.28**	ns	ns
Sapi	Sakri		ns	ns	ns	ns	ns	0.15*
Topi	Sakri	+	ns	ns	0.01	ns	0.01	ns
Upit	Sakri	+	ns	ns	ns	0.21**	ns	ns
Dili	Sakri	+	ns	ns	0.01	ns	0.01	ns

Djalan	Mayun	+	0.01	0.34**	0.01	0.28**	ns	ns
Toko	Mayun	+	0.01	0.29**	0.02	0.25**	ns	ns
Topi	Mayun	+	ns	−0.15*	0.04	ns	−0.01	−0.15*
Dili	Mayun	+	ns	0.19**	0.01	0.25**	−0.01	ns
Titin	Mayun	+	−0.01	ns	ns	ns	−0.01	ns

+: Actor is higher ranking than receipient.
ns: Not significant
*: Significant at 5% level.
**: Significant at 1% level.

Table 22.1 summarizes the results for neighbourship scores and grooming frequencies. For each dyad the grooming given to the specialist as well as the grooming received from the specialist are analysed. In 16 dyadic relationships among other group members and the specialists Wilcoxon rank sum tests revealed a significant increase of neighbourship scores during their specialist phase. A decrease occurred in only six dyads. In 14 of these relationships even an increase of grooming received by the specialist occurred. A decrease occurred in only one of the dyads. When analysing the data for threatening, chasing, and providing alliances it appeared that for most of the dyads an insufficient amount of data was available. No significant reductions of threatening or chasing could be found. Concerning the alliances it appeared that in only one of the dyads a significant effect could be found: Djalan increased the number of alliances that he formed with Mayun (Wilcoxon rank sum test, $P \leq 0.05$). The results of the Kendall τ correlations over all trials carried out with the subgroup a specialist belonged to are summarized in Table 22.1 too. Significant positive correlations between joint presence and neighbourship scores were found in 11 dyads, negative ones in only three dyads. Correlations between joint presence and grooming given to the specialist are significantly positive in 11 dyads and significantly negative in one dyad. Correlations between joint presence and grooming received from the specialist are significantly positive in three dyads and significantly negative in two dyads.

Table 22.1 indicates that in seven dyadic relationships, namely Ukui-Junus, Topi-Sanah, Rini-Saja, Tjat-Saja, Djalan-Sakri, Djalan-Mayun, and Toko-Mayun, statistical analyses revealed significant changes in both association measures. In the dyad Rini-Saja not only the grooming frequencies directed towards the specialist Saja increased, but her grooming of Rini as well. Six of these seven animals that improved their social interactions with one of the specialists were also those who gained the most benefit in the respective replicates, as measured by in earlier analyses. Topi is the only exception. I conclude that in six dyads the changes in

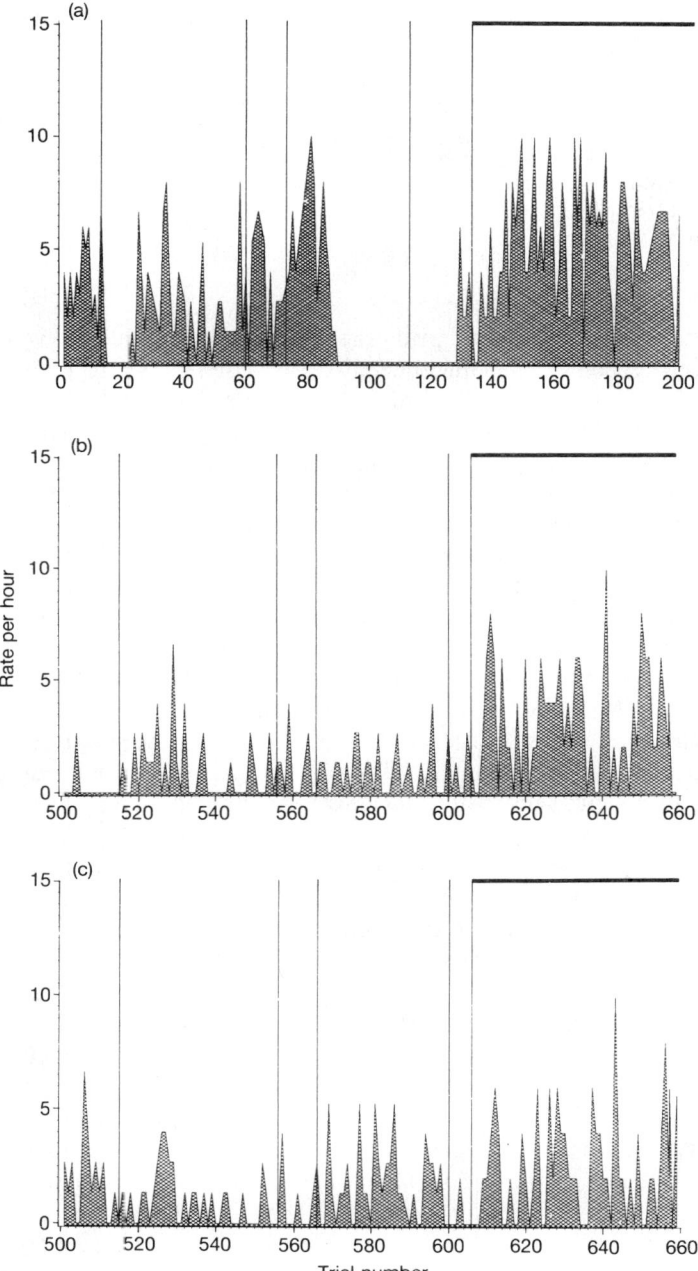

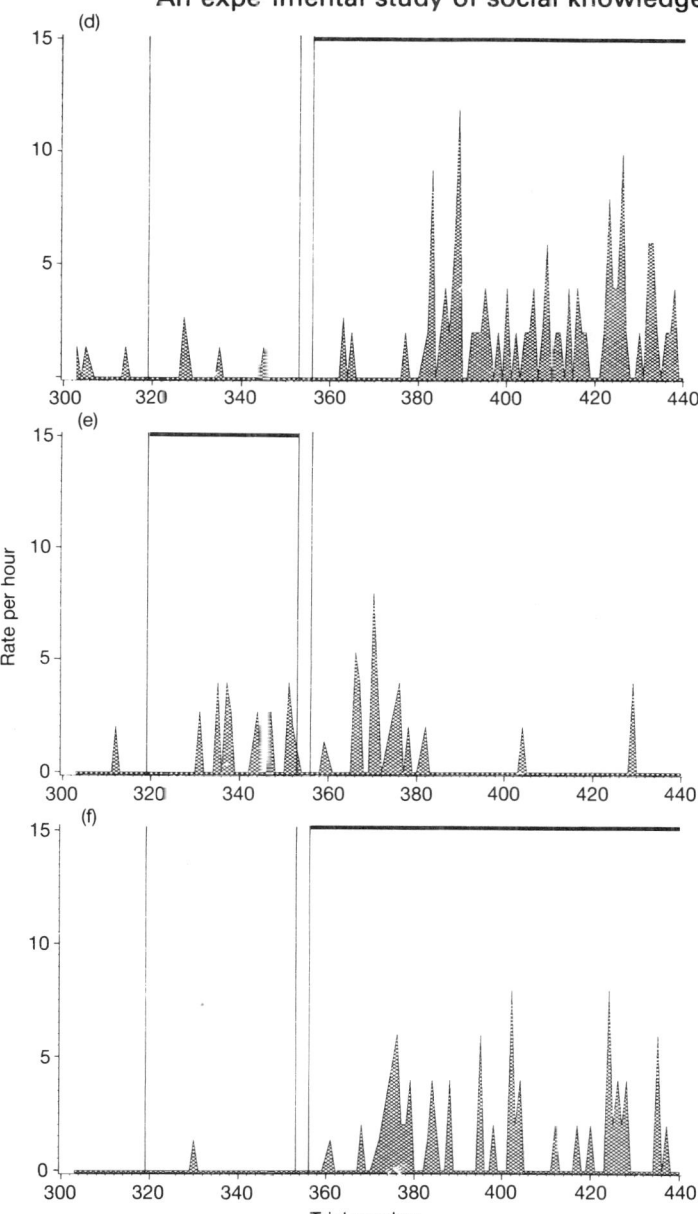

Fig. 22.3 Neighbouring scores in each trial as observed in the dyads where significant changes in social interactions happened in reaction to the experimental conditions: (a) UK–JN; (b) RN–SJ; (c) TJ–SJ; (d) DA–MY; (e) DA–SK; (f) TK–MY. Vertical lines indicate the beginning or the end of a replicate. A black horizontal bar indicates the specialist's replicate.

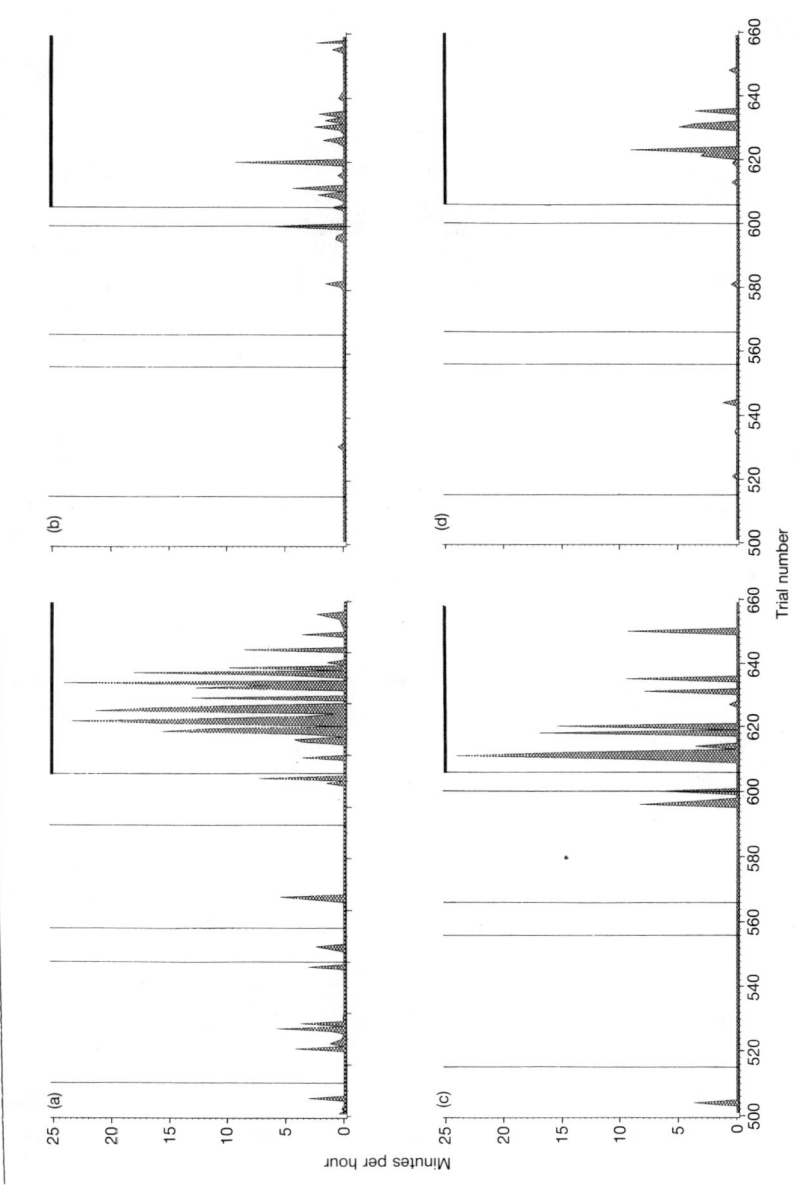

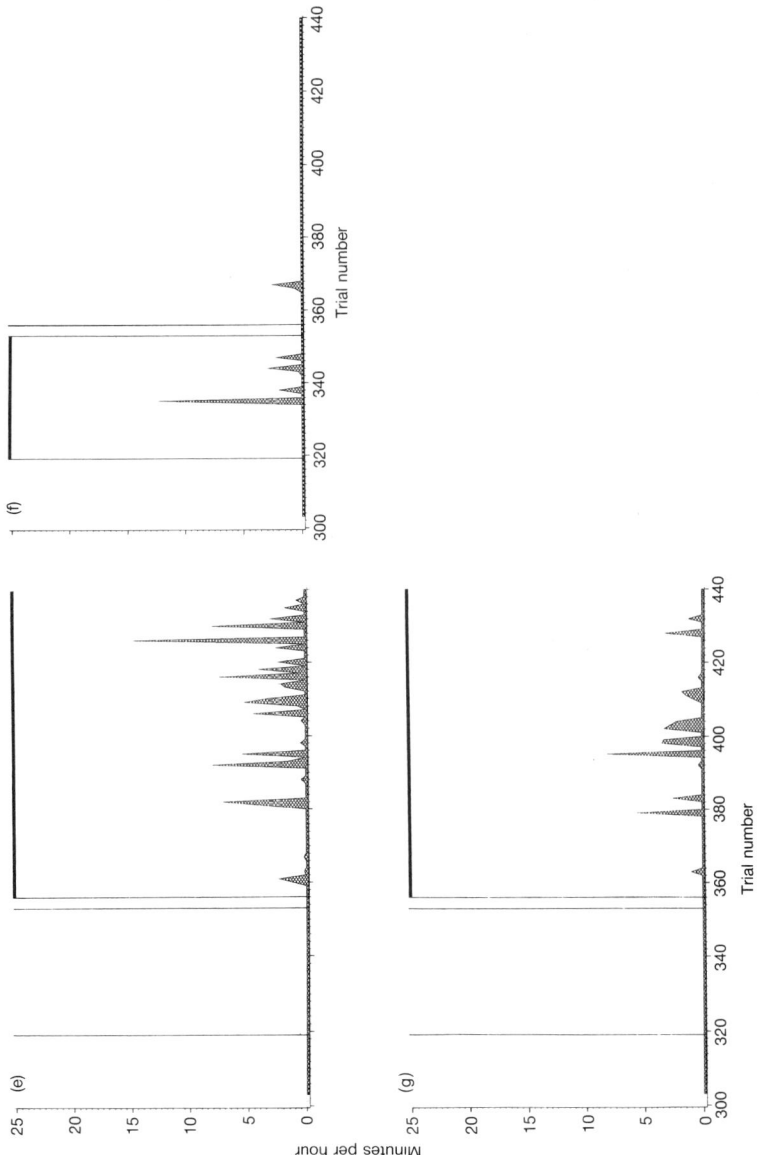

Fig. 22.4 Grooming rates per hour in each trial as observed in the dyads where significant changes in social interactions happened in reaction to the experimental conditions: (a) UK–JN; (b) RN–SJ; (c) SJ–RN; (d) TJ–SJ; (e) DA–MY; (f) DA–SK; (g) TK–MY. Vertical lines indicate the beginning or the end of a replicate. A black horizontal bar indicates the specialist's replicate.

behavioural treatment of the specialists is particularly clearly explicable by a differential amount of benefit that was gained.

Figures 22.3 and 22.4 show how grooming rates and neighbourship scores changed in the course of the trials that were carried out with one subgroup.

As mentioned above, in the dyad Djalan-Mayun an increase of alliances provided by Djalan to Mayun could be observed. It is Djalan whose behaviour is the most interesting. First, he began to increase his friendly interactions with Sakri, the first of the specialists established in his subgroup. When Mayun became specialist next, Djalan switched to the new popcorn producer and began to 'flatter' Mayun. If we consult Figs 22.3 and 22.4 we can see how Djalan's grooming and neighbouring in fact depended on the specialist status of both Sakri and Mayun.

In one of the subgroups I had the occasion to carry out an additional trial about 6 months after the last one. The last regular trial with subgroup 1 was carried out in June 27th 1984 and the additional one took place on December 11th 1984. It was quite impressive to see how Junus, who was the last of the three specialists of his subgroup, immediately approached the apparatus and manipulated the three levers as if the last trial had taken place some days ago while the other animals held back with lever manipulations of their own. No chasing was observed. Junus performed 93.5 per cent of the total and 100 per cent of the successful lever manipulations. Only Ketut, who was taken away from his subgroup for the last 20 trials, was responsible for 6.5 per cent of the unsuccessful manipulation. Obviously, the low ranking Junus was still kept in mind as a food producer by the other subgroup members even 6 months after the last regular trial. We conclude that knowledge on individual competences of other group members is not subject to quick extinction, but is remembered for a long period.

4. Discussion

If we try to summarize the different results and to find a coherent interpretation we must refer to the main questions of this study:
(1) Is it possible to establish low ranking group members as specialized food producers;
(2) are other, mainly higher ranking group members able to adapt their behaviour in order to gain benefits;
(3) do changes occur even in social contexts outside the feeding phase of the trials?

1. In fact all animals trained as specialists succeeded in operating the apparatus regularly and non-specialists ceased to pull levers on their own.

The latter might simply be explained by extinction: lever manipulations of non-specialists dropped because these were not rewarded for their incorrect manipulations.

2. Further analysis showed that non-specialists that gained benefit from one of the specialists could learn to refrain from displacing and chasing the specialist in the course of a replicate. This finding suggests that non-specialists realized that the specialist's presence at the feeding site was necessary for subsequent food release. The study gives evidence that non-specialists seem to be able to anticipate the specialist's later actions, as suggested by the increase of following and passing that can be shown for some of the dyads. Results on following and passing allow us to conclude that the non-specialists are able to foresee that the specialist that is on his way to the apparatus will activate the food release soon.

3. The most interesting finding however is that besides these short term adaptations, the specialists were treated differently in a social context outside the experimental situation by some of their partners as a consequence of their activity at the apparatus. In seven of the 55 dyads a significant change in social associations occurred. It was shown that six of the seven partners that treated the specialist differently in a social context were those of their respective replicates that gained the largest amount of benefit from them. This makes it very improbable that other reasons than gaining benefit from the specialist induced the increase of sociopositive interaction. One could interpret the increase of neighbourships as an extended anticipation of the specialist's role from the feeding to the social phases of the trials, as if his partners assumed that the specialist has other skills in other contexts as well, and thus maintaining spatial proximity is favourable in terms of later benefits. Increased grooming, however, cannot be interpreted as anticipation of food production, but suggests that non-specialists try to improve their bonding to an individual from which benefits may be gained. In conclusion: monkeys are able to assess exceptional capabilities of others. Surprisingly, this awareness seems somehow to be restricted to a certain location and probably to other characteristics of the situation. Furthermore, the knowledge can be remembered for a considerable amount of time.

The aim of this study was to inquire how monkeys conceptualize other individuals; what then is relevant in other studies dealing with questions on cognition? Different studies have revealed quite surprising results concerning cognitive capabilities of monkeys. Turning first to conceptualizing social relationships, Bachmann and Kummer (1980) showed that hamadryas baboon males (*Papio hamadryas*) were more likely not to challenge the female in possession of another male when the female preferred her present owner as measured in choice trials. Hamadryas males thus seem to be able to assess social preferences of females but the experimental procedure does not exclude the possibility that the rivals read the relationships from the behaviour of only one group member. Cheney and Seyfarth (1980, Chapters 6 and 19) presented a field study that reveals another interesting result. Vervet monkeys (*Cercopithecus aethiops*) are able to perceive their group as hierarchically structured into different social units. By replaying the screams of juveniles to their mothers, the authors showed that mothers respond to the scream replays by orientating towards the loud speaker with significantly shorter latency and longer duration than did non-mothers, and playback increased the probability that controls looked at the mother. Cheney and Seyfarth conclude that vervets are able to distinguish specific mother-infant relationships. Dasser (Chapter 7) was able to demonstrate that longtailed macaques (*Macaca fascicularis*) distinguish among mother-infant and non-mother-infant relations too. This suggests that monkeys have a rich understanding of their own and of other individuals' relationships. This suggestion is supported by the fact that social behaviour such as grooming is not distributed randomly among group members but according to dominance rank, personal preference and kin relationship (Stammbach 1978; Stammbach and Kummer 1982; Seyfarth 1977; Kurland 1977; Silk *et al.* 1981).

However, do monkeys have knowledge on individual capabilities and skills? Beck (1972) describes the development of co-operative tool use in captive hamadryas baboons. A male learned to get food with a tool. His female partner learned to give him the tool, thus demonstrating that she had some kind of knowledge on the skills of her partner. Sigg (1980) found evidence in captive hamadryas baboons (*Papio hamadryas*) that they can judge the ecological competence of others. He informed only one female of a caged one-male-group about the location of artificial waterholes. The other group members seemed to trust informants that were experienced in ecological tasks more than they trusted less experienced ones. According to Menzel (Chapter 12) young chimpanzees are able to judge whether other individuals were furnished with information on the location of food that was hidden in the enclosure in front of their eyes. Thus, considering both the results of this and of the above discussed studies we must conclude that monkeys possess a considerable amount of knowledge not only on social status, but also on capabilities of their group members as shown in the present study. This knowledge can be applied in such a way that they may adapt their behaviour according to special characteristics of individuals in a flexible manner.

23

Invention and social transmission: new data from wild vervet monkeys
MARC D. HAUSER

One important aspect of primate social intelligence according to Jolly's analysis is an individual's capacity to benefit from the discoveries of others, through social learning. In this chapter Hauser uses new observations on tool use in vervet monkeys for a detailed dissection of the variety of mental processes potentially involved in innovation and social transmission.

Introduction

Few would question the point that inventing novel solutions to either old or new problems counts as a form of intelligent behaviour. Moreover, discussions about which solutions count as intelligent should not be based on whether the environmental context is social or ecological. According to the social intellect hypothesis (see Chapters 1–5), however, the social domain represents greater complexity than the ecological domain. Thus, it is argued, social intelligence is cognitively more demanding than ecological intelligence.

There are two central issues addressed in this chapter. First, what are the important conditions leading to the invention of a novel solution in the ecological domain? Secondly, what are the processes underlying the transmission of such novel solutions? As mentioned above, the issue of invention is clearly relevant to hypotheses about the nature of intelligence. The relevance of transmission mechanisms to these hypotheses is, perhaps, less obvious. I argue that social learning, as one mechanism of transmission is directly relevant to our concerns about social intelligence. Once solutions to environmental problems have emerged, social learning allows these solutions to be transmitted to other individuals who have not yet figured the problem out. That is, social learning allows many individuals to 'become intelligent' (see Chapter 3). What the social intellect hypothesis then postulates is that the problem of learning how to replicate the inventor's behaviour represents an intelligent mechanism within the social domain. More specifically, complex mental processes underly the ability to imitate another's behaviour. As Bruner (1972) has pointed out, true imitation involves first translating one's perception of someone else's behaviour into a mental picture, and secondly, making use of this picture to guide the production of a similar behavioural act (see Chapter 5).

The empirical foundation for this discussion is an interesting case of technological invention and subsequent transmission in a group of free-ranging vervet monkeys (*Cercopithecus aethiops pygerythrus*). A descriptive account of the observations is first provided. I then discuss the case in greater detail using a framework developed by Galef (in press) that clearly sets up the necessary criteria for describing a particular phenomenon as invention or socially mediated learning. Two important points emerge out of this detailed case study analysis. First, a group may be homogeneous with regard to the expression of a trait, and yet individuals may differ with regard to underlying mechanisms for acquisition. Secondly, there are complex mental processes underlying some of the mechanisms for transmission and acquisition of a novel trait. Whether these mechanisms first evolved in the context of ecological or social problems remains an open question.

The case of invention involved (i) the modification of an *Acacia tortilis* tree pod into a tool for more efficient access to the exudate from the tree and (ii) subsequent digestion of the pod which had been immersed in the exudate. This observation is particularly interesting for at least three reasons. First, most cases of invention and transmission in nonhuman primates have been described for provisioned animals; most of these cases involve apes rather than monkeys (apes: McGrew and Tutin 1978; McGrew 1983; Boesch and Boesch 1983; Takasaki 1983; monkeys: Itani and Nishimura 1973; Camberfort 1981; Scheurer and Thierry 1985). Thus, the observations I present provide potentially new information on how technological inventions arise in the absence of provisioning and how such an invention spreads in a free-ranging monkey group.

Secondly, the modified *Acacia* pod was first used or invented by the oldest individual in the group. If the oldest individual was responsible for the invention, then this differs from several previously reported incidences of technological invention. In most cases, the newly observed behaviour or technology was first recorded for a young animal and was then gradually passed on to other members of the population (e.g. potato washing in Japanese macaques; Itani and Nishimura 1973). Conversely, if the oldest individual was the first to use the technique, but did not invent it, then this represents an extraordinary case where knowledge gained early in life was stored for a long interval of time without being used.

A final point is that the innovated trait most probably emerged in response to the drought. If this is true, then technological inventions may be one way for an organism to adapt quickly to changing ecological conditions and constraints. It is possible then that ecological pressures provide 'fuel' for such inventions. Moreover, if experience plays a role in the invention of new behavioural traits, then one might expect older members of a population to be more likely to be 'responsible' for such innovations than younger individuals.

Case study description

Observations on six groups of free-ranging vervet monkeys were collected from August 1983 to June 1985 in Amboseli National Park, Kenya. Since 1977, three of the six study groups (A, B, and C) have been observed intensively and in 1983, three additional groups (2, 3, and 4) were added for intensive observations.

Vervet monkeys in Amboseli occupy and defend small territories (mean size = 0.37 km^2, standard deviation = 0.10; Lee 1981; Cheney et al. in press). Group size varies considerably and since 1977 there have been as many as 35 individuals and as few as six in each group. Even though the entire study area is only 3.5 km^2, there are striking ecological differences between the groups (Hauser 1987; Cheney et al. in press). For instance, both groups A and 2 lack surface water throughout most of the year. Thus, they depend primarily on the water contents of the plants in their habitat and a viscous exudate produced by *Acacia tortilis* trees. Competition between individuals over exudate is intense (Wrangham 1981). In addition to intergroup contrasts in water availability, there are marked differences in the distribution of food resources both within and between groups.

During the drought of 1983–1984, there were very few resources of high quality in the six vervet study groups. This was particularly true for groups 2, 4, and A. As a result, the vervets in these latter three groups were forced to feed on food items that typically comprised a small portion of the diet (Lee 1981, 1986; Wrangham 1981; Wrangham and Waterman 1981).

In September 1983, the oldest and highest ranking adult female (BA) in group A was observed taking dry *tortilis* pods (only 14 per cent water compared with 67 per cent for ripe pods) and dipping them into the well of a *tortilis* tree containing a viscous exudate. After the pods had remained in the exudate for approximately 2 minutes, BA removed and consumed them. From previous records, this feeding behaviour had never been observed (R. M. Seyfarth and D. L. Cheney, personal communication), even though continuous observations on this group have been gathered since 1977, and the group was studied for 18 months during 1975 and 1976 (Klein 1978).

Approximately 8 days after BA invented the technique, the next member of group A was observed feeding on pods dipped in exudate. During the interval of time between invention and transmission, group A was observed for at least 3 hours per day and during all times of the day from 06.30 to 18.00. Table 23.1A lists the animals observed using this modified feeding technique in addition to their age and dominance rank. In section B of Table 23.1 the order of transmission of the technique is presented schematically. These data are based on both continuous focal samples of adult females ($n = 4$) and *ad libitum* observations.

The order of transmission corresponds primarily with genetic relatedness within the group and high rank. The first three individuals to feed on pods and exudate were BA's offspring who were all high ranking. The animals acquiring the trait next were unrelated, but acquired it as a function of rank; the third ranking female (YV) fed on the pods and exudate before the fourth ranking female (ES). The alpha male, TT, was the last to acquire the trait. The other two males and the lowest ranking juvenile were never observed feeding on the pods and exudate in combination.

What benefits might vervets obtain from this pod-exudate technique? There are at least two potential benefits. Vervets typically consume the tree's exudate by dipping their hands into the wells and then licking them off. Since the wells are narrow, use of pods may facilitate access and removal of the exudate. Since group A lacks surface water throughout most of the year, increased access to other sources of water would be beneficial. Additionally, since the dry pods are extremely hard, the exudate most likely softens the seeds and thereby facilitates processing time. Unfortunately, since time spent feeding on dry pods with and without exudate was not recorded, it is not possible to directly quantify this benefit. However, because the quality and abundance of food in group A was poor during this period of time, any additional source of food would have been beneficial.

Figure 23.1 presents a more quantitative assessment of the adult females' feeding behaviour. In September 1983, during the drought, *tortilis* pods comprised a high portion of the diet for all animals in group A. Out of the total time spent feeding on pods, approximately 30 per cent involved the use of pods and exudate. In October and November 1983, pods comprised a lower proportion of the diet. However, out of the total time spent feeding on pods, most of the time involved pods dipped in exudate.

In 1984, when the rains were good, the proportion of pods in the diet was comparable to 1983. However, in 1984 there were no observations (both from focal and *ad libitum* observations) of animals feeding on pods dipped in exudate. The rains provided both surface water and a diversity of plant types that most likely alleviated the necessity to feed on dry pods and perhaps more importantly, dry pods dipped in exudate.

Invention and social transmission 331

Table 23.1 *Demographic data on individuals in group A and acquisition of pod-exudate technique*

A. Demography of group A in 1983.

Animal	Sex	Age	Dominance rank*	Used technique?
TT	Male	10	1	Yes
KM	Male	10	2	No
KJ	Male	8	3	No
BA	Female	14	1	Yes
LS**	Female	8	2	Yes
YV	Female	7	3	Yes
ES	Female	7	4	Yes
BN**	Juvenile male	4	1	Yes
GO**	Juvenile male	2	2	Yes
WP	Juvenile female	2	3	No

In 1984, the members of Group A were: TT, BA, LS, ES, BN, GO, and an infant born to BA. Dominance ranks stayed the same.

*A dominance rank of 1 indicates the highest ranking individual. Dominance ranks have been organized according to age/sex categories: adult male, adult females and juveniles.

**Offspring of BA.

B. Order of acquisition of pod-exudate technique.

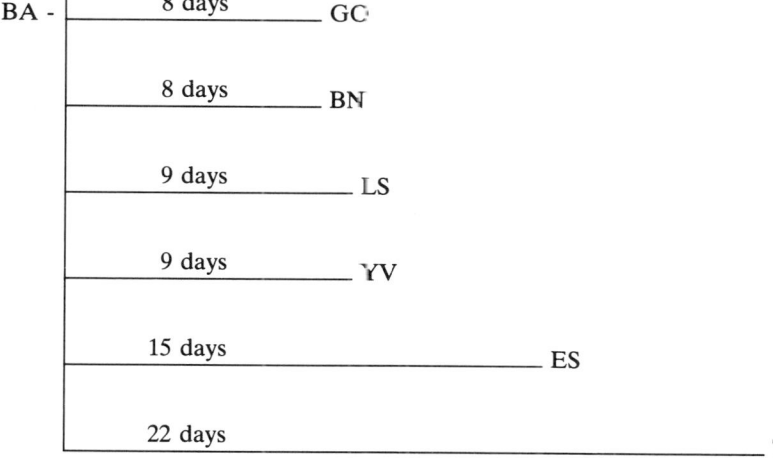

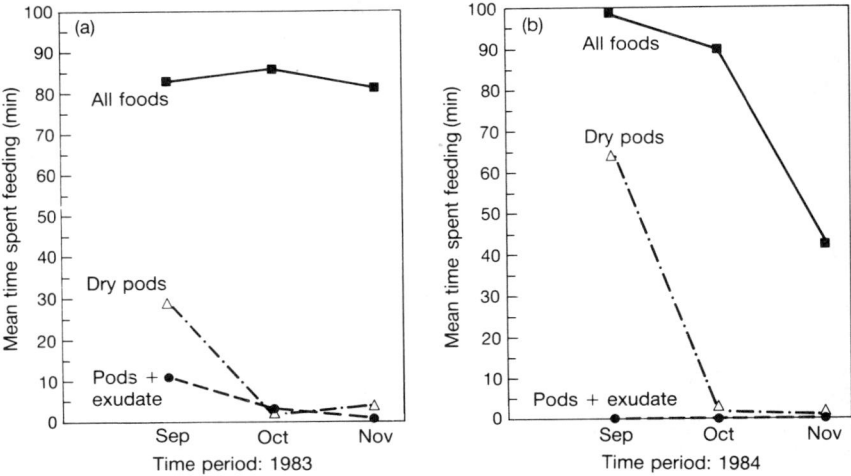

Fig. 23.1 (a) The mean time spent feeding in minutes by adult females ($n = 4$) in group A during 1983. *Closed squares* indicate time spent feeding on all food items in the vervets' diet. *Open triangles* indicate time spend feeding on dry *A. tortilis* pods. *Closed circles* indicate time spent feeding on dry *tortilis* pods dipped in exudate. (b) The mean time spend feeding in minutes by adult females ($n = 4$) in group A during 1984.

How was BA's modified feeding technique transmitted through the group?

The case study described above suggests that transmission may have been mediated by social factors. However, the literature on transmission of novel traits is replete with mechanisms that are primarily the result of social interactions. As a result, it is important for our understanding of animal intelligence to distinguish between these different mechanisms. The discussion below provides a more detailed analysis of the vervet case. The emphasis is on the kind of data necessary to support a particular hypothesis concerning the acquisition and/or transmission of a new trait. To do this, several processes related (either semantically or in terms of underlying mechanism) to the notion of imitation are examined following the arguments and presentation by Galef (in press).

Galef's article is particularly useful for four reasons. First, it is based on a comprehensive review of the literature on imitation, not just in primates, but in vertebrates generally. Secondly, it distinguishes phenomena that are often confused. Thirdly, it provides both the necessary and sufficient criteria for demonstrating that a particular process of transmission or acquisition has occurred (see also Palameta and Lefebvre 1985; Payne 1985). Fourthly, it suggests the many ways in which a trait could be transmitted (see also Boyd and Richerson 1985, in press, for general models of transmission). By concentrating on the criteria set out by Galef, a more rigorous treatment of the vervet case study is possible.

Social learning

Social learning has been used interchangeably with imitation (Thorndike 1911) and observational learning (Hall 1963a, b). These terms refer to situations where a *naïve* individual(s) *acquires* a given behaviour as a result of *social interactions* with others *exhibiting* the behaviour (Galef in press). The emphasized words require careful attention with regard to placing a phenomenon under the category of social learning.

'*Naive*' Let us consider the vervets' case. When BA was first observed taking pods and dipping them in exudate, none of the other group members were in visual proximity, and when GO and BN (BA's sons) first observed BA, she was already eating the dipped pods. In other words, they did not see the process or preparation of the dipped pods, but rather their ingestion. Were GO and BN naive? On the one hand they were naive with regard to the actual technique of gathering pods, placing them in the exudate, waiting for the exudate to be at least partially absorbed, and then ingesting them. However, if the entire technique is decomposed into its relevant components, then GO and BN cannot be considered entirely naive. They had eaten both pods and exudate, and even gone through the motor processes of picking up dry pods on the ground and carrying them up into the tree to feed on. These observations suggest that in many cases of natural social learning, successful transmission may well depend upon the pre-existence of certain relevant experiences. This idea of pre-existing experiences or mental structures has been expounded upon by Bruner (1972) in light of the problem of imitation. In essence, he argues that the individual holds a mental structure of experiential building blocks that serve to guide and potentially facilitate the process of imitation.

'*Acquisition*' The term 'acquires' is also not straightforward. When GO and BN first ingested the pods and exudate, they consumed pods that BA had left in the well. The second time that GO ate the pods and exudate, he picked up dry pods from the base of the tree and then placed them in the exudate for approximately 10 seconds. He then looked at the pods and replaced them in the exudate for approximately 2 minutes more and then ingested them. The third time that GO was seen feeding on pods and exudate, he exhibited the same actions as BA; that is, he left the pods in the exudate for approximately 2 minutes and then ingested them. In contrast to GO, when LS (BA's daughter) fed on pods and exudate for the first time, she replicated the technique exhibited by BA. At what stage can we claim that GO 'acquired' the technique? In one sense, GO's initial use of pods and exudate only involved obtaining the product or reward from the technique without the preparation. In the second observation, GO prepared the pods and exudate, but withdrew the pods prematurely. This initial withdrawal and subsequent replacement suggests that GO recognized the purpose of the preparation phase and therefore knew that the process was not complete. The final observation clearly shows that GO acquired the technique.

These observations show that 'acquisition' is not a simple all-or-nothing affair. Rather, acquisition can be achieved through a number of progressive steps. Moreover, it is useful to ask not merely about acquisition of the behaviour, but about acquisition of understanding. For instance, GO's second attempt to use pods and exudate showed that he did not completely acquire the technique. However, it did suggest that he understood the association between a pod's time spent submerged in exudate and the desired product. In this sense, GO acquired the association between preparatory phase and product.

'*Social interaction*' Considering the first individuals to exhibit the new technique, GO and BN, their first exposure to pods and exudate was not in the form of a social interaction with BA. As mentioned above, they simply fed on what BA had left behind. However, in the second incidence, they both used the technique after having sat next to BA while she ate pods and exudate. In contrast to GO and BN, the other adult females (LS, ES, and YV) used the technique immediately after BA. If one considers the initial behaviour by GO and BN, one would conclude that social interaction does not completely explain acquisition. Rather, GO and BN may have 'deduced' the preparation from having received, fortuitously, the product of such preparation. With regard to the adult females in the group, it seems much more likely that they acquired the technique through social learning. Each female exhibited the preparatory phases and ultimate ingestion after watching an individual that knew the technique.

'*Exhibition*' The observations indicate that it is not always necessary for the full act to be 'exhibited' in front of the learner for transmission to be achieved. For example, consider the observation that all animals in the group were capable of feeding on pods and feeding on exudate (two different preparatory techniques). What behavioural components of BA's pod-exudate technique would be required for individuals to express it? Gathering dry pods and carrying them up into the tree? Sitting in front of a well with dry pods inside? Eating pods that appear moist? It is possible that

Invention and social transmission   335

only one of these behavioural acts would be sufficient to achieve successful transmission. In fact, if transmission occurred on the basis of only a partial demonstration, one could argue that the learner showed remarkable intelligence. Such intelligence, which would enable an individual to 'fill in the gaps' or make inferences, will be returned to later on.

Social facilitation/enhancement and local enhancement

To illustrate the idea of social facilitation (Zajonc 1965) or social enhancement, Galef quotes the definition used by Clayton (1978):

. . . an increase in the frequency or intensity of responses or the initiation of particular responses already in an animal's repertoire, when shown in the presence of others engaged in the same behaviour at the same time.

Clayton goes on to point out that social factors will have differential effects depending on whether the individual has the behaviour in its repertoire or whether the behaviour is acquired *de novo*.

Clayton's definition can also be dissected into its components and applied to the vervets' case. As with social learning, it is difficult to assess what 'particular responses already in an animal's repertoire' means. It is clear that the individuals who acquired the feeding technique were capable of performing each of the behavioural components of the technique; that is, all of the individuals could pick up dry pods, carry dry pods up into the tree, drink exudate and so on. However, what makes BA's technique novel is the piecing together of various preparatory behaviours used for gathering different foods. Thus, the individuals who learned the pod-exudate technique did not have this preparatory behaviour in their repertoire, *sensu strictu*.

Another relevant part of Clayton's definition concerns the idea that the behaviour emerges 'when shown in the presence of others engaged in the same behaviour at the same time'. LS, who was the third individual to use the technique, actually sat, watched, and waited for BA to finish eating dipped pods, and then performed the entire behaviour. LS's performance would seem to fall reasonably well into a framework of social enhancement. In contrast with LS, GO's performance does not appear to fit with the social enhancement hypothesis. GO's first experience with pods immersed in exudate appeared to be in terms of a final product rather than with the preparatory technique. During the second observation of GO's use of pods and exudate, his execution of the technique did not mirror that of BA. Thus, an *imprecise* copy of BA's behaviour emerged at approximately the same time as it was demonstrated. By the third observation, GO had refined his technique and produced a precise copy of BA's behaviour.

A closely related concept, local enhancement, was defined by Thorpe (1956) as

... apparent imitation resulting from directing the animal's attention to a particular object or to a particular part of the environment.

When BA first fed on the dipped pods, GO and BN were the only other animals in the tree. Although BA did not directly attract their attention either behaviourally or vocally, her high rank and relatedness may have served as indirect attractants. However, even if BA had attracted their attention, they would only have observed ingestion of the pods and exudate. From such observations, GO and BN could either have concluded that

(1) BA had found dried pods, left them in the exudate and then ingested them; or,
(2) that she found pods in the exudate and consumed them.

Unfortunately, the data collected are insufficiently detailed to differentiate between these alternatives, particularly since GO and BN may have seen BA using the technique prior to when they first used it (see below). In the discussion, however, the possibility that BA found the pods already dipped in exudate is entertained as a possible explanation for the emergence of the technique.

Contagious behaviour and vicarious instigation/observational conditioning

Contagious behaviour is said to occur when performance of a behaviour (usually thought to be innate) by one individual acts as a 'releasing mechanism' for the performance of the same behaviour in other members of a social group (Thorpe 1963). If GO and BN's use of the pods and exudate is used as the first incident, then one could argue that BA's ingestion of dipped pods served as a releaser to eat. However, because only one individual can use a well with exudate at a given time, there was a delay between BA's feeding behaviour and that of GO and BN. It is not clear from the definition above whether there can be a delay in the releaser's effects. If so, then GO and BN's ingestion of dipped pods on the first observation would count as a case of contagious behaviour. However, the second observation of GO can not be explained as contagious behaviour. As previously mentioned, when GO ingested dipped pods for the second time he used the preparatory technique and did so in the absence of BA. Furthermore, the preparatory technique and ingestion occurred at a different tree from the one where GO first observed BA feeding on dipped pods.

Vicarious investigation (Berger 1962), which is perhaps more appropriately called observational conditioning (Minkeka *et al.* 1984; Cook *et al.* 1985; Galef in press), occurs when an

... observer responds emotionally to a performer's unconditioned emotional response ... (Berger 1962).

The critical point here is that the observer must somehow perceive or deduce the performer's emotional or motivational state (e.g. hunger). Although it is unlikely that data from a field situation would ever be suitable for testing the idea of observational conditioning, the concept is an appealing one in terms of social intelligence, and is considered below.

Consider the following scenarios:
1. Animal A sees BA picking up dry pods at the base of the tree.
2. Animal B sees BA drop dry pods into a well with exudate.
3. Animal C sees BA retrieve moist pods from an exudate well.

In scenarios 1 and 3, it seems likely that both animals A and C would infer that BA was preparing to feed. In scenario 2, it seems less likely that animal B would infer that BA was preparing to feed.

In which of the scenarios would the individual be most likely to acquire the new technique? This question is important for the following reasons. In scenario 1, if A was satiated, it would most probably fail to continue its observations of BA. This is because most individuals would probably perceive BA's pod gathering behaviour as leading to ingestion and not further preparation. In scenario 2, animal B would most probably infer that BA dropped the pods as a result of satiation, rather than as a part of the preparation phase. Finally, in scenario 3, animal C might perceive BA's ingestion of moist pods as novel given that pods during this period of the year are typically dry. Of all the scenarios, it seems the third is the only that would present a problem for the naive (with regard to the technique) observer. If BA never repeated the technique again, scenario 3 is perhaps the only one that would provide sufficient information to deduce it.

Matched-dependent behaviour

Miller and Dollard (1941) suggested the idea of matched-dependent behaviour to explain, in part, their observations of aggression in children. In general, they found that a performer's behaviour acts as a reinforcer for an observer who matches the behaviour demonstrated. Furthermore, matched-dependent behaviour can produce homogeneity among the population of observers. However, the behaviour will only become an integrated part of the group's repertoire if it is independent of the conditions under which it was initially demonstrated.

A number of aspects of the vervets' behaviour fit the definition of matched-dependent behaviour. First, it seems likely that BA's behaviour acted as a reinforcer for when GO and BN performed the pod-exudate technique. There is evidence from field studies of nonhuman primates that young animals feed on what their mothers feed on and that this contributes to the acquisition of their diet (Watts 1985). Secondly, most members of group A acquired the feeding technique and its expression mirrored that of BA. And thirdly, some members of group A (e.g. LS and ES) used the technique in the absence of BA and at a different tree from the one where it was initially performed.

As with the previous analyses, deciding whether a trait has been transmitted through matched-dependent behaviour is not straightforward. For example, according to Miller and Dollard's (1941) view, a behaviour becomes incorporated into the group's repertoire when the initial conditions in which the trait was expressed are absent. What, however, would it mean for the initial conditions to be absent? In organisms that live in a clearly defined home range (most Old World monkeys, such as vervet monkeys), details of the environment gradually become associated with past experiences. For example, in vervet monkeys, areas where predation has occurred will often be avoided for weeks or months. Thus, even if BA was absent from the tree where the technique was first expressed, the tree, in and of itself, would be part of the 'initial condition'. This means that for field observations, it will be difficult to claim that the behaviour became incorporated in the absence of initial conditions. Only through experiments can such conditions be controlled for.

Summary

The dissection of the case study raises two points. First, there has been a tendency to assume that homogeneity in a trait's expression is the result of only one underlying mechanism of transmission. However, as argued above, it is quite possible for different mechanisms to express a trait in the same way. In the vervets' case, one animal may have deduced the technique from obtaining the end product whereas another animal may have acquired the technique from direct observations. The important point is that both animals ultimately learned the technique and were capable of reproducing it in its original form (i.e. as first exhibited or demonstrated by BA).

Secondly, some of the mechanism of transmission appear to rely on fairly complex mental processes. For example, imitation consists of creating a mental picture of some novel behaviour and then using this picture to reproduce the behaviour. Such mental processes form the foundation for intelligent behaviour, whether in the social or ecological domain.

Discussion

When ecological conditions change, an organism is often confronted with the task of solving new problems (Kohler 1925; Hall and Schaller 1964). The ability to solve problems will be constrained by various selective pressures and the species' history. For some species, there exists the possibility of inventing behavioural solutions, either by modifying some aspect of the ecological environment or by manipulating individuals within the social environment. It is interesting to consider how such solutions might arise. This discussion, which raises more questions than it answers, begins by addressing the origins of innovations with an emphasis on the role of memory and past experience. I then discuss the process of transmission, focusing in particular on social factors. One of the arguments to emerge from this discussion is that individuals may be able to comprehend means-ends relationships from the 'inventor's' behaviour without having the entire technique or skill demonstrated. Similarly, experience with the inventor's product could be sufficient to deduce the processes or preparations that lead to such products. Once the ability to make inferences evolves, social learning becomes a powerful transmission mechanism.

What causes an innovation?

Throughout this presentation it has been assumed that, since vervets in Amboseli were never seen feeding on pods dipped in exudate, BA's behaviour represents a case of invention. Is this assumption valid? As Nishida (1987) points out, studies of primate behaviour are based on limited sampling periods. Thus observational bias can confound one's conclusions about cases of invention and transmission. This confound is certainly a possibility in the work presented here.

BA was the oldest member of her group and when the project began in 1977, she was already an adult. Thus, it is possible that BA had acquired the feeding technique prior to 1977 and simply failed to express it since observations began. This is also the case for LS, who was an infant at the start of the study. Whether or not BA is credited with the innovation, the important point is that the use of pods in combination with exudate seems to present a response to ecological pressures. Even if one argues that BA learned the feeding technique when she was younger, we are still left with the question of why it wasn't used since 1977.

Based on records of rainfall in Amboseli, 1976 was the last time when rainfall was as low as in 1983–84. If the use of pods and exudate is a feeding technique that developed in response to severe ecological conditions, then its expression would not be expected during times when ecological pressures are relaxed, from 1977 to 1982. Conversely, if one assumes that BA is responsible for the innovation, then it seems likely that the trait emerged in response to ecological pressures (i.e. the drought of 1983–84). If the latter is true, there are at least two relevant points. In species where the potential for innovations has evolved, ecological pressures may initiate their expression (Hall and Schaller 1964). In addition, given a change in ecological constraints, one might expect older individuals (who have experienced various fluctuations in the environment) to be 'responsible' for such innovations. It has often been suggested, on the basis of anecdotal evidence, that older members of a population are sources of knowledge and in times of stress can lead their groups out of potential bottlenecks (e.g. baboons: Kummer 1971; elephants: C. Moss, personal communication.

In the case described here, it is clear that BA would have accumulated the greatest amount of knowledge concerning fluctuations in the availability of resources. The ability to combine previous experiences over a long period of time could provide the foundation upon which novel solutions are invented. This suggests that older individuals are an asset and can potentially provide solutions to problems that younger individuals would fail to solve due to a relatively impoverished data base of knowledge (Nishida 1987).

There are at least two possible explanations for how the technique might have been invented: deduction or prediction. The first suggests that dry pods were found immersed in exudate, and from this observation, the technique necessary to obtain such a product was deduced. It is unlikely that this top-down approach would be dependent on memory. The fundamental property of this type of inventiveness is the ability to deduce the building blocks of a completed product. This explanation is analogous to that offered previously for GO. Recall that GO found pods dipped in exudate and from this may have solved the technique without observing BA. In light of recent experiments by Sherry and Galef (1984), this type of top-down approach to problem solving is particularly relevant. They show that black-capped chickadees acquire the ability to open milk bottles through either social learning or finding open bottles with milk inside. These results suggest that previous observations of the milk-bottle-opening technique in great tits (Fisher and Hinde 1949) may not have been entirely due to social learning, as some have claimed (Mainardi 1980).

The second mechanism for invention, prediction, assumed that an individual would have had to infer that the combination of two food items (and two preparatory techniques) provides greater benefits than either food item alone. In terms of underlying mechanisms, this second case clearly implies more complex cognitive abilities than the first. In the first explanation, the combined food items are ingested and then, the technique is determined on the assumption or prediction that it will provide a beneficial return.

These two explanations raise the important point that animals of different taxa may be able to solve problems of comparable difficulty, and yet, the underlying mechanisms for such situations may differ. This is particularly relevant in terms of the recent experiment by Epstein and his colleagues (1984) which shows that pigeons can solve the same banana-retrieving problem that chimpanzees can solve (Kohler 1925). Even if pigeons could solve every problem that human beings could solve, this would not necessarily imply anything substantive about their intelligences. What we need to demonstrate are the steps that lead to a problem's solution.

What factors lead to an invention's survival or extinction?

Although social transmission can potentially expedite the integration of a new trait, a number of evolutionary mechanisms can stall its integration and expression. For example, in the vervets' case, since group size is small the trait has a high probability of disappearing by chance alone. Regardless of how one interprets the use of pods and exudate by BA (i.e. as original invention or transmission of a previously acquired technique) there was at least a seven year period where the technique had not been observed. If, as I have argued, this technique arose in response to the drought, then we are left with the following question: What if the drought of 1983–84 had not occurred and BA had died before another drought? Given that BA was the only animal with any knowledge of the feeding technique, the trait would have disappeared. Thus, if technological innovations arise in non-human animals primarily as the result of ecological pressures, it is quite possible that a trait will be innovated and then lost due to a relaxation in pressures. This leaves open the possibility that what we see as cyclic (though not necessarily periodic) patterns of social transmission and evolution, may in fact represent cases of independent innovation. Nishida (1987) suggests that the potato washing and placer-mining techniques of Japanese macaques might also be the result of independent innovation and are equally susceptible to extinction.

A trait can also survive by transmission to other groups. For the study population in Amboseli, transmission could occur as the result of cross-group observation or migration. Since group A is surrounded by four groups, individuals approaching the territorial boundaries during a bout of pod-exudate feeding could directly observe the technique. Conversely, since males, but not females, leave their natal groups at reproductive maturity (or earlier if an older brother is leaving; Cheney and Seyfarth 1983), such migration could serve as a direct mechanism for transmission. If adult males are less likely to acquire the trait than adult females or juveniles, then one might expect transmission to proceed more rapidly by cross-group observation than by migration and subsequent demonstration. This type of prediction could be tested with models for cultural evolution developed by Cavalli-Sforza and Feldman (1981), and Boyd and Richerson (1985).

In Galef's (in press) review article, he concludes that

> It is somewhat surprising that almost 100 years of study of social learning in animals has failed to produce a clear answer to the question of whether animals can in fact learn to do an act from seeing it done, whether they can, in Thorndike's sense, truly imitate . . . [The]majority of attempted demonstrations of imitation have failed to provide convincing evidence of the phenomenon.

This criticism includes work on imitation in non-human primates. Galef does, however, suggest that true imitation can be demonstrated if the imitation of motor patterns is used as a prerequisite. This was the procedure used by Dawson and Foss (1965) in their study of imitation in budgies. Galef considers this work to be exemplary.

Applying the motor pattern prerequisite to the current work suggests that at least some animals acquired the pod-exudate technique by true imitation. Once again, consider GO's behaviour. In both the second and third observation of GO's performance, he displayed the pod-exudate behaviour both while occupying a different tree from where the technique had originally been performed and during a period of time when BA was not demonstrating the technique. The second performance was an imprecise copy of the model. The third performance represented a precise copy. In both cases, GO appeared to perform the motor patterns by referring to a mental image rather than direct visual imput from the model's behaviour. This suggests true imitation of the form described by Thorndike and endorsed by Galef.

The final issue to be raised is the role of deduction in the transmission of a trait. As previously mentioned, GO may have acquired the feeding technique without a great deal of social interaction. This is not to say that socially-mediated factors were unimportant. However, GO's acquisition of the technique is likely to have been the result of independent problem solving abilities rather than socially transferred information. The important point is that in social groups, it will often be difficult to untangle the factors that led to a trait's transmission. Frequently, an individual will come into contact with only a particular aspect of a trait (e.g. dry pods lying next to an exudate well) and this will be sufficient to determine all of the properties of the trait (e.g. the preparation of the dipped pods and their ingestion). This suggests that in some animals, those we might want to call more intelligent, problem solving is not a laborious series of causal associations (i.e. If I do A, then I will get B, and if I do B, then I will get C, and so on) that lead to a solution. Rather, it is a process of making long-range predictions or inferences about the outcome of a potential solution. Once this ability evolves, it enables individuals to solve problems in either the social or ecological domain.

Acknowledgements

I would like to thank the Ministry of Wildlife and Tourism, the Office of the President, and the National Council for Research and Technology of the Republic of Kenya for permission to work in Amboseli National Park. I also would like to thank the park wardens and rangers, who were extremely helpful with logistical aspects of my research. Jim Else and Maria Buteyo of the Institute for Primate Research, Kenya were invaluable. To my assistant, Bernard Musyoka Nzuma, I extend special thanks for herculean help and stories in the field. Increased terseness in writing is due to Lisa Halko Hauser. For comments on various drafts of this manuscript I thank Robert Bailey, Robert Boyd, Nick Blurton Jones, Dorothy Cheney, Walter Goldschmidt, and Robert Seyfarth. This research has been supported primarily by an NSF grant (BNS 82-15039) to R. Seyfarth and D. Cheney. Additional support was obtained by the author from the Wenner-Gren Foundation, Sigma Xi, and the University of California, Los Angeles.

Taking stock

24

The experiential context of intellect
JOHN H. CROOK

When we asked Crook to contribute to this book, we did not think it appropriate to dictate a particular area; as a founder of modern socioecology who has never lost sight of the importance of the evolution of human intellect, Crook's contribution is too broad to be easily pigeon-holed. We simply explained our plan and left it up to him, and we were not disappointed. This chapter shows that the computational skills of intellect, on which the book focuses, are only a narrow part of the full range of mental processes, and argues that other—experiential—aspects of consciousness cannot safely be neglected in the future.

Introduction: performance and experience

As comparative psychology developed in the 1930s it gradually became clear that animals such as rats and pigeons were capable of solving computational problems that could only imply a complex interior process. Cognitive abilities of the order uncovered, for example by Tolman (1948), showed that a merely externalistic or behaviouristic account making simplistic assumptions about mediating processes based on elementary building bricks such as 'reflexes' were quite inadequate to the task. That animals had 'maps in the head' and plausibly an awareness of their positions on them became an inevitable conclusion upon which the current interest in a cognitive ethology is based.

In recent years three trends in cognitive ethology have emerged. The first is the realization that complex social organization in groups of relatively constant membership often correlates with social relationships that are elaborate. Since social organization itself is the expression of behavioural adaptations of individuals in a community to the prevailing ecological environment, complex social interaction becomes viewed as an aspect of particular ecological adaptations (Crook 1970a, b; Clutton-Brock and Harvey 1977a, b; Emlen 1978). The complexity of relationships is the expression of competitive and collaborative interaction patterns of individuals driven by basic sociobiological strategies. In solving problems of choice in the face of alternative interaction possibilites higher mental activities of individuals such as learning, insight and insight-learning (Thorpe 1956) play an important role.

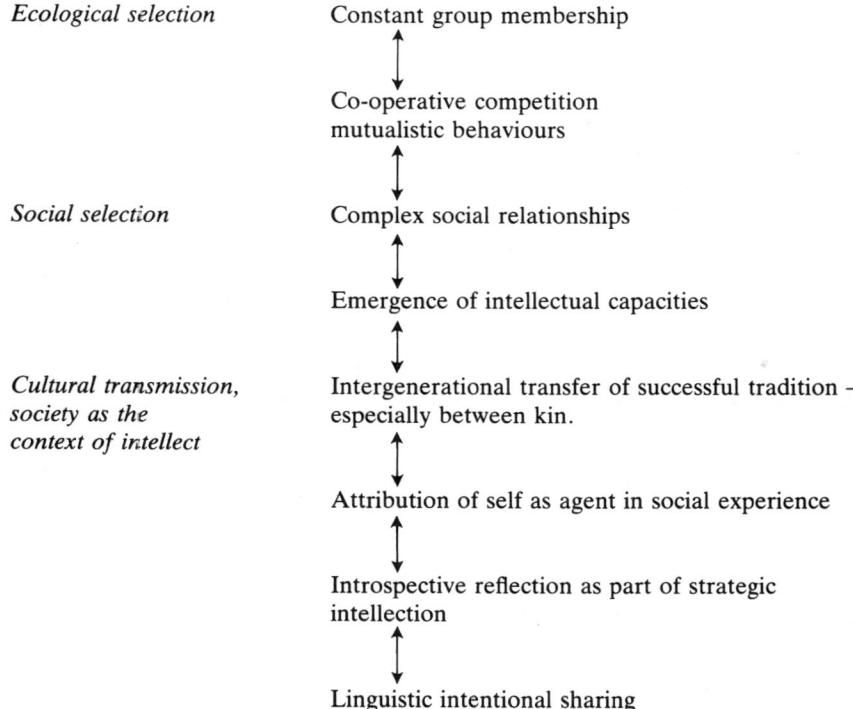

Fig. 24.1 Co-emergent evolution of social complexity and intellectual functions: a diagrammatic synopsis of recent perspectives. Many processes of change are dialectical in the sense that an effect may exert feedback on its precipitatory stimulus, thus maintaining a process of change.

The second trend follows from the first: as individuals adapted to environments with increasing social complexity so, it is argued, did the capacity for higher mental functions become enlarged through differential natural selection. In particular, the ability to compute solutions for multifactorial situational dilemmas became extended as social complexity deepened. As increased learning capacity gave rise to an intergenerational transfer of social skills, so the selection of the genetic basis for intellectual operations became enhanced (Chapter 1–4). With such a view we gain a useful new perspective on the evolutionary origins of human intellect itself

(Fig. 24.1). Part of the excitement of this approach is that successful performances in given social situations can be objectively measured and comparative, functional and causal studies of intellect in animals become valuable within an extended evolutionary paradigm (Griffin 1982).

The third approach centres upon what Donald Griffin (1981) aptly terms the 'question of animal awareness'. Are animals aware of situations? In what way are they aware? Which animals are aware of themselves as social agents? What is the form of any self-consciousness? At once we plunge into a dilemma caused by the use of 'experience' words (Crook 1983); for how can we ever find a reliable basis for inferring the nature of experience in sentient beings other than ourselves? Griffin's approach has indeed not been welcomed in all quarters because of the major philosophical issues the entire set of problems poses. To my mind, however, this issue remains a fundamental one upon the resolution of which an understanding of mind depends. It is at once clear that while an analysis of intellect using a performative approach can proceed much in the fashion of normal science, the analysis of awareness necessarily involves complexities of inference that traditional behaviouristic or ethological thinkers will shy away from.

Models of mind tend to be couched in the terms of the most complex machinery of control known to the modellers. Descartes was fascinated by pipes and clockwork; contemporary thinkers turn naturally to the computer for inspiration (Sayre 1976; Boden 1977; but see also Searle 1980; Russell 1984). A moment's reflection suggests an oddness about this procedure that is quickly found to be strongly characteristic of European thought in general: the tendency to reify the subjective. Mind is treated in terms of performance and modelled therefore with a vocabulary derived from an engineer's understanding of performing machines, yet, to us as human subjects, mind is also a matter of experience. Western models using a vocabulary of experience are less fashionable in scientific circles and, until recently (Hallowell 1956; Chance and Mead, Chapter 4; Griffin 1976, 1982; Humphrey, Chapter 2, 1980, 1983; Crook 1980; Woodgush et al. 1981) beyond the pale of evolutionary thought.

Performance words and experience words rarely assemble together in model building because the interface between the implications of their meanings is a grey area for the most part inaccessible to a traditional scientific thinker. This is particularly clear when the emergence of consciousness in animals is addressed. Discussions of consciousness anchored in the dictionary meaning of that term involve experience words, whereas discussions of behaviour, from which inferences to consciousness might plausibly be made, involve performance words (Crook 1983). These vocabularies are highly distinctive, mutually exclusive, and miscible only with difficulty.

In this chapter I want to attempt a top-down approach to cognitive evolution within the context of an extrapolational theory of mind (Crook 1980, p. 10). In causal studies, a top-down approach means looking at the design features of elaborate performance and then inferring the sorts of components that could process the performance: the bottom-up approach is to look at the physiological components and infer from a knowledge of them how the whole system must work. In understanding the mind top-down approaches are provoking new research in areas where bottom-up analyses have hit sterile ground. In an evolutionary context we need to understand the human mind better and then look back into its evolutionary history, rather than supposing that we can infer the structure of mind from the evolution of animal behaviour. Furthermore, as soon as we actually look at the operations of our personal minds, we see that performance, intellectual operations and experiencing are all interlinked into a complex whole. To abstract the intellect for analysis without a concomitant study of the effects of emotive and conscious states on performance and cognitive capacity is clearly psychologically naive.

Mind as a model of reality

A number of authors have developed Craik's (1943) original notion that the brain maintains a scheme or representation of processes that maintain physical and social stability (Fisher and Cleveland 1956; Sokolov 1960; Mackay 1965; McFarland 1966; O'Keefe and Nadel 1978; Inglis 1980; Oatley 1985). Representational systems depend upon feedback that allows the organism to monitor the effects of activity and to compare behavioural outcomes with those predicted when action was initiated. Wherever a change has occurred the monitoring process detects dissonance and incorporates change so that the representation remains updated and stable. We may distinguish at least three levels of representation which are assumed to be operational in an organism such as a social mammal (Crook 1987). These are the somatic model of body condition, positioning and movement, the soma-situational model that locates the body in its topographic relationship with environment and the socio-focal which, in addition to subsuming and depending upon the other two, places the organism within its social relationships. In the human subject all three are closely integrated in a single experiential system, but we are usually only conscious of particular aspects of the whole at any one time. The question naturally arises as to what controls the focus of awareness that directs attention to priorities for action from moment to moment.

The tripartite representation of a person's situation in the world amounts to a model of that individual's 'reality'. Indeed the system actually *is* the reality so far as the experiencing subject is concerned—it symbolizes whatever relationships obtain between the in-here and the out-there aspects of a personal situation. Normally, this being in an experiential world is unquestioning—reality is simply assumed in the act of experiencing. The natural orderliness of experience is emphasised by cases in which degrees of disorder intrude upon the conscious world. Susan Blackmore (1982, 1984) has studied the 'out of body' experience and suggests the following psychological interpretation of this and other mental states.

1. The brain maintains models of reality by monitoring ongoing experience by reference to recent memory. Longer term memory is called upon to stabilise the model when it is severely disturbed (3).
2. Unstable models are rejected as being inconsistent with expectations of stability and only stable models are used as bases for action.
3. Where the brain is disturbed by drugs, illness, emotional imbalance, or death it seeks to maintain stability by drawing upon memory to replace parts of a representational system that are highly distorted. Under extreme conditions, such as the Out-of-Body Experience, reality is attributed to representations that are predominantly based in memory but assembled often in bizarre ways.
4. Restoring defective models by building a preferable replica from memory may in extreme cases be the only way to restore the input control essential if an organism is to continue with survival-related behaviour.

In this view 'reality' is thus a psychological attribution by the agent, projecting a particular status on a representation believed to symbolise an 'actual' state of affairs. This 'reality' is never a matter of external 'fact'.

Where we experience ourselves to be within this 'reality' is a function of the current focus of attention. This focus is itself dependent upon the nature of a current issue which requires intentional resolution: there is a state in experience calling for change by design. As an issue is resolved and another becomes salient so attention becomes focused upon the latter. The criteria for changes in attentive focus seem to be hierarchically organised: thus, a dramatic change in a social situation may commonly take precedence in attention over postural discomforts or bodily states such as hunger. We may suppose that there are sets of rules defining the criteria for attention to differing sorts of input and these rules will be a complex interrelationship of sociobiological tactics, acquired responses and graded reactivity responsive to given cultural situations. Dealing with issues is what attention is commonly about and it will be intentionally directed.

The intentionality of animal behaviour has become an important talking point in discussions of consciousness. In its antique philosophical sense the term intentional means that one process is 'about' another. Intentionality is a property of 'aboutness' forming a key part of a system (Dennett 1979). It seems to me that in this general philosophical sense any living system, in that its existence depends on a relationship with environment if only for energy assimilation, must necessarily show intentionality; the degree of subjective awareness of intentionality is a problem of metacognition rather than of the cognitive process itself. The term also carries however, the more limited sense of 'having an intention'—i.e. being goal oriented in some way. According to the viewpoint we are developing, whenever a deflection from an expected value (a *sollwert*) representing a desirable condition of stability arises the individual faces an issue, and intentionally sets about resolving it and moving towards the value set for stability in that particular regard.

Self representation as identity

Some recent studies have argued that a sense of self arises as an imputation that the experiencer is the agent of a particular behaviour or affective state. This seems to be the way in which a sense of self arises in ontogeny (Lewis and Brooks 1979). Phylogenetically a comparable process in evolutionary time seems plausible. As social situations became complex the role of an individual in relation to two or more others required understanding. Humphrey (1983) has stressed the point that to predict another's response to self may have necessitated an ability to empathize with the feelings of another. The observation of another's behaviour calls forth feelings in self which are empathic, thus allowing the self to predict the other's behaviour on a basis of what self (in the same situation) might try to do. Crook (1980) emphasized the significance of empathy and the necessities of distinguishing between empathic feeling and self's feelings in monitoring responses in behaviour involving reciprocal altruism—and particularly in the detection of cheating (Trivers 1971). It would for example be cognitively confusing if an individual could not distinguish clearly between feelings attributed to another through empathy and those arising as a consequence of its own state; the ability to predict an action that would enhance its own condition would then be impaired (see further below). Colvin (1982) has stressed that in choosing to select one of two or more individuals as an ally a monkey must be able to understand the different responses of affiliation or rejection which others make to him. Jane Goodall's work (1986) and that of Frans de Waal (1982 and Chapter 10) on chimpanzees show how subtle a sense of self in social situations these animals have.

Among humans the being aware of self's awareness and representing it metacognitively to others in language means that self-knowledge—treating the experience of self as an agent in experiential change—can become explicit, whereas in non-verbal animals we can only infer it from outcomes in complex social situations. This sharing of explicit information about the experience of self becomes the basis of propositional statements whereby the wants or needs of individuals can be stated for others to comprehend.

Generally speaking, theories of identity stress the defensive nature of the processes whereby a human sense of self is achieved. As an individual grows away from dependence upon mother he/she experiences various hurtful or frightening situations; to these he/she adapts by establishing means whereby self-esteem can be maintained and confidence assaulting social events coped with satisfactorily. Many humans are subjected to harsh treatment or inadequate loving from parents and as a consequence come to introject their parents' attitudes as their own—believing for example, that 'I am bad' because I have been told so many times 'You are a bad girl/boy'. As an adult such introjections tend to rule personal experiencing by predetermining defensively the outcome of what might have been novel personal experiences (Sullivan 1955; Erikson 1968; Swenson 1973; Guntrip 1983; Main 1980). Self-punishment replaces other-punishment, object relationships become distorted and perceptions of those who become personally close distorted along channels established in infancy. A common tendency to avoid ancient unbearable experiences produces schizoid personal dispositions whereby individuals split away from confrontation with the real through avoidance, fantasy, or depressive aggression. Becoming a person with self-awareness is fraught with culs-de-sac of socially induced illusion filled with models of 'reality' that relate badly to occurrences actually happening in the world.

We come now therefore to a key point in relation to models of mind that focus upon the computational skills of intellect. The above arguments make it clear that for human subjects the facing of a personal issue requiring an intelligent appraisal of self in social setting involves more than a supra-developed arithmetical ability to work out best plans of action. The process is throughout embedded in feeling states of varying degrees of confidence, suspicion, fear, self-doubt, defensive assertion, or tentative objectively interested exploration. These states of mind undoubtedly influence the course of decision making in subtle and little analysed ways—in spite of an extensive literature on the psychodynamics of the subject. The experiential context of intellect becomes in this viewpoint a vital component in the determination of personal action.

Dimensions of personal awareness

What actually is awareness and what is it for? Why should not all these computational procedures run off as it were automatically without this curious property we call consciousness? Attempts to answer this question have not got very far and range from dualism to agnosticism. Today, however, most thinkers are not inclined to deny the question significance.

It is clear that not all representational processes are conscious ones. In proprioceptive modelling, which allows postural precision in movement, an actual awareness of the process is not usually present. In the learning of skills the task is often clearly experienced element by element, but gradually, as in learning to ride a bicycle or play a piano, the elementary skills become unconscious and awareness is now more focused on style or manner—the minutiae of performance execution need no longer receive conscious attention. Lower order skills can now be run off as required with subtle nuances formerly unobtainable. 'Flow' in performance commonly results from an awareness of effective skill in relation to an exacting task—worry about detail is replaced by a joyous flow of performance (Czikszentmihalyi 1975).

The difficulties we have in conceiving the nature and functions of awareness perhaps arise from our tendency to reify experiences so that we turn the concept of awareness into a *property* we have. Experience words should perhaps be examined as much from the perspective of the basis that underlies them (ontology) as from a perspective examining only the objects focused upon (phenomenology) in experience (Crook 1987). In such an extended perspective 'being' is seen as the starting point and not as a property of something else. It may be more meaningful to assert that the core of the concept 'I' is the process of representation of cognitive materials rather than that I have such a property as awareness. Thus, when certain forms of cognitive processing arise the reality modelling constitutes consciousness itself.

I have suggested elsewhere (Crook 1987) that the essential feature in making a representational process conscious is the temporal continuum of the dissonance detection that constantly updates the model and uses memory to control for coherence. The process involves a feed-forward process whereby an expectation is created that predicts the relationships between components of the model in immediately future time. The continuity of the match-mismatch comparison involves the creation of cognitive patterning against an equally cognitive background. Within this figure-ground relationship issues arise and receive attention within the psychological 'time' defined by the process itself. The temporal continuity of this process is felt by the subject as 'experience' and this awareness, I submit, constitutes the subjective basis (ontology), of phenomenal sensing. To the question—why should this process constitute awareness—the answer, unfortunately, seems to be: it is so.

Duval and Wicklund (1972) have made a useful distinction between objective self consciousness and subjective-self-consciousness. OSC involves the reification of 'self as agent' for example, in a conversation about personal behaviour. It is experienced as a felt concept that is evaluated by others and compared by self with others. Such objective-self-consciousness takes itself as the object of awareness and as the target for introspection. The empirical research of these authors points to a degree of attendant anxiety wherever OSC is present—a stress due to implicit comparison with others in social evaluations which may or may not support self-esteem. The effect is often subtle, as when a subject reports a more anxious condition when doing a simple task in a room with a mirror than when in a room without one. SSC by contrast is an awareness of process—walking, hammering nails, singing to oneself, without the concomitant presence of the idea of self as performer. The non-social, purely personal, aspect of SSC accounts for its more relaxed character and value in renewal from social anxieties. Crook (1980) suggests that when a subject is in flow (Czikzentmihalyi 1975), performing an activity joyously without boredom or anxiety, the process is essentially one involving SSC—which is why flowing activity is felt to be relaxing, energizing and self-confirming.

It becomes clear, therefore, that while representational processes are going on in both cases, OSC and SSC are very different experiential states determined by the nature of the focus of awareness. At this point therefore we need to enquire into the range and variety of conscious states in order to attempt a more comprehensive mapping of the dimensions of awareness and the way in which these relate to the computational functions of intellect. The association between OSC and socially-based anxiety on the one hand, and SSC and a more relaxed experiencing is for example found again in recent research on stress. Both the western method of autogenic training created by J. H. Schultz in Germany early this century (Rosa 1976) and many oriental meditation techniques clearly involve the transition from OSC to SSC. This shift can, for example, be clearly observed and experienced in Zen retreats of about 5 days or more duration. There is an increasing dis-identification from the anxieties of OSC leading to an open awareness of process; a capacity to be acutely aware of circumstances, both external and internal to the body, which are normally precluded by defensive self-concern. The many approaches to Buddhist meditation have the cultivation of this open awareness as a prime object (Sekida 1975; Berzin 1978; Hopkins 1981; Lati Rimpochay et al. 1983; Kelsang Gyatso 1982; Crook, in prep.).

The range of subjective experience can be usefully envisaged as the possibility of placing awareness anywhere on a circular surface defined by two crossed dimensions (Fig. 24.2). In this simple model the first of these dimensions defines experience that ranges from being highly intentional (having a purpose or goal in mind) to being highly attentional (focused on here and now experience without goal orientation). The second dimension

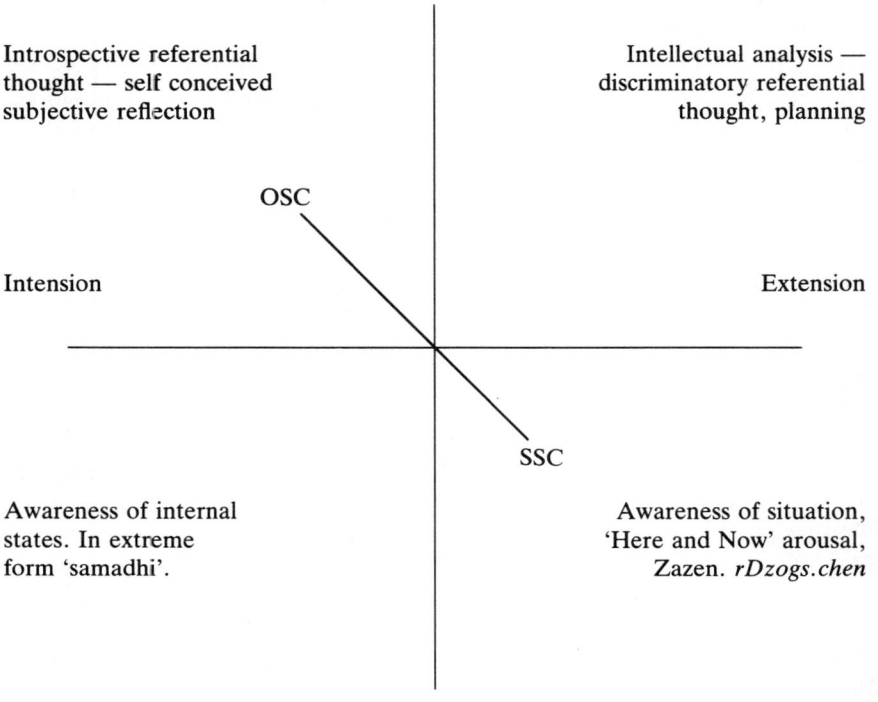

Fig. 24.2 The range of states of *human awareness* as an interaction on three dimensions. See text for explanation.

expresses experience that at one extreme is totally concerned with environment (Extension) and, at the other, focused on interior awareness (Intension). When these two dimensions are crossed we can define any locus on the surface so formed in terms of degrees of Intention-Attention and Extension-Intension. Broadly described, four quadrants arise:
Top left Subject-related thought
Top right Object-related thought
Bottom right Situational awareness
Bottom left Inner state awareness

The vertical dimension calls for futher comment, for it will be clear that high levels of attention often accompany goal-orientated thought and action. What is implied here is, however, the relative degree to which the subject is orientated towards a desirable situation in future time (intention-expectation, etc.) or, conversely, orientated towards the quality of on-going present moment experiencing in which focus on alternative conditions is absent. The dimension defines what is essentially a difference between concern about a desirable or non-desirable state other than in the 'Now', and a non-judgemental observation of the actual passing moment.

During the course of any activity plottable on such a surface the subject may or may not be consciously or unconciously self-regarding. Thus, in the top right position a subject, although explicitly or implicitly utilizing self-conception as the basis for planning, may or may not be synchronously self-conscious in the sense of feeling regarded in whatever way by another. Similarly, an introspecting subject (top-left) may or may not feel regarded by another at the same time. It thus appears that the OSC–SSC distinction cuts across the two axes of these conditions creating a third dimension. In the bottom left and right states, the increasing loss of intentionality with heightened single-pointedness of attention suggests that the presence of this third-dimension becomes weaker as the activity deepens in the downward direction. The result is thus, perhaps, not a geometrical sphere with three dimensions of equal potency, but a shape more like a hot-air balloon with the lower surface open!

The prime aspect of cognitive ethology which is the concern of this book focuses on the significance of top-right processes—objectifying intellect; the evolution of which is perceived as closely correlated with the emergence of social complexity calling for social calculation and behavioural calibration in relation to others. Humphrey (1980) talks about 'nature's psychologists' and Reynolds (1976) offers the term *Homo sociologicus* as a replacement for *Homo sapiens*. Figure 24.2 provides a general map of conscious states within which varying degrees of social concern manifest themselves. Most Westerners will be more familiar with the top half than with the bottom; a result of educational bias under cultural pressure. The figure represents in a simple way wider capacities of mind than the merely intellectual; and in an evolutionary perspective we may examine whether the less purpose-orientated and less socially concerned states of mind also have survival value. To see what adaptive significance these may have had, it is useful to conceive of their functions in an imaginary hunter-gathering society early in human ancestry.

The positioning of consciousness in top right is to be actively involved in planning action in the world or reviewing past action with a view to improvements. Top left states are often highly objective self conscious conditions which take the self and its experiences as the topic of intellectual examination and relate them to ideas of personal social effectiveness. These introspective concerns are doubtless significant in producing understanding of feelings in relation to others. Trivers (1971) has argued that individuals acting in reciprocally altruistic exchanges with others need to guard against failures in returns of benefit that would arise when a partner cheats. It is particularly here that an ability to read another's intentions from his/her non-verbal expression in posture and gestural style through an empathic induction of the other's motivating conditioning within oneself, can provide valuable information about a partner's intentions. However, when an individual experiences 'intuitively' what an other feels, there is the risk of confusing empathetically induced feelings with one's own. An introspective ability, firmly distinguishing self-states from empathic ones, does much to prevent the appearance of such errors in social awareness and has an important function in preventing the exploitation of self by others. It also allows an estimate of the strength of one's own motivation for comparison with values of motivational strength perceived in others, thus permitting calibration of responsiveness in balancing hawk versus dove strategies in competitive situations. Individuals are also enabled to play mental games with varying outcomes, on a basis of visualizing alternative situations in which motivational strength in self and others are set at differing values. Mental games of this kind are valuable rehearsals for real life social contests.

Bottom left states arise when the inward regard continues, but without a focus on any problematic issue; direct attention on the inward state deepens without attendant thought processes. In meditation training such states are induced by closing the eyes and focusing attention initially on a repetitive activity which cuts out thought. A typical example is counting the breath up to ten or twenty-one and then starting again, or repeating a phrase or mantra, or simply watching the breathing come and go. After a length of time (which varies with the pressure of thoughts demanding attention) the mind settles into a relative silence which deepens into trance and finally a 'consciousness without an object' known in Sanskrit as 'samadhi'. This inward trance, as in Transcendental Meditation and Autogenic Training, has been found to initiate a 'deep relaxation' particularly beneficial to people prone to anxiety (review in Crook 1980, pp. 338–351; West 1979). Since this condition is not sleep (EEG differences confirm this), arousal from it can be virtually immediate and an unconscious monitoring of situation seems to be present throughout—although this may vary with the depth of samadhi.

In our primitive hunter-gatherer the ability to relax in this way during pauses in long treks in potentially dangerous country would have been valuable, but there may also have been other benefits. As the mind slips into inward meditative trance, a variety of conditions may arise in which disinhibited unconscious material from memory appears out of context superimposing dream on a semi-alert awareness. This material may often be distorted into fantasy which however retains its own logic (as Freud and Jung emphasize in their respective analytical systems). As this logic works its way through, partly repressed personal issues may be resolved in symbolic form—a process also thought by some to be a key function of dreaming. As this repressed material is often based in childhood memories too painful for more direct experiencing, the resolution of such issues may become a major factor in clarifying awareness of self—although not necessarily at an explicit intellectually comprehensible level. Such symbolic processing may, nonetheless, involve intellectual functions.

Conditions of trance on the border between top and bottom left also allow peculiarly direct empathic apperceptions of mental and physical states in others—such apperceptions again being intuitive or 'heart-felt' rather than intellectually accountable. It is in trance states of this kind that the shaman, the *Hla.ba* of Ladakh, for example (Brauen 1980), can perceive the problems of his/her patient and often evoke a healing state through charismatic influence on individuals sharing the same culture. Furthermore, on the borderline between top and bottom right, a shaman may develop a comparable intuition into conditions governing the environmental state—such as the onset of rain, or irrational awareness of potential benevolence or threat. To enter into such apperceptive states the shaman commonly uses a combination of high and low sounds, drum and bell, which help to empty the mind of a rationalising consciousness thereby rendering it intuitive in the manner described.

Such states of mind are well known in subsistence economies where non-technologically supported awareness of environmental situations and changes is important. Current western educational bias precludes their development by favouring the hyperdevelopment of intellectual reasoning.

Bottom right states are induced in meditative training when the subject keeps his/her eyes open either gazing at a wall, into the sky, into flowing water or out over a vast landscape. The gaze is kept steady and unfocused. In this condition the concerns of ego-related thinking recede and ultimately a direct apperception of situation arises in which self is no longer experienced—all that remains is the world. As Douglas Harding (1965) has put it, one reaches a 'headless' condition and he describes his method as the 'headless way'. In this condition, there is an acute sensitivity to the world around one; as might be suggested in a training session—you can hear a woodlouse crawling on the carpet behind you.

In a hunter gathering economy conditions of comparable attentiveness in the here and now would have high value when a hunter, weapon poised, and quietly moving in a landscape, heard a sound. At once, a totally focused here and now condition arises in which attention is wide open to the slightest situational change which might presage either the appearance of prey or danger. This openness is quite unreflective for purposive intentionality has receded out of awareness. Thought has been gotten out of the way so that it may not interfere with the Zen-like immediacy of the archer's spontaneous shot or sudden protective response (Herrigel, 1953; Legget 1978). To rediscover such focused attention the reader has merely to travel to the Serengeti game reserve in East Africa and walk away from his landrover among the trees.

In meditative training the practiced induction of bottom right states leads to experiences in which the boundary between self and situation dissolve—in the 'headless' state there is simply the environment within which a body sits—awareness reaches from horizon to horizon as one undivided experience, Self, and 'reality' appear as one. In the Tibetan practice of *rDzogs.chen* this undivided openness (unbounded-self-environment-continuum) is called *rig.pa*—a term meaning space; and indeed unlimited spaciousness is the prime quality of mental experiencing in this mode.

The model implies that the stressful states induced in the upper intentional modes in the figure will be relieved by recourse to the private concerns involving low intentionality in the lower half. The mind left in a socially neutral condition might then oscillate between periods spent in the two halves, perhaps with a developing preference for states of subjective self consciousness (lower half) over the referential states of the upper half. The system implies a periodic shift between preferred conditions. David McFarland's state space interpretations of cybernetic motivational systems (McFarland and Houston 1981) suggest that individuals will move towards *sollwerten* of optimum comfort. However, within this tendency, endogenous need states (hunger, thirst, sex) arise giving effect to motivational deflections from a maintenance level. Similar patterns of change are likely to influence the cognitive states we have been considering. The relationship between *sollwerten* of personal rest and social action is likely to be expressed by the periodic appearance of intentional behaviour and action ultimately serving sociobiological strategies favouring outcomes maintaining inclusive fitness. In strongly cultural animals the attainment of traditional goals is observable and measurable in the outcomes of behaviour, while any sociobiological outcomes can only be assessed retrospectively in terms of changes in reproductive fitness.

These arguments suggest that the range of states of awareness available to human subjects are as a whole part of the mental adaptation of the species to the environment of evolutionary origin. The capacity to utilize all four quadrants effectively may have been of major adaptive significance for early Man.

The context of intellect

The reason for placing this discussion within this book may perhaps now be clearer. Intellect is essentially concerned with goal oriented problem solving and, in human subjects, it is commonly experienced as internalized speech or verbal thought. Intellectual operations (as in problem solving by dogs, cats, etc.) are, however, clearly not dependent upon the verbal medium nor do human subjects necessarily clearly conceptualize the frame or contents of considerations involved in reaching a conclusion or a decision. Intellectual operations may in human subjects be either implicit or verbally explicit in speech or thought. Explicit intellectual operations may use pictorial or symbolic representations besides or in addition to verbal ones—or use media to which only gifted individuals have access—as in mathematics or musical composition. It seems, however, that all these intellectual operations are based essentially in a purposive intentionality which is orientated towards a time other than the present. The intention may be to reach a particular goal or a situation other than that conceived as an undesirable or inadequate one; alternatively rehearsal of the past may be employed to gain understanding of the present or of a general issue. As Freud said 'Denken ist probieren'—thought is rehearsal. The function of intellect is to rehearse action in scenarios that will produce a future state or lead to a comprehension of what has happened—and which could happen again. In all cases the experiential context of intellect lies predominantly in the upper half of Fig. 24.2. Introspection is essentially an intellect focused inwards upon its own nature and social status.

An understanding of intellect in this perspective is thus only half an understanding of mind since the two lower quadrants have yet to be examined. Ethologically, it is clearly easier to focus on problems of the evolution of problem solving than on problems of awareness, but that does not mean that questions concerning the origin of these other states can be ignored. They may be difficult, but as Don Griffin repeatedly points out, they are not likely to go away. Furthermore, these bottom half conditions may be fundamental to human well being and require effective understanding.

In Western cultures both the educational system and the pattern of commercial life in a consumer economy promote a hyperdevelopment of the highly intentional (goal orientated) modes of mind with the intellect as their prime mechanism. The average Westerner is, thus, almost continuously embedded mentally in a project, his own or his company's, or in relating to the projects of others. We have seen that such a mentality is predominantly in a state of objective self consciousness (OSC) which, in that it involves implicit comparisons with others, constitutes a continuous concern with degrees of self-esteem and social coping. There is, thus, a continuing level of stress, varying in intensity according to personal dispositions derived largely from childhood conditioning, which, for many, constitutes a sort of socially constructed cage. It is this background stress that contributes greatly to feelings of personal alienation in western cultures (see also Czikzentmihaly 1975), and generates an intentionality that seeks resolution.

There is indeed a striking contrast between the mental experience of the average Westerner and that of an individual living in a traditional subsistence economy where self-sufficiency, rather than maximization of gain, is the only possible objective. In the traditional village culture of Ladakh, for example, reciprocity replaces exploitation and social security depends on an extensive system of intra- and inter-family support within a population of reciprocally supportive lineages (Crook and Osmaston 1988). In Ladakhi culture a present-centredness in task performance is a normal attitude of mind; and amongst the religious virtuosi, whose life style in monasteries and hermitages generates the attribution to them of charisma, extensive meditative training has led to a considerable disidentification from social concern through a concomitant emergence of a non-egoic sense of participation in a universal process. The result is a population of people whose mental health seems remarkably developed (although threatened by contemporary developments especially commercial monetization). It is, of course, important not to idealize a tradition which has perhaps been unusually fortunate in its social history, but the point remains that contrasting balances in the degree to which goal orientated obsessions are softened by present centred awareness arise in situations of differing socio-economic structure. It is clear that in the West the possibility of managing society in ways more beneficial to individuals remains an option open to policy makers and planners alike.

The role of intellect in fostering circumstances where the non-intellectual apperceptive functions of mind can function more freely may, thus, be highly significant—but we cannot achieve advances here without understanding that intellect is itself only one aspect of the experience of mankind. Creativity, rather than a defensive responsiveness, often results from periods of reflective withdrawal during which present centred activity is enhanced. An understanding of social circumstances that favour an optimum balance between purposive action, present awareness, and creative reflection should clearly be an important objective in social psychological research and one to which ethological study on the origins of higher mental activities can make a major contribution.

25

The evolution of purpose
ALISON JOLLY

In this chapter, one of the originators of the Machiavellian Intelligence hypothesis has the last word. We simply gave Jolly all the other chapters, asking her to survey the 'state of the art' displayed in them and make an audit of the achievements of the 20 years since her own pioneering paper.

Introduction

It is almost as if the human brain were specifically designed to misunderstand Darwinism,

Richard Dawkins writes in *The Blind Watchmaker* (1986). Absolutely right. The human mind is intrinsically teleological. We can't help believing that moving, complex, and beautiful objects must have a purpose. The world, and even the universe are full of beautiful, complex, moving parts, intricately linked by electromagnetism and gravity, by the chemical interactions of Gaia and by tasting good to each other, and by sexual desire.

Teleology, however, may be an extreme example of the phenomenon Humphrey describes in Chapter 2. That is, teleology may be a mistaken view which comes easily to the mind of a social primate. Consciousness implies the process of setting goals, either goals far in the distant future, or as near-term feedback that monitors the ongoing situation. Intelligence is designed by selection to actively look for and impose patterns on an inchoate world, in the service of achievable goals. Surely social intelligence would be most useful to an evolving primate if couched in terms of what other animals *want*. The kind of mind which believes the world, and evolution itself, has purpose, could have started long ago by being particularly sensitive to the desires of other *Aegyptopithecus*.

The blind process of natural selection (whether operating fast or slow or in hiccups,) has produced creatures which want things, creatures aware of their own goals, and aware of the goals of others. We may be over-generalizing when we impute purpose to the universe at large, but we know it is true that purpose exists among ourselves and our social companions.

Intention and adaptation

Daniel Dennett (1983) claims that there is a close parallel between 'the intentional stance', and the Panglossian assumption of optimality in evolution (Voltaire 1759). I agree, and agree with his approval of both views, but I would rather rephrase it, because Dennett doesn't mean any of the words the way most of us do.

The 'intentional stance' (Chapter 14) isn't about intentions as purposes; it is about 'aboutness'. The feedback loop of a thermostat is 'about' room temperature in this sense. So is the *sollwert* of a fly's eye. A *sollwert* represents the way the world should look. This lets the fly sort out the way the world spins across its compound eye when the fly itself circles round Grandfather's moustache from the way the world spins when Grandfather sneezes at the fly. This 'aboutness' property should really be called just feedback, or at most monitoring or reference.

The metaphorical term 'intention' is a bid to link such self-reference with the everyday meaning of conscious purpose. Dennett feels there is a true continuum between the two.

My bid is supported by the fact that the logic of the explanation is fundamentally the same whether one is explaining the most vaunted conscious purpose or the free floating rationale of some simple feedback device—or the evolutionary design process itself (D. C. Dennett, personal communication).

Intention in the everyday sense is a conscious intention. The earliest such intentions might have been simple desires—'hunger-eat', or 'bite', or 'suck' that could serve like a thermostat strip to monitor actions on the environment. [This takes us back to Konrad Lorenz (1975) the founder of modern ethology. He distinguished appetitive behaviour from the consummatory acts which are, in a sense, goals.] Adding a second layer of feedback with built-in time delays could make the monitoring device self-monitoring. Setting a future value might allow a variety of 'appetitive' intervening acts. At that level of complexity most people might agree we are talking of common-or-kitchen-garden intentionality. At what stage in the tree of life such self-monitoring became conscious awareness is an argument several millenia long which will not be solved in this paragraph. I am going to assume that you and I are both conscious, and that if you have read so far, you agree consciousness has evolved.

Crook (Chapter 24) emphasizes the fundamentally temporal nature of consciousness.

It is an essentially temporal continuum of dissonants that constantly updates the mental model, and uses memory to control for coherence. This involves a feed forward process whereby an expectation is created that predicts the relationships between components of the model in immediately future time . . . The temporal continuity of this process is felt by the subject as experience.

One of Crook's main axes of consciousness is temporal, stretching from an almost immediate receptivity and awareness forward to earnest scheming toward a future goal. Note that it takes something like Zen training or intense physical exhilaration for most of us to approach pure awareness without the contamination of planning. Even then, the feed forward monitoring of consciousness maintains some slight dissonance if we are conscious at all.

In this picture of evolution, then, consciousness does not have to spring fully formed from nothing. Like the vertebrate eye, there are plausible intermediate steps, each useful in themselves: first feed-back, then more complex *sollwert*, then self-monitoring with temporal dissonance, and finally increasing the time and number of steps between goal state and present state to the point of imagining alternate routes to the goal, while discarding unproductive routes. Chance and Mead (Chapter 4) had the original insight that deferring immediate action while calculating alternatives is very important in primate social relations. It is clearly one of the chief functions of developed consciousness.

Humphrey (Chapter 2) describes the importance of farther goals even in chess-playing computer programs:

Weak players grow short bushy [decision] trees, looking a short way ahead at a mess of poorly differentiated possibilities; strong players prune the tree much more efficiently and . . . construct long thin trees, looking much deeper into a few critical variations. This pruning is the heart of the problem . . . Which branches are critical and which can safely be cut off? (from *New Scientist* **66**, 119, 1975).

Stephen Jay Gould (1987) points out that when evolution produces long thin trees, we get a false impression of purpose. For instance, the evolution of the horse is now known to have been a mass of bushy branches, mostly a variety of many-toed smallish animals that browsed on forest leaves. As it happens, only one lineage survives—an aberrant group of grass-eaters who thundered about the plains of North America, tearing up divots of the savannah with their hooves. By sheer luck, a few of them reached the Old World, whence their still later descendents came home to America as the Spaniards' war machines. Early palaeontologists, who saw a kind of inevitability in the horse lineage, were working from an incomplete fossil record that could all be arranged in a straight line from *Eohippus* to the Cortez chargers as though Nature planned it that way.

When Dennett compares the intentional stance with optimality in evolutionary theory, he is saying that in some sense one assumes the animal intends its behaviour to do the best thing for it in the circumstances, and in some sense the species is adapted as best it can be to its ecological niche. In any one case of course, that assumption may be wrong—the courting antelope may not see the lion, or the population may be stranded on a lesser adaptive peak in the genetic landscape. As a general guideline, though, we use the optimality assumption all the time. We presume when a person does something she means to do it, as well as assuming that an animal is well-fitted to her place in nature. We picture both as a thrusting mind (or species) choosing or finding a course through the limitations of the environment. Sometimes, indeed, those choices may change the environment, which in turn may change the species. (Aside to fellow feminists. What was I supposed to say there: personal she, but animal he?)

Scientists of different temperament then pull different stories out of different (or even the same) sets of data. Some, the adaptationists, delight in finding functional explanations for observed facts. Others prefer to debunk such explanations by pointing out how many apparent adaptations are side effects of overriding developmental patterns, as the spandrels between the arches that support a Byzantine dome result inevitably from the arch and dome construction (Gould and Lewontin 1979). However, even these arguments suppose that arch and dome have an evolutionary 'purpose'. All Darwinians come back to adaptation at some stage of the discussion.

Even Aristotle (350 BC), codifier of teleogy, came close to a Darwinian formulation, only to reject it.

> Why should not even bodily parts like teeth have developed in the necessary course of nature—sharp front teeth, suited for tearing, the molars broad and suited for grinding down the food? May they not have been produced not to some end, but by coincidence? And may it not be so with all bodily parts supposedly having some inherent end or purpose? Those organic structures then which came into the world as if they had been produced to some end, survived because they had been automatically organized in a fitting way; all others, like the man-faced offspring of oxen in the theory of Empedocles, have perished and continue to perish. (*Physica* 198).

Aristotle rejects this supposition, however, because organic life is 'generated in definite ways', not by luck or chance. He is clear that this generation may not have involved conscious deliberation:

> As we pass gradually down the scale, it becomes apparent that there is adaptation to ends also in the growth of plants which, for example, put forth leaves to shelter

their fruit. Hence, if it is both by nature and to an end that the swallow builds its nest, and the spider spins its web and that plants put forth leaves for the benefit of the fruit and send their roots down rather than up for nourishment, it is evident that there is such a factor [as an "end"] in natural processes and beings (*Physica* 199).

Darwin and Wallace showed us how the mechanism Aristotle rejected could, indeed, operate to produce animals fitted for an end. They saw how it could operate in relatively small steps over evolutionary time. The current interest in punctuated equilibrium remains firmly Darwinian though supposing that a few macromutations, like the man-faced offspring of oxen, have also found a niche to survive.

Desire and belief

One criterion of intention seems to be belief statements: Vervet Tom believes that vervet Sam has not seen the leopard, for a complex example (Dennett, Chapter 14). It seems to me that desires are probably simpler than factual beliefs, and older both phylogenetically and ontogenetically. The desire for food or sleep or safety is a very small step from hunger, fatigue or fear: a first step into the future. Crystallizing this into complex mental statements, or beliefs can stepwise generate more and more elaborate complexity. 'Ripe figs . . . This way to the figs . . . This way not that one is the shortest to the fig tree . . . I know this is the shortest way . . . I believe (but am not quite sure) this is the shortest way . . . She (the matriarch) knows the shortest way; I'll follow her . . . She may even know some fruit tree I don't; I'll follow her.'

In this sense, intention in the colloquial sense precedes knowledge.

It also seems likely that perception of others' emotions or desires could have long preceded understanding other's states of knowledge—and, indeed, must have for social life to function among birds and mammals. (Griffin adds: 'Or social insects!') This understanding, and response to other's desires underlies both the transactional nature of society and the extrapolation to desires in nature. As Humphrey says (Chapter 2), primitive and not so primitive people commonly attempt to bargain with nature. Even to a primate, a generalization from 'the lion wants to eat me' to 'the river wants to drown me' might have helped survival (Jolly 1983), and anyone who has lived through an English winter can see the point of building Stonehenge to make the sun come back.

Much normal primate social life does not require that an animal make any clear attribution of states of knowledge outside the immediate situation. Even quite sophisticated social behaviour can fall into this immediate class, for instance, Kummer's (Chapter 9) seminal description of tripartite relationships in hamadryas baboons. If a female threatens another animal from beside her male, all three can react to the whole situation.

Still, the threatened animal must know or deduce that the male will defend that female if the threat is to work. It would be useful to take the next step and to remember the relationship, in case of temptation to punish the female if ever she's caught alone—she may well go back to enlist stronger aid. Kummer's (Chapter 9) account of young males using infants to buffer threat and reduce stress led him to point out how often these three-way relationships have components of mother-infant behaviour—in the one case females acting 'infantile', in the other, males acting 'maternal'. Two other examples seem more innovative: when a male threatened a young female, who did not respond, an older female rushed over and groomed him—what the young one (probably her daughter) 'should' have done. Once, a mother ignored her infant's screams when it fell down a crevice, so another female pulled it out and pressed it to the mother's breast. Three-part interactions demand a background of social knowledge.

Harcourt (Chapter 11) reviews and analyses all the current data on primate two- three-, and more part alliances and interaction. He shows that primates react differently to close and distant kin, to greater and smaller threats to kin, to dominance status of the threatener, to whether their kin or friends are threatened or threatening, and to more distant friends and kin of the antagonist. Much of this can be put in emotional, rather than cognitive terms: close kin are loved more, threats to self mean scared more, threats to kin mean scared for loved one, etc. Apparent long-range strategies, such as a dominant female breaking up grooming between two subordinates, may have a fairly simple immediate basis—'I feel cranky today and I can't stand those two just sitting there grooming' or else 'Why aren't they grooming me?' Even if one looks for the simplest possible explanations, it is clear that the alliances take a large number of social factors into simultaneous account; and it is easy (and sometimes even correct) to empathize all the way from our own vantage point—if the mother whose daughter transgresses a social code can see herself in Kummer's baboon so Harcourt's last figure of baboons currying favour with dominant mothers' offspring is a lurid insight into the origins of juvenile human snobbery. Harcourt makes a strong case that the evidence so far suggests more complex social alliances in primates than nonprimates, but we need much more data on social carnivores—especially the toothed whales—and may be even on parrots or ants.

Seyfarth and Cheney (Chapter 6 and 19) have gone far to decipher the mental skills which underly these alliances. Yes, vervet monkeys do know that a particular infant belongs to a particular mother, and that a particular neighbour's grunt belongs to his own troop range, not another's. Vervets can extrapolate from a 'wrr' to a 'chutter'. That is, Cheney and Seyfarth framed one vervet, playing its 'wrr' that should warn of the arrival of another troop eight times running with no other troop actually there. When they then played its 'chutter' which has the same (or very similar) meaning, the other vervets transferred the habituation, or disbelief, to the new call. They did not habituate if another individual's voice gave the new call, and did not habituate if the same individual gave a call with a different meaning.

Cheney and Seyfarth point out that recognizing close bonds between other individuals could result from simple associative learning, but at some point it must have become more efficient to form some more abstract concept of relationships into which particular animals fit. Perhaps such an abstract conception of kinship is revealed by the fact that vervets are more likely to threaten each other if close kin have previously fought.

Verena Dasser's experiments in the laboratory amply confirm that long-tailed macaques form such social categories (Chapter 7). Her two subjects could distinguish slides of familiar mother-offspring pairs from other pairs of monkeys, even otherwise related animals. The offspring in transfer tests ranged in age from new black infant to full-grown adult—it seemed to be the actual relationship which the monkeys picked out, not other cues. One monkey also learned to distinguish adult-sister pairs from other kinds of grouping. It thus seems clear that a catarrhine can deal in social categories, beyond the immediate associations of specific animals.

The deciphering of chimpanzee abstraction and empathy in the laboratory is principally due to the work of David Premack. He has revealed (or perhaps he and an extremely intelligent chimp named Sarah have revealed) that a chimpanzee can solve such problems as 'knife is to apple as scissors are to paper' or 'key is to padlock as can opener is to tin can' or the causal sequences 'blank paper—pencil—marked paper' but 'marked paper—eraser—blank paper' (Premack 1983).

Sarah can, as well, choose appropriate endings for a video tape of one of her trainers attempting to solve a problem. For instance, a favourite trainer is shown trying to reach a suspended banana. Sarah chose the picture of his reaching his goal, having piled up blocks to stand on. For a disliked trainer, though, she chose the man prone and strewn with concrete blocks. She was exposed to tests either with a kind trainer who showed her which of two cups held concealed food, or with a liar who indicated the empty cup, or, if she showed him the right answer, ate the food himself. Sarah learned to lie, actively, in return by indicating the wrong cup, though she sometimes gave vent to her feelings by hurling things at the liar (Premack and Woodruff 1978).

This degree of social sophistication takes account of others' wishes. Premack has also demonstrated that chimpanzees can deal with others' limitations of knowledge. He blindfolded a trainer who was wearing the key to a concealed food box on a chain round his neck. Younger chimpanzees learned to lead him along by the hand instead of just indicating where they wanted to go. Sarah, ever superior, simply pulled down the blindfold. (Premack and Premack 1982).

The complexity, and advantage of alliances as outlined by Harcourt, and their psychology as outlined by Cheney, Seyfarth, Dasser, and Premack have led us from desire to knowledge. Smuts (1985) describes the case of a baboon male taking 'revenge' after 24 hours by wounding a female who had persistently interrupted his consortship with another over the previous 3 days, and of a mother who searched and called with agitation whenever the troop passed through the grove where she had lost the corpse of her dead baby, weeks after the loss. These incidents could be attributed to blind hate or sorrow triggered again by the sight of the relevant individual or place, but it now seems as reasonable to propose the baboons remembered, and knew, what happened before.

Deception

When social knowledge is working well, the group functions like a well-oiled machine. Milton (Chapter 21) quite rightly stresses that co-operation has been as important as competition in primate society, but competitive deception is a tool that lets us dissect situations because some animals do not know what is happening. Then we can be much more sure what they should know.

Whiten and Byrne's catalogue, and classification of tactical deception (1988 and Chapter 16) will set standards for a long time to come. Of course, it is the fate of most such classifications for the edges to blur, and the categories to be superseded by others, but this one should provoke its share of thought and controversy before wearing out.

As Whiten and Byrne say, the more elaborate examples demand that the deceiver imagine what the target knows. For instance Kummer's baboon who spent 20 minutes inching behind a rock, so that her overlord could see only her back, while her concealed hands groomed a subadult male, must have had some notion of what the dominant male could and could not watch.

The multiple steps of third- and perhaps fourth-order deception not only show states of knowledge of others' intentions, but steps of contingency within the agent's own intentions. At the least it is clear that the animals are fooling each other on purpose.

The phylogenetic conclusions are less clear. Ristau (in press) makes a convincing case that piping plovers monitor their 'targets', triangulate the predator's approach, remember previously threatening targets, and in short act as though they were playing 'distracting by leading away', much like a clever chimp. Although I am certain that lemurs are far less subtle than baboons (or plovers), I am not sure that triadic behaviour has been looked for or that it won't turn up. They may turn out, though, to be too naive to deceive. The suggestion that apes are too clever to be fooled by baboon–type deceptions, made by both de Waal (1986), and Whiten and Byrne, must be separately considered.

La Frenière (Chapter 18) raises a different kind of classification by considering what capacities would forestall another's deception. He quotes Ekman and Friesen's (1975) distinction between deception cues, which can inform a receiver that deception is in progress, and actual leakage of the concealed information. Total inhibition of information gives clues that deception is happening. As de Waal pointed out, baboons and macaques can just blankly ignore threats as though they weren't perceived, and get away with it—chimps don't.

The shift in strategy sounds useful to primates. In La Frenière's game, young children progress from identifying with the adult's goals under 4-years-old, to inhibition, and finally in some children near 6, to providing false cues—like Sarah showing the wrong box to the 'lying' trainer. The ontogeny of human deception has tantalizing resemblences to the contrasts in primate deception.

With the analysis of deception, then, we have clear indication of some primate beliefs about states of knowledge, and the use of those beliefs in planned behaviour towards a goal.

Intelligent foraging

However, what about the traditional picture of environmentally-driven evolution of intelligence? All these primates must eat, and some get eaten. They spend far more of the day in foraging than socializing *per se*.

Of course, two good reasons for evolving intelligence are better than one. There is no reason why intelligence and brain size should not reflect both environmental competence and social sophistication. Milton (Chapter 21) points out some crucial correlations between brain size and aspects of diet. The larger brains of frugivores than folivores, and the still larger brain of hunter-gatherer hominids implies that something in the diet demands or allows brain growth. She makes the very important point that allowing could be as, or more important than demanding. High energy foods may be necessary to support the metabolic cost of a large brain.

I think we just do not know the necessary background to decide the relative importance of each factor. This book begins to provide the background, delineating gross categories of social sophistication. We can just about define what we mean by apes being really more clever socially than baboons, and baboons than lemurs—a contrast which will be obvious to most zoo-goers using humans' long-evolved social intuition. We haven't yet measures to be sure whether frugivores are actually more clever socially than folivores.

On the dietary side, we have a good many descriptions of some species search for embedded foods (cf. Jolly 1985). We have also a theoretical Piagetian justification for regarding embedded foods as mentally taxing (Parker and Gibson 1979). However, we don't have comparative work on the sophistication of mental maps. It takes heroic effort to address the 'travelling salesman problem' of optimal visits to food trees. It is still difficult to prove, mathematically, that an intensively studied troop must be using some mental map to cover their home range (Robinson 1986).

Sigg and Stolba (1981) tackled the mental map of hamadryas baboons. They showed that merely following topographical features, such as wooded ravines, could not explain the band's daily march to a diversity of waterholes. The well-used 'street segments' of the routes could be flexibly connected by a wide variety of transitions. Well-used waterholes were approached from a wide variety of directions. Bands which were in less familiar area stuck to more rigid routes, while regular visitors seem to know more sites off the direct line of travel. Finally, as bands approached the mid-day waterholes, they accelerated their travelling speed. All these observations imply a mental map, and purposeful visiting of goals within it.

One further set of observations suggests that the group goal was common knowledge to members. The social groups often split and travelled out of each others' sight, rejoining later.

In 5 cases the separate parties were pursued by 2 observers: the parties followed quite different routes, eg one part crossed a hill chain while the other part went around it. Nevertheless they later met at the same rich feeding site, despite other feeding sites situated closer to the route of one party . . . Optical or acoustic communication was unlikely—rather the events support the hypothesis that the baboons know in advance where to go.

We do not yet have comparative data on folivores and frugivores in a given forest, which would tell us how many individual tress each species visits, and which trees are approached from how many directions.

Besides the huge commitment in human time that such studies demand, there is a problem of tautology. You almost have to assume that the monkeys are optimizing their use of the habitat. Even if your surveys show that an apparently suitable food tree is ignored, it might have an individually unpleasant chemical content (Glander 1982). Or you might be working to the wrong time scale. Milton (1980), in her classic study of Barro Colorado howlers, found that her troops did not minimize travel time, but moved around more than they seemed to need to for the immediate food distribution. She reasoned that it might, in fact, save energy in the long run, because they would have explored a little each day, and be prepared to shift to newly fruiting and leafing trees as their current ones went out of season. Furthermore, there may be a tradeoff between physical energy and memory. Mackinnon (1978) argues that gibbons and macaques with low locomotion costs are explorers, covering much of the home range daily, while orangs are mappers, going straight to trees likely to be in fruit.

Of course, it could always be that the particular troops one is studying are not optimizing their foraging style. They could be sick, or subordinate, or new in their territory—or just dumb. One study which suggests (and measures) suboptimal use is Pollock's (1977) work on indri. He compared a pair which were new to their territory with an older, long-established family. The two groups were similar in time spent at major food sources. However, when the older family visited minor food sources (less frequently visited quadrats) they had longer feeding bouts than the young pair, who just travelled on through. That is, says Pollock, the long-established animals knew their own territories' potential resources in greater detail.

Perhaps the most helpful approach would be experimental. Either one could study newly released folivores and frugivores as they explored the potential of a new forest or one might put out new food sources in the ranges of known troops, and see what helped or constrained the wild animals in discovering new foods. There is another problem here—fruits are not very likely to be toxic; leaves often are. Whitehead (1986) describes food learning by young howler monkeys—tasting fruits freely, but leaves only after observing an adult eat them. Russell (personal communication) even suggests the *Lepilemur* confine themselves to particular trees learned in youth. Of course, *Lepilemur* is no one's candidate for the most intelligent primate.

There is one final problem in ascribing intelligence to 'either—or'. Clumped, patchy foods are the sort that demand mapping—they are also the sort that encourage co-operation and social life. Thus, mapping and social life should correlate with each other as well as with brain size. Further, the highest energy foods are the patchiest. Thus, even granting the argument that high energy matters for high metabolism, and this in turn for a costly brain, energy also correlates both with social *nous* and ranging complexity.

Wynn (Chapter 20) takes on one of the correlations for our own hominid lineage. His paper is a bombshell to the older 'Tools makyth Man' view—and especially impressive from a scientist like Wynn with a vested interest in tool making. Few can argue with him, if he says stone pebble tools are within the intellectual capacity of modern chimpanzees, but were made by far more encephalized *Homo*. He cogently argues that at every stage *Homo* had far greater cranial capacity than necessary for his material culture. In fact, only in the 20th century are our tools beginning to actually challenge our mental capacities, evolved long since. Wynn throws the question of the cause of human brain size back into the realm of the invisible: either the social relationships or the lifestyle which produced technology, not the technology itself.

Stammbach (Chapter 22) and Hauser (Chapter 23) describe yet another aspect of the interrelationship of society and technology—one in the laboratory, one in the wild. Stammbach trained low-ranking long-tailed macaques to operate a complex lever sequence to dispense popcorn. Higher-ranking group members learned not to chase the 'specialist' away from the dispenser. In some cases they groomed and were affiliative toward the specialist—especially those group members who benefited most from taking the popcorn. They all remembered their roles 6 months later. Stammbach compares this to the trusted animals who chose the day range in Sigg and Stolba's hamadryas baboons.

Of course, we then run into Dennett's observation that there is relatively little private information, and probably not many specialist skills in most cercopithecine troops. Hamadryas, and chimpanzees and spider monkeys, with their fission-fusion lifestyles more often have new information to conceal or impart. Chimpanzees, as Jane Goodall reports, do develop individual prowess at termite fishing. However, most vervets are unlikely to discover a popcorn dispenser, and if they do it may not be the low ranking animal who becomes a specialist.

Which said, Hauser (Chapter 23), in fact, describes a case of a vervet novel skill. An old female either invented, or remembered over more than 10 years, the technique of soaking dry *Acacia tortillis* pods in acacia exudate. The pods absorb water from inaccessible tree wells which also softens the dry hard seeds. Hauser discusses the means of social learning—the old female's son and daughter obtained pods soaked by their mother, but other adult females simply learned by observing or even observing only part of the 'inventors' behaviour without having the entire technique demonstrated. This, as Hauser points out, puts a premium on the ability to infer the intentions of another animal. This technique appeared in the drought year of 1983, and did not continue when the rains returned in 1984. It is just possible that it is not an invention, but a tradition, passed on by the old female to her offspring, used only when in need.

To conclude this section, it is clear that the environment makes demands that interact in every way with social relationships. Cheney and Seyfarth, like Wynn, have ventured a comparison between the mental skills required to deal with each. They conclude (Chapter 19) that vervets are highly sophisticated in their social categories, but don't make obvious connections between events in the physical world—even events that should matter to them such as linking a python-track with a python. They report (personal communication) that a mother leopard with two cubs is now simply stalking the troops, eating one vervet after another, cafeteria style. The vervet troops seem unable to change their ranging behaviour to avoid the leopards or to learn any other means of dealing with the situation.

Peter Smith (Chapter 8) also takes on a direct comparison between social and manipulative understanding in growing humans. Infants on the whole recognize that their mother is permanent, and can be searched for if she disappears, younger than they will search for disappearing objects. The surprise to me is that the difference in timing is not overwhelming, or true in all cases. Surely the survival value of recognizing or missing the mother is far more important to any young primate than any other single object. In fact, the survival value of the continuing existence of one's great-aunt must outweigh that of pink rattles or bits of fruit. Similarly, memory for friendships and dominance hierarchies matters in a primate play group at an age when adults still mediate the demanding aspects of the physical environment, such as how to find one's dinner. Smith's useful review of the well developed social skills of young children should not be extrapolated causally as a phylogenetic sequence—rather, it should help us unravel *juvenile* primate strategies.

At the other end of the phylogenetic scale, what has happened to my original argument that lemurs' social intelligence far outstrips their intelligence towards gadgetry? (Chapter 3).

Since the early study, lemur society has proved to be far more complex than I guessed in 1965. *Lemur catta* and *Propithecus verreauxi* recognize and respond differentially to males outside their own group; their society is a neighbourhood, not just a single troop (Richard 1978, 1985; Mertl Millhollen 1977). During the birth season about a quarter of the *Lemur catta* males migrate to new troops. They rarely move alone. More often they are in pairs, trios, or even a gang of five (Jones 1983; personal observation). In my early study I observed male-female friendships within troops—perhaps these continue the bonding of migration. Matrilines are also bonded. Daughters rank under their mothers in inverse order of age like the matrilines of baboons or macaques (Taylor and Sussman 1985). Thus ringtailed lemurs, like higher primates, live in a complex web of individual relationships.

Meanwhile lemurs continue to seem uninterested in technology. Their diet is more varied than I first reported. Diurnal *Lemur catta* occasionally eat insects and even catch birds (Glander *et al.* 1985; Jolly and Oliver 1985). They do not, however, have noticeably complex feeding habits and still do not extract hidden foods or peel bananas. Since we do not yet have a measure of ranging strategy we can say nothing about lemurs' cognitive maps. Lemurs do form learning sets when properly trained. On the whole though, there have been no great revelations of subtlety toward the environment. Thus, if one accepts the 1965 argument that lemurs' social skills reflect an original primate bias toward social intelligence, then the studies of the intervening 20 years reinforce that point of view.

So far, the observations of social skills still seem to take the lead over mental maps or technology, but this book should provoke more analysis of both and of the intelligence which both demand.

Pretence and play

Whiten and Byrne (1988), in their catalogue of deception, point out that the highest primate social acuity is either a property of chimpanzees, or of chimp-watchers, or both. The kind of people who are most interested in primate social subtlety go out and associate with chimpanzees.

Jane Goodall (1986) has recently summarized 25 years of work in the Gombe Stream Reserve. She frames her many discoveries in all spheres of behaviour and ecology with opening and closing discussions of the chimpanzee mind. She gives many examples of chimp deception, distraction, alliances, and calculation of potential audiences in the wild. She does add, though, that she suspects the sophistication of de Waal's chimpanzees (Chapter 10) has been increased by captivity. Goodall has not seen, for instance, mediation by a female chimpanzee, who attracts one male to groom her, entices him to follow to where his rival can groom her other side, and then slips away leaving the two males grooming each other.

However, the step beyond both deception and social skill in representing others' awareness must be imagination. Imagination is picturing an idea that one knows oneself is not, or may not be true. When a baboon looks suddenly into the distance, or a vervet gives a leopard alarm during inter-troop conflict, or a juvenile baboon gives an accusatory scream at a dupe which never attacked it, perhaps these animals are 'pretending' a false situation. De Waal's (Chapter 10) chimpanzees seem to have actual mutual pretence. After a fight, one may 'find something' very interesting or even groomable in the grass. The rival who needs reconciliation earnestly examines the same spot, until the antagonists are finally sitting together without tension. Such pretence can also appear in play. Goodall describes 4-year-old Wunda who

> watched intently from a safe distance as her mother, using a long stick, fished for fierce driver ants from a branch overhanging the nest. Presently Wunda picked a tiny twig, perched herself on a low branch of a sapling in the same attitude as her mother, and poked her little tool down—into an imaginary nest? And when she

subsequently withdrew it, how do we know it was not swarming with a record, though non-existent catch? (p. 591).

Savage-Rumbaugh (Chapter 17), at last, gives us a corpus of such observations to add to Cathy Hayes' (1951) classic account of Vicki and her imaginary pull-toy. Austin making noises on a pipe to frighten Sherman in the dark, and the sly Matata inciting Sherman and Austin to display at Savage-Rumbaugh, and then putting a hose into Savage-Rumbaugh's hands to discipline the two males are social deception of high order. However, Kanzi's invisible objects go far beyond deception. Pretend eating or spitting out a non-existent 'object' and signing 'bad' is the realm of true symbolic play. He may give an alarm call to stop a game he is losing like the vervets in a territorial dispute. Kanzi does far more than the vervets. He stares in the direction he has called, then asks to leave where he is and go off in that direction with the human group, which starts a whole new activity. Given the rest of Savage-Rumbaugh's account, when Sherman and Austin attacked and made 'waa-barks' at an empty cage, after seeing a video tape of King Kong in such a cage, it is reasonable to join Savage-Rumbaugh in the richer, not the kill-joy interpretation.

Conclusion

This book is not going to 'solve' debates over the relative importance of social intelligence or object oriented intelligence. I think it strengthens the case for the importance of social intelligence, but then I would do so, wouldn't I?

The more lasting effects may be first, to define levels of understanding of other's states of knowledge and one's own, and secondly, to show how this social knowledge interacts with knowledge about the environment. The new observations reported here, as well as the new categorizing, should provoke a deluge of further information.

Whiten and Byrne's three-stage rocket of social intelligence (Chapter 1) could perhaps be elaborated as a progression from intuition of others' desires, to knowledge of others' states of knowledge, to the kind of semi-detached view of oneself which allows pretence. Note that I retain the adaptationist's bias toward small, gradual evolutionary steps. I do not think that full self-consciousness switched on in the proto-pongid—rather that we should tease out the aspects of social intelligence that served as useful precursors, as the light-sensitive pigment-spot preceded the vertebrate eye. I do fully agree with Humphrey's developed argument (1983) that human consciousness evolved in large part as a model to predict others' actions—probably beginning with more attention and awareness to others than to oneself. I would add that so did those earlier abilities we sum up as social uses of intelligence.

This book, with its accounts of social intelligence, begins to offer a taxonomy of purpose. We may never be sure just how widely to believe in conscious awareness among animals, (Griffin 1984), but we may ascribe to others at least some of the sensations which enter our consciousness. Lemurs surely have desires and goals in the simplest sense. Perhaps so do sea urchins and amoebas, but lemurs surely do. Lemurs differently respond to other individuals as kin, friends, or menaces, and to some extent seem to predict others' behaviour, but they haven't been noted (yet) to indulge in complex three- and four-part social interactions. The cercopithecines who do juggle three- and four-part behaviours are calculating others' desires. When we see them engage in deception, we can also be sure that they can calculate others' knowledge. They sometimes even show that they guess others' knowledge of their own knowledge. Chimpanzees and pygmy chimpanzees and gorillas do all this and more. They go beyond deception to pretence, which I take as something like self-deception with a mental frame that adds 'this is not true' (Bateson 1973).

What is the upshot? Well, we are. We are, with our biologists and philosophers who sometimes seriously argue this is the best of all possible worlds. I have tried to show that this point of view is almost built into the socially conscious primate. We are inclined to believe the world must be this way on purpose.

As we grow up to accept that the purpose is our own, a further step is thrust upon us in the immediate future. Nature evolved our sense of purpose blindly, but we now have it. That sense of purpose has given us dominion over the biosphere. We, who were not created, have become creators. We are ignorant and fallible, and like any other animals our minds work best in the short term and the bodily scale. We do not think clearly in megatons or picture generations much beyond our grandchildren. Nonetheless, the earth has become our garden; it behoves us to cultivate it with wisdom.

References

Abramovitch, R. (1976). The relation of attention and proximity to dominance in preschool children. In *The social structure of attention* (ed. M. R. A. Chance and R. R. Larson). Wiley, London.
Abramovitch, R. (1980). Attention structures in hierarchically organised groups. In *Dominance relations: an ethological view of human conflict and social interaction* (ed. D. R. Omark, F. F. Strayer, and D. G. Freedman). Garland STPM Press, New York.
Ainsworth, M. D. S., Blehar, M. C., Waters, E., and Wall, S. (1978). *Patterns of attachment: a psychological study of the strange situation*. Lawrence Erlbaum, New York.
Alexander, R. D. (1977). Natural selection and the analysis of human sociality. *The changing scenes in natural science* **12**, 283–337.
Allee, W. C. (1942). Social dominance and subordination among vertebrates. *Biological Symposium* **8**, 139–63.
Altmann, J. (1974). Observational study of behavior: sampling methods. *Behaviour* **49**, 227–67.
Altmann, S. A. (1962). A field study of the sociobiology of rhesus monkeys, Macaca mulatta. *Annals of the New York Academy of Science* **102**, 338–435.
Altmann, S. A. (1965). Sociobiology of rhesus monkeys. II: Stochastics of social communication. *Journal of Theoretical Biology* **8**, 490–522.
Altmann, S. A. (1967). The structure of primate social communication. In *Social communication among primates* (ed. S. Altmann), pp. 325–62. University of Chicago Press, Chicago, Illinois.
Altmann, S. A. and Altmann J. (1970). *Baboon ecology*. University of Chicago Press, Chicago.
Anderson, F. J. and Willis, F. N. (1976). Glancing at others in preschool children in relation to dominance. *Psychological Record* **26**, 467–52.
Anderson, J. R. (1978). Arguments concerning representations for mental imagery. *Psychological Review* **85**, 249–77.
Andrew, R. J. (1962). Evolution of intelligence and vocal mimicking. *Science* **137**, 585.
Aristotle (c. 350 BC) *Physica* (trans. Richard Hope, 1961). University of Nebraska Press, Lincoln, Nebraska.
Armstrong, E. (1983). Metabolism and relative brain size. *Science* **220**, 1302–4.
Armstrong, E. (1985). Allometric considerations of the adult mammalian brain with special emphasis on primates. In *Size and scaling in primate biology* (ed. W. L. Jungers), pp. 115–46. Plenum Press, New York.
Asdell, S. A. (1946). *Patterns of mammalian reproduction*. Comstock Publication, New York.
Augspurger, C. A. (1978). Reproductive consequences of flowering synchrony in Hybanthus prunifolus (Violaceae) and other shrub species of Panama. Unpublished PhD thesis, University of Michigan.

References

Axelrod, R. and Hamilton, W. D. (1981). The evolution of cooperation. *Science* **211**, 1390–6.

Bachmann, C. and Kummer, H. (1980). Male assessment of female choice in hamadryas baboons. *Behavioural Ecology and Sociobiology* **6**, 315–21.

Baddeley, A. D. and Hitch, G. (1974). Working memory. In *The psychology of learning and motivation* (ed. G. H. Bower), **8**, 47–90.

Baldwin, J. D. and Baldwin, J. (1972). The ecology and behavior of squirrel monkeys (*Saimiri oerstedi*) in a natural forest in Western Panama. *Folia Primatologica* **18**, 161–84.

Baldwin, J. M. (1897, 1973). *Social and ethical interpretations in mental development*. Arno, New York.

Barlow, H. B. (1985). Discussion point. In *Animal intelligence* (ed. L. Weiskrantz), p. 64. Clarendon Press, Oxford.

Barnard, C. J., Brown, C. A. J., Houston, A. I., and McNamara, J. M. (1985). Risk-sensitive foraging in common shrews: an interruption model and the effects of mean and variance in reward rate. *Behavioural Ecology and Sociobiology* **18**, 139–46.

Bartholomew, G. A. and Birdsell, J. B. (1953). Ecology and the protohominids. *American Anthropologist* **55**, 481–98.

Barton, R. (1987). Allogrooming as mutualism in diurnal lemurs. *Primates* **28** (4), 539–42.

Bateson, G. (1973). *Steps to an ecology of mind*. Paladin, St. Albans.

Bateson, P. P. G. (1976). Specificity and the origins of behavior. *Advances in the Study of Behavior* **6**, 1–20.

Bateson, P. P. G. (1980). Optimal outbreeding and the development of sexual preferences in Japanese quail. *Zeitschrift für Tierpsychologie* **53**, 231–44.

Bauchot, R. and Stephan, H. (1969). Encephalisation et niveau evolutif chez les simiens. *Mammalia* **33**, 225–75.

Beach, F. A. (1947a). Hormones and mating behaviour in vertebrates. *Proceedings of the Laurence Hormone Conference* **1**, 27–64.

Beach, F. A. (1974b). Evolutionary changes in the physiological control of mating behaviour in mammals. *Psychological Review* **54**, 297–315.

Beck, B. B. (1972). Tool use in captive Hamadryas baboons. *Primates* **13**, 276–96.

Beck, B. B. (1974). Baboons, chimpanzees, and tools. *Journal of Human Evolution* **3**, 509–16.

Bell, S. (1970). The development of the concept of the object and its relationship to infant–mother attachment. *Child Development* **41**, 291–312.

Beninger, R. J., Kendall, S. B., and Vanderwolf, C. H. (1974). The ability of rats to discriminate their own behaviours. *Canadian Journal of Psychology* **28**, 79–91.

Bennett, J. (1964). *Rationality*. Routledge and Kegan Paul.

Bennett, J. (1976). *Linguistic behaviour*. Cambridge University Press, Cambridge.

Bennett, J. (1983). Cognitive ethology: theory or poetry? *The Behavioral and Brain Sciences* **6**, 356–8.

Berger, S. M. (1962). Conditioning through vicarious instigation. *Psychological Review* **69**, 450–66.

Berman, C. M. (1980). Early agonistic experience and rank acquisition among free-ranging infant rhesus monkeys. *International Journal of Primatology* **1**, 153–70.
Bertram, B. C. R. (1976). Kin selection in lions and evolution. In *Growing points in ethology* (ed. P. P. G. Bateson and R. A. Hinde), pp. 281–301. Cambridge University Press, Cambridge.
Berzin, A. (ed.) (1978). *The mahamudra eliminating the darkness of ignorance by the ninth karmapa wang-ch'ug dorje*. Library of Tibetan Works and Archives, Dharamsala.
Bickerston, D. (1986). More than nature needs? A reply to Premack. *Cognition* **23**, 73–79.
Birch, H. G. (1945). The relation of previous experience to insightful problem solving. *Journal of Comparative Psychology* **38**, 295.
Blackmore, S. J. (1982). *Beyond the body*. Heinemann, London.
Blackmore, S. J. (1984). A psychological theory of the out of body experience. *Journal of Parapsychology* **48**, 201–18.
Boden, M. A. (1977). *Artificial intelligence and natural man*. Harvester, Brighton.
Boesch, C., and Boesch, H. (1984). Mental map in wild chimpanzees: an analysis of hammer transports for nut cracking. *Primates* **25**, 160–70.
Bowlby, J. (1969). *Attachment and loss: volume 1, attachment*. Hogarth Press, London.
Boyd, H. (1953). On encounters between wild white-fronted geese in winter flocks. *Behaviour* **37**, 291–319.
Boyd, R. and Richerson, P. J. (1985). *Culture and the evolutionary process*. Chicago University Press, Chicago.
Boyd, R. and Richerson, P. J. (in press). An evolutionary model of social learning. In *Comparative social learning* (ed. T. Zentall and B. G. Galef Jr). Erlbaum, Hillsdale, N.J.
Bramblett, C. A. (1970). Coalitions among gelada baboons. *Primates* **11**, 327–33.
Brauen, M. (1980). *Feste in Ladakh*. Gratz.
Bretherton, I. (1985). Attachment theory: retrospect and prospect. In *Growing points of attachment theory and research*, Monographs of the Society for Research in Child Development, **50**. (ed. I. Bretheron and E. Waters), pp. 3–35.
Bronson, W. C. (1974). Competence and the growth of personality. In *The growth of competence* (ed. K. Connolly and J. Bruner). Academic Press, London.
Brooks-Gunn, J. and Lewis, M. (1978). Early social knowledge: the development of knowledge about others. In *Issues in childhood social development* (ed. H. McGurk). Methuen, London.
Brooks-Gunn, J. and Lewis, M. (1981). Infant social perception: responses to pictures of parents and strangers. *Developmental Psychology* **17**, 647–9.
Bruner, J. S. (1972). Nature and uses of immaturity. *American Psychologist* **27**, 687–708.
Bryant, P. E. and Kopytynska, H. (1976). Spontaneous measurement by young children. *Nature* **260**, 773.
Bryant, P. E. and Trabasso, T. (1971). Transitive inferences and memory in young children. *Nature* **240**, 456–8.

Bunn, H. T. (1981). Archaeological evidence for meat-eating by Plio-Pleistocene hominids from Koobi Fora and Olduvai Gorge. *Nature* **291**, 574–80.
Busse, C. D. (1978). Do chimpanzees hunt cooperatively. *American Naturalist* **112**, 767–70.
Butterworth, G., and Cochran, E. (1980). Towards a mechanism of joint visual attention in human infancy. *International Journal of Behavioural Development* **3**, 253–72.
Bygott, J. D., Bertram, B. C. R., and Hanby, J. P. (1979). Male lions in large coalitions gain reproductive advantages. *Nature* **282**, 839–41.
Byrne, R. W. (in press). Social expertise and verbal explanation: the diagnosis of intelligence from behaviour. In *Expertise and explanation: the knowledge-language interface* (ed. C. Ellis). Ellis Horwood.
Byrne, R. W. and Whiten, A. (1987). A thinking primate's guide to deception. *New Scientist* **116**, No. 1589, 54–7.
Byrne, R. W. and Whiten, A. (1988). Towards the next generation in data quality: a new survey of primate tactical deception. *The Behavioral and Brain Sciences* **11**, No. 2.
Camberfort, J. (1981). A comparative study of culturally transmitted patterns of feeding habits in the chacma baboon *Papio ursinus* and the vervet monkey *Cercopithecus aethiops*. *Folia Primatologica* **36**, 234–65.
Cargile, J. (1970). A note on 'iterated knowings'. *Analysis* **30**, 151–5.
Carpenter, C. R. (1934). A field study of the behaviour and social relations of howling monkeys (*Alouatta palliata*). *Comparative Psychology Monographs* **10**, 48–168.
Carpenter, C. R. (1935). Behaviour of red spider monkeys in Panama. *Journal of Mammology* **16**, 171–80.
Carpenter, C. R. (1942a). Sexual behaviour of free-ranging rhesus monkeys (*Macaca mulatta*). I. Specimens, procedures, and behavioural characteristics oestrus. II. Periodicity of estrus, homosexual, autoerotic and non-comformist behaviour. *Journal of Comparative Psychology* **23**(1).
Carpenter, C. R. (1942b). Societies of monkeys and apes. *Biological Symposium* **8**, 177–204.
Cavalli-Sforza, L. L. and Feldman, M. W. (1981). *Cultural transmission and evolution*. Princeton University Press, Princeton.
Cerella, J. (1979). Visual classes and natural categories in the pigeon. *Journal of Experimental Psychology: Human Perception and Performance* **5**, 68–77.
Chance, M. R. A. (1961). The nature and special features of the instinctive social bond of primates. *Viking Fund Publications in Anthropology* **31**, 17–33.
Chance, M. R. A. (1980a). An ethological assessment of emotion. In *Emotion, theory research and experience, 1: Theories of emotion* (ed. R. Plutchick and H. Kellerman). Academic Press, London.
Chance, M. R. A. (1980b). The social structure of attention and the operation of intelligence. In *The exercise of intelligence: the biosocial preconditions for the operation of intelligence* (ed. E. Sunderland and M. T. Smith). Garland, New York.
Chance, M. R. A. and Jolly, C. (1970). *Social groups of monkeys, apes and men*. Jonathan Cape, London.

Chance, M. R. A. and Mead, A. P. (1953). Social behavior and primate evolution. *Symposia of the Society for Experimental Biology, Evolution* **7**, 395–439.

Chance, M. R. A. and Larsen, R. R. (1976). *The social structure of attention*. John Wiley, Chichester.

Chapais, B. (1983a). Adaptive aspects of social relationships among adult rhesus monkeys. In *Primate social relationships* (ed. R. A. Hinde), pp. 286–9. Blackwell Scientific Publications, Oxford.

Chapais, B. (1983b). Dominance, relatedness and the structure of female relationships in rhesus monkeys. In *Primate social relationships* (ed. R. A. Hinde), pp. 208–19 Blackwell Scientific Publications, Oxford.

Chapais, B. (1983c). Structure of birth season relationships among adult male and female rhesus monkeys. In *Primate social relationships* (ed. R. A. Hinde), pp. 200–8. Blackwell Scientific Publications, Oxford.

Chapais, B. (1985). An experimental analysis of a mother-daughter rank reversal in Japanese macaques (*Macaca fuscata*). *Primates* **26**, 407–23.

Chapais, B. and Schulman, S. (1980). An evolutionary model of female dominance relations in primates. *Journal of Theoretical Biology* **82**, 47–89.

Charlesworth, W. R. (1983). An ethological approach to cognition. In *Recent advances in cognitive-development theory* (ed. C. Brainerd). Springer-Verlag, New York.

Charlesworth, W. R. and Kreutzer, M. A. (1973). Facial expressions of infants and children. In *Darwin and facial expressions, a century of research in review* (ed. P. Ekman). Academic Press, New York.

Cheney, D. L. (1977). The acquisition of rank and the development of reciprocal alliances among free-ranging immature baboons. *Behavioural Ecology and Sociobiology* **2**, 303–18.

Cheney, D. L. (1978). Interactions of immature male and female baboons with adult females. *Animal Behaviour* **26**, 389–408.

Cheney, D. L. (1983). Extrafamilial alliances among vervet monkeys. In *Primate social relationships* (ed. R. A. Hinde), pp. 278–86. Blackwell Scientific Publications, Oxford.

Cheney, D. L. and Seyfarth, R. M. (1980). Vocal recognition in free-ranging vervet monkeys. *Animal Behaviour* **28**, 362–7.

Cheney, D. L. and Seyfarth, R. M. (1981). Selective forces affecting the predator alarm calls of vervet monkeys. *Behaviour* **76**, 25–61.

Cheney, D. L. and Seyfarth, R. M. (1982a). How vervet monkeys perceive their grunts. *Animal Behaviour* **30**, 739–51.

Cheney, D. L. and Seyfarth, R. M. (1982b). Recognition of individuals within and between groups of free-ranging vervet monkeys. *American Zoologist* **22**, 519–29.

Cheney, D. L. and Seyfarth, R. M. (1983). Non-random dispersal in free-ranging vervet monkeys: social and genetic consequences. *American Naturalist* **122**, 392–412.

Cheney, D. L. and Seyfarth, R. M. (1985a). The social and non-social world of non-human primates. In *Social relationships and cognitive development* (ed. R. A. Hinde, A-N. Perret-Clermont, and J. Stevenson-Hinde), pp. 23–44. Clarendon Press, Oxford.

Cheney, D. L. and Seyfarth, R. M. (1985b). Vervet monkey alarm calls: manipulation through shared information? *Behaviour* **94**, 150–66.

Cheney, D. L. and Seyfarth, R. M. (1986). The recognition of social alliances by vervet monkeys. *Animal Behaviour* **34**, 1722–31.

Cheney, D. L. and Seyfarth, R. M. (1988). Assessment of meaning and the detection of unreliable signals by vervet monkeys. *Animal Behaviour* (in press).

Cheney, D. L., Lee, P. C., and Seyfarth, R. M. (1981). Behavioural correlates of non-random mortality among free-ranging adult female vervet monkeys. *Behavioural Ecology and Sociobiology* **9**, 153–61.

Cheney, D. L., Seyfarth, R. M., and Smuts, B. B. (1986). Social relationships and social cognition in nonhuman primates. *Science* **234**, 1361–6.

Cheney, D. L., Seyfarth, R. M., Andelman, S. J., and Lee, P. C. (in press). Reproductive success in vervet monkeys. In *Reproductive success* (ed. T. H. Clutton-Brock). Chicago University Press, Chicago.

Cherniak, C. (1981). Minimal rationality. *Mind* **99**, 161–83.

Chevalier-Skolnikoff, S. (1986). An exploration of the ontogeny of deception in human beings and nonhuman primates. In *Deception: perspectives on human and nonhuman deceit* (ed. R. W. Mitchell and N. S. Thompson). State University of New York Press, Albany.

Chomsky, N. (1967). The general properties of language. In *Brain mechanisms underlying speech and language* (ed. F. L. Darley), pp. 73–88. Grune and Stratton, New York.

Clayton, D. A. (1978). Socially facilitated behavior. *Quarterly Review of Biology* **53**, 373–91.

Clutton-Brock, T. H. (ed.) (1977). *Primate ecology: studies of feeding and ranging behaviour in lemurs, monkeys and apes*. Academic Press, London.

Clutton-Brock, T. H. and Harvey, P. H. (1976). Evolutionary rules and primate societies. In *Growing points in ethology* (ed. P. P. G. Bateson and R. A. Hinde), pp. 195–237. Cambridge University Press. Cambridge.

Clutton-Brock, T. H. and Harvey, P. H. (1977a). Species differences in feeding and ranging behaviour in primates. In *Primate ecology: studies of feeding and ranging behaviour in lemurs, monkeys and apes* (ed. T. H. Clutton-Brock), pp. 557–84. Academic Press, London.

Clutton-Brock, T. H. and Harvey, P. H. (1977b). Primate ecology and social organisation. *Journal of Zoology* **183**, 1–39.

Clutton-Brock. T. H. and Harvey, P. H. (1980). Primates, brains and ecology. *Journal of Zoology* **190**, 309–23.

Colvin, J. (1982). Social integration and emigration of immature male rhesus macaques. Unpublished PhD thesis, University of Cambridge.

Colvin, J. (1983). Rank influences rhesus male peer relationships. In *Primate social relationships* (ed. R. A. Hinde), pp. 57–64. Blackwell Scientific Publications, Oxford.

Cook, M., Mineka, S., Wokenstein, B., and Laitsch, K. (1985). Observational conditioning of snake fear in unrelated rhesus monkeys. *Journal of American Psychology* **94**, 591–610.

Corsaro, W. A. (1981). Friendship in the nursery school: social organisation in a peer environment. In *The development of children's friendships* (ed. S. R. Asher and J. M. Gottman). Cambridge University Press, Cambridge.

Craik, K. J. W. (1943). *The nature of explanation*. Cambridge University Press, Cambridge.
Crook, J. H. (1970a). The socio-ecology of primates. In *Social behaviour in birds and mammals* (ed. J. H. Crook). Academic Press, London.
Crook, J. H. (1970b). Social organisation and the environment: aspects of contemporary social ethology. *Animal Behaviour* **18**, 197–209.
Crook, J. H. (1980). *The evolution of human consciousness*. Oxford University Press, Oxford.
Crook, J. H. (1983). On attributing consciousness to animals. *Nature* **303**, 11–14.
Crook, J. H. (1987). The nature of conscious awareness. In *Mindwaves* (ed. C. Blakemore and S. Greenfield). Blackwell, Oxford.
Crook, J. H. and Osmaston, H. (eds.) (1988) *Himalayan Buddhist Villages*. Aris and Phillips, Warminster.
Crook, J. H. (in press). A model of mind for Western Zen. For the proceedings of the Conference on 'Eastern Approaches to Self and Mind'. Cardiff, 1986.
Czikszentmihalyi, M. (1975). *Beyond boredom and anxiety—the experience of play in work and games*. Jossey-Bank, San Francisco.
Darling, F. F. (1937). *A herd of red deer*. Oxford University Press, Oxford.
Darwin, C. (1875). *The origin of species*. John Murray, London.
Darwin, C. (1965). *The expression of the emotions in man and animals*. University of Chicago Press, Chicago.
Dasser, V. (1985). Cognitive complexity in primate social relationships. In *Social relationships and cognitive development* (ed. R. A. Hinde, A-N. Perret-Clermont, and J. Stevenson-Hinde), pp. 9–22. Clarendon Press, Oxford.
Dasser, V. (1987a). A social concept in Java monkeys. *Animal Behaviour* **36**, 225–30.
Dasser, V. (1987b). Slides of group members as representations of the real animals (*Macaca fascicularis*). *Ethology* **76**, 65–73.
Datta, S. B. (1983a). Patterns of agonistic interference. In *Primate social relationships* (ed. R. A. Hinde), pp. 289–97. Blackwell Scientific Publications, Oxford.
Datta, S. B. (1983b). Relative power and the acquisition of rank. In *Primate social relationships* (ed. R. A. Hinde), pp. 93–103. Blackwell Scientific Publications, Oxford.
Datta, S. B. (1983c). Relative power and the maintenance of dominance. In *Primate social relationships* (ed. R. A. Hinde), pp. 103–12. Blackwell Scientific Publications, Oxford.
Datta, S. B. (1986). The role of alliances in the acquisition of rank. In *Primate ontogeny, cognition and social behaviour* (ed. J. G. Else and P. C. Lee), pp. 219–26. Cambridge University Press, Cambridge.
Daugherty, J. W. D., and Keller, C. M. (1982). Taskonomy: a practical approach to knowledge structures. *American Ethnologist* **5**, 763–74.
Davidson, D. (1974). Philosophy as psychology. In *Philosophy of psychology* (ed. S. C. Brown). Macmillan, London.
Davies, N. B. and Houston, A. I. (1984). Territory economics. In *Behavioural ecology*, 2nd edn (ed. J. R. Krebs and N. B. Davies), pp. 148–69. Blackwell Scientific Publications, Oxford.

Dawkins, R. (1976). Hierarchical organisation: a candidate principle for ethology. In *Growing points in ethology* (ed. P. P. G. Bateson and R. A. Hinde), pp. 7–54. Cambridge University Press, Cambridge.

Dawkins, R. (1986). *The blind watchmaker*. W. W. Norton, New York.

Dawkins, R. and Krebs, J. R. (1978). Animal signals: information or manipulation? In *Behavioural ecology: an evolutionary approach* (ed. J. R. Krebs and N. B. Davies), pp. 282–314. Blackwell, Oxford.

Delfosse, P. and Smith, P. K. (1979). Memory for companions in preschool children. *Journal of Experimental Child Psychology* **27**, 459–66.

Dennett, D. C. (1969). *Content and consciousness*. Humanities Press.

Dennett, D. C. (1971) Intentional systems. *Journal of Philosophy* **68**, 68–87.

Dennett, D. C. (1976). Conditions of personhood. In *The identities of persons* (ed. A. O. Rorty). University of California Press.

Dennett, D. C. (1978*a*). *Brainstorms*. Harvester, Brighton.

Dennett, D. C. (1978*b*). Why not the whole iguana? *The Behavioural and Brain Sciences*, **1**, 103–4.

Dennett, D. C. (1981*a*). Making sense of ourselves. *Philosophical Topics* **12**, 63–81.

Dennett, D. C. (1981*b*). Three kinds of intentional psychology. In *Reductionism, time and reality* (ed. R. Healey). Cambridge University Press, Cambridge.

Dennett, D. C. (1981*c*). True believers: the intentional strategy and why it works. In *Scientific explanation* (ed. A. Heath). Oxford University Press, Oxford.

Dennett, D. C. (1982). How to study consciousness empirically: or, nothing comes to mind. *Synthese* **53**, 159–80.

Dennett, D. C. (1983). Intentional systems in cognitive ethology: the 'Panglossian paradigm' defended. *The Behavioral and Brain Sciences* **6**, 343–90.

Dennett, D. (in press, *a*). Cognitive ethology: hunting for bargains or a wild goose chase? In (title not determined) (ed. D. McFarland). Open University Press.

Dennett, D. (in press, *b*). Out of the armchair and into the field. In *Poetics today*. Israel.

De Vore, I. (1962). The social behavior and organization of baboon troops. Unpublished PhD thesis. University of Chicago.

De Vore, I., and Hall, K. R. L. (1965). Baboon ecology. In *Primates behavior* (ed. I. De Vore). Holt, Rinehart and Winston, New York.

De Vries, R. (1970). The development or role-taking as reflected by behavior of bright, average, and retarded children in a social guessing game. *Child Development* **41**, 759–70.

de Waal, F. (1977). The organisation of agonistic relations within two captive groups of Java monkeys. *Zeitschrift für Tierpsychologie* **44**, 225–82.

de Waal, F. (1982). *Chimpanzee politics*. Jonathan Cape, London.

de Waal, F. (1986). Deception in the natural communication of chimpanzees. In *Deception: perspectives on human and non-human deceit* (ed. R. W. Mitchell and N. S. Thompson) pp. 221–4. State University of New York Press, Albany.

Dickinson, A. (1980). *Contemporary animal learning theory*. Cambridge University Press, Cambridge.

Dodge, K. A., Pettit, G. S., McClaskey, C. L. and Brown, M. M. (1986). Social competence in children. *Monographs for the Society for Research in Child Development* **51**, 1–80.

Dodge, K. A., McClaskey, C. L., and Feldman, E. (1985). A situational approach to the assessment of social competence in children. *Journal of Consulting and Clinical Psychology* **53**, 344–53.
Dohl, J. (1968). Uber die fahigheit einer schimpansin, umwege mit selbstandigen zwischenzielen zu uberblicken. *Zeitschrift für Tierpsychologie* **25**, 89–103.
Douglas, R. J. and Marcellus, D. (1975). The ascent of man: deductions based on a multivariate analysis of the brain. *Brain, Behaviour and Evolution* **11**, 179–213.
Dover Wilson, J. (1951) *What happens in Hamlet?* 3rd edn. Cambridge University Press, Cambridge.
Dretske, F. (1981). *Knowledge and the flow of information*. Bradford/MIT Press.
Dunn, J. and Kendrick, C. (1982). *Siblings: love, envy, and understanding*. Grant McIntrye, London.
Duval, S. and Wicklund, R. A. (1972). *A theory of object self-awareness*. Academic Press, New York.
Echstein, P. (1949). Patterns of mammalian sexual cycle. *Acta anatomica* **7**, 389–410.
Edelman, M. S. and Omark, D. R. (1973). Dominance hierarchies in young children. *Social Science Information* **12**, 103–10.
Edwards, C. P. and Lewis, M. (1979). Young children's concepts of social relations: social functions and social objects. In *The child and its family* (ed. M. Lewis and L. A. Rosenblum), pp. 245–66. Plenum Press, New York.
Ehardt, C. L. and Bernstein, I. S. (1986). Matrilineal overthrows in rhesus monkey groups. *International Journal of Primatology* **7**, 157–81.
Eimas, P., Siqueland, P., Jusczyk, P., and Vigorito, J. (1971). Speech perception in infants. *Science* **171**, 303–6.
Eisenberg, J. F. and Wilson, D. E. (1978). Relative brain size and feeding strategies in the Chiroptera. *Evolution* **32**, 740–51.
Ekman, P. (1985). *Telling lies: clues to deceit in the market place, politics and marriage*. Norton, New York.
Ekman, P., Friesen, W. V. (1969). Nonverbal leakage and clues to deception. *Psychiatry* **32**, 88–106.
Ekman, P. and Friesen, W. V. (1974). Detecting deception from the body or face. *Journal of Personality and Social Psychology* **29**, 288–98.
Ekman, P. and Friesen, W. V. (1975). *Unmasking the face*. Prentice–Hall, New Jersey.
Emlen, S. T. (1978). Cooperative breeding. In *Behavioural ecology—an evolutionary approach* (ed. J. R. Krebs and N. B. Davies. Blackwell, Oxford.
Epstein, R., Kirshit, C. E., Lanza, R. P. and Rubin, L. C. (1984). 'Insight' in the pigeon: antecedents and determinants of an intelligence performance. *Nature* **308**, 61–2.
Erikson, E. (1968). *Identity, youth and crisis*. Faber and Faber, London.
Ettlinger, G. (1983). A comparative evaluation of the cognitive skills of the chimpanzee and the monkey. *International Journal of Neuroscience* **22**, 7–20.
Fagen (1976). Modelling how and why play works. In *Play: its role in development and evolution* (ed. J. S. Brunner, A. Jolly, and K. Sylva), pp. 96–116. Basic Books, New York.

Fairbanks, L. (1975). Communication of food quality in captive *Macaca nemestrina* and free ranging *Ateles geoffroyi*. *Primates* **16**, 181–90.
Fairbanks, L. A. (1980). Relationships among adult females in captive vervet monkeys: testing a model of rank-related attractiveness. *Animal Behaviour* **28**, 853–9.
Falk, D. (1980). Language, handedness, and primate brains: did the australopithecines sign? *American Anthropologist* **82**, 72–8.
Feinman, S. (1982). Social referencing in infancy. *Merrill-Palmer Quarterly* **28**, 445–70.
Feiring, C., Lewis, M. and Starr, M. D. (1984). Indirect effects and infants reaction to strangers. *Developmental Psychology* **20**, 485–91.
Feldman, R., Jenkins, L., and Popoola, O. (1979). Detection of deception in adults and children via facial expression. *Child Development* **50**, 350–5.
Fellows, B. J. (1967). Chance stimulus sequences for discrimination tasks. *Psychological Bulletin* **67**, 87–92.
Fisher, J. and Hinde, R. A. (1949). The opening of milk bottles by birds. *British Birds* **42**, 347–57.
Fisher, S., and Cleveland, S. E. (1956). *Body image and personality*. Van Nostrand, New York.
Flavell, J. H. (1978). The development of knowledge about visual perception. In *Nebraska Symposium on Motivation, 25* (ed. C. B. Keasy). University of Nebraska Press, Lincoln.
Flavell, J. H., Shipstead, S. G., and Croft, K. (1978). Young children's knowledge about visual perception: hiding objects from others. *Child Development* **49**, 1208–11.
Flavell, J. H., Flavell, E. R., and Green, F. L. (forthcoming). Young children's knowledge about the apparent-real and pretend-real distinctions.
Fletcher, H. J. (1965). The delayed-response problem. In *Behaviour of nonhuman primates*, Vol. 1 (ed. A. M. Schrier, H. F. Harlow, and F. Stollnitz), pp. 129–65. Academic Press, New York.
Fodor, J. A. (1983). *The modularity of mind*. The MIT Press, Cambridge, MA.
Frank, L. G. (1986). Social organisation of the spotted hyena (*Crocuta crocuta*). II. Dominance and reproduction. *Animal Behaviour* **34**, 1510–27.
French, G. M. (1965). Associative problems. In *Behavior of non-human primates* (ed. A. M. Schrier, H. F. Harlow, and F. Stollnitz). Academic Press, London.
Freud, S. (1925/1959). Fragment of an analysis of a case of hysteria. *Collected Papers*, 3. Basic Books, New York.
Frost, G. T. (1980). Tool behavior and the origins of laterality. *Journal of Human Evolution* **9**, 447–59.
Galef, B. G., Jr (1976). Social transmission of acquired behavior: a discussion of tradition and social learning in vertebrates. In *Advances in the study of behavior*, Vol. 6 (ed. J. S. Rosenblatt, R. A. Hinde, E. Shaw, and C. G. Beer), pp. 77–100. Academic Press, New York.
Galef, B. J., Jr (in press). Imitation in animals: history, definition and interpretation of data from the psychological laboratory. In *Comparative social learning* (ed. T. Zentall and B. G. Galef Jr). Erlbaum, Hilldale, N.J.
Gallup, G. G. (1970). Chimpanzees: self-recognition. *Science* **167**, 86–87.

Gamble, C. (1982). Interaction and alliance in palaeolithic society. *Man* **17**, 92–107.
Gatewood, J. (1985). Actions speak louder than words. In *Directions in Cognitive Anthropology* (ed. J. W. D. Daugherty), pp. 199–220. University of Illinois Press, Urbana.
Gellerman, L. W. (1933). Chance orders of alternating stimuli in visual discrimination experiments. *Journal of Genetic Psychology* **42**, 206–8.
Gelman, P. and Spelke, E. (1981). The development of thoughts about animate and inanimate objects. In *Social cognitive development* (ed. J. H. Flavell and L. Ross), pp. 43–66. Cambridge University Press, Cambridge.
Gibb, J. (1956). Food, feeding habits and territory of the rock pipit, Anthus spinoletta. *Ibis* **98**, 506–30.
Gibbs, C. A. (1969). Leadership. In *The handbook of social psychology*, 2nd edn (ed. G. Lindzey and E. Aronson), pp. 205–82. Addison-Wesley, Reading, Massachusetts.
Gillan, D. J. (1981). Reasoning in the chimpanzee. II. Transitive inference. *Journal of Experimental Psychology: Animal Behaviour Processes* **7**, 150–64.
Gillan, D. J., Premack, D., and Woodruff, G. (1981). Reasoning in the chimpanzee. 1. Analogical reasoning. *Journal of Experimental Psychology: Animal Behaviour Proceedings* **7**, 1–17.
Glander, K. E. (1975) Habitat and resource utilisation: an ecological view of social organisation in Mantled Howler Monkeys. Unpublished PhD thesis, University of Chicago.
Glander, K. E. (1978). Howler Monkey feeding behavior and plan secondary compounds: a study of strategies. In *Ecology of arboreal folivores* (ed. G. G. Montgomery), pp. 561–74. Smithsonian Press, Washington D.C.
Glander, K. E. (1982). The impact of plant secondary compounds on primate feeding behavior. *American Journal of Physical Anthropology* **25**, 1–18.
Glander, K. E., Freed, B. Z. and Ganzhorn, J. V. (1985). Meat eating and predation in captive-born semi-free-ranging *Lemur fulvus* and caged *Lemur macaco*. *Zoo Biology* **4**, 361–5.
Glickman, S. E. and Scroges, R. W. (1966). Curiosity in zoo animals. *Behaviour* **26**, 151–8.
Goffman, E. (1972). *The presentation of self in everyday life*. Penguin, New York.
Goodhall, J. (1963). Feeding behaviour of wild chimpanzees: a preliminary report. *Symposia of the Zoological Society of London*, **10**, 39–47.
Goodall, J. (1964). Tool using and aimed throwing in a community of free-living chimpanzees. *Nature* **201**, 1264–6.
Goodall, J. (1965). Chimpanzees of the Gombe Stream Reserve. In *Primate Behavior* (ed. I. de Vore), pp. 425–47. Holt, Rinehart and Winston, New York.
Goodall, J. (1968). The behaviour of free-living chimpanzees in Gombe Stream Reserve. *Animal Behaviour Monographs* **1**, 161–311.
Goodall, J. van (1971). *In the shadow of man*. Collins, London.
Goodall, J. (1973). Cultural elements in a chimpanzee community. In *Precultural primate behaviour* (ed. E. W. Menzel Jr), pp. 144–84. S. Karger, Switzerland.
Goodall, J. (1986). *The chimpanzees of Gombe*. Belknap, Harvard.
Goodall, J., Bandora, A., Bergmann, E., Busse, C., Matama, H., Mpongo, E., Pierce, A. and Riss, D. (1979). Intercommunity interactions in the chimpanzee

population of the Gombe National Park. In *The great apes* (ed. D. Hamburg and E. R. McCown), pp. 13–54. Benjamin Cummings, Menlo Park.

Goldberg, S. (1977). Social competence in infancy: a model of parent-infant interactions. *Merrill-Palmer Quarterly* **23**, 163–77.

Goldman, R. and Goldman, J. (1982). *Children's sexual thinking*. Routledge and Kegan Paul, London.

Gottman, J. M. (1986). Merging social cognition and social behavior. *Monographs of the Society for Research in Child Development* **51**, 81–5.

Gould, J. L. and Gould, C. G. (1982). The insect mind: physics or metaphysics. In *Animal mind—human mind*. Dahlem Life Sciences Workshop, report 21 (ed. D. R. Griffin). Springer-Verlag, Berlin.

Gould, S. (1975). Allometry in primates, with emphasis on scaling and the evolution of the brain. In *Approaches to primate paleobiology* (ed. Szalay). Karger, Basel.

Gould, S. J. (1987). Life's little joke. *Natural History* **96**, 16–25.

Gould, S. J. and Lewontin, R. C. (1979). The spandrels of San Marco and the Panglossian Paradigm: a critique of the adaptationist programme. *Proceedings of the Royal Society, B* **205**, 581–98.

Gouzoules, S., Gouzoules, H. and Marler, P. (1984). Rhesus monkey (*Macaca mulatta*) screams: representational signalling in the recruitment of agonistic aid. *Animal Behaviour* **32**, 182–93.

Gowlett, J. (1984). Mental abilities in early man: a look at some hard evidence. In *Hominid evolution and community ecology: prehistoric human adaptation in biological perspective* (ed. R. Foley), pp. 167–92. Academic Press, London.

Green, S. (1975). Dialects in Japanese monkeys: vocal learning and cultural transmission in locale-specific vocal behaviour? *Zeitschrift für Tierpsychologie* **38**, 304–14.

Grice, H. P. (1957). Meaning. *Philosophical Review* **66**, 377–88.

Grice, H. P. (1969). Utterer's meaning and intentions. *Philosophical Review* **78**, 147–77.

Griffin, D. R. (1978). Prospects for a cognitive ethology. *The Behavioral Brain Sciences* **1**, 527–38.

Griffin, D. R. (1981). *The question of animal awareness*. Rockefeller University Press, New York.

Griffin, D. R. (ed.) (1982). *Animal mind—human mind*, Dahlem Life Sciences Workshop, report 21. Springer Verlag, Berlin.

Griffin, D. R. (1984). *Animal thinking*. Harvard University Press, Cambridge, Mass.

Griffin, D. R. (1985). The cognitive dimensions of animal communication. In *Experimental behavioral ecology and sociobiology* (ed. B. Holldobler and M. Lindauer), pp. 471–82. Springer-Verlag, Berlin.

Gunnar, M. R. (1980). Control, warning signals, and distress in infancy. *Developmental Psychology* **16**, 281–9.

Guntrip, H. (1983). *Schizoid phenomena, object relations and the self*, International Psychoanalytic Library, No. 77. Hogarth, London.

Gurche, J. A. (1982). Early primate brain evolution. In *Primate brain evolution* (ed. E. Armstrong and D. Falk), pp. 227–46. Plenum Press, New York.

Hailman, J. P. (1977). *Optical signals*. Indiana University Press, Bloomington.

Hall, K. R. L. (1963a). Tool using performance as indications of behavioural adaptability. *Current Anthropology* **4**, 479.
Hall, K. R. L. (1963b). Observational learning in monkeys and apes. *British Journal of Psychology* **54**, 201–26.
Hall, K. R. L. and De Vore, I. (1965). Baboon social behaviour. In *Primate Behavior* (ed. I. De Vore). Holt, Rinehart and Winston, New York.
Hall, K. R. L. and Schaller, G. B. (1964). Tool-using behavior of the California sea otter. *Journal of Mammology* **45**, 287–298.
Hallowell, A. I. (1956). The structural and functional dimensions of human existence. *Quarterly Review of Biology* **31**, 88–101.
Handel, M. D. (1982). Intelligence and deception. *Journal of Strategic Studies* **5**, 122–54.
Harcourt, A. H. (1987). Cooperation as a competitive strategy in primates and birds. In *Animal societies: theories and facts* (ed. Y. Ito and J. L. Brown), pp. 141–57. Japan Scientific Societies Press, Tokyo.
Harcourt, A. H. and Stewart, K. J. (1987). The influence of help in contests on dominance rank in primates: hints from gorillas. *Animal Behaviour* **35**, 182–90.
Harding, D. (1965). *On having no head: a contribution to Zen in the West*. Harper and Row, New York.
Harlow, H. F. (1965). Total social isolation: effects on macaque monkey behaviour. *Science* **148**, 666.
Harlow, H. F. and Harlow, M. K. (1963). A study of animal affection. In *Primate social behavior* (ed. Charles H. Southwick), pp. 174–84. Van Nostrand, Princeton, New Jersey.
Harlow, H. F., Harlow, M. K. and Meyer, D. R. (1950). Learning motivated by a manipulation drive. *Journal of Experimental Psychology* **40**, 228–34.
Harris, J. W. K. (1983). Cultural beginnings: plio-pleistocene archaeological occurrences from the Afar, Ethiopia. *The African Archaeological Review* **1**, 3–31.
Harris, P. L., Donnelly, K., Guy, G. R. and Pitt-Watson, R. (1986). Children's understanding of the distinction between real and apparent emotion. *Child Development* **37**, 895–909.
Hartman, C. G. (1930–2). Studies in the reproduction of the monkey, Macacus rhesus, with special reference to menstruation and pregnancy. *Contributions to Embryology, Carnegie Institution* **134**, 22–3.
Hartshorne, H. and May, M. A. (1928). *Studies in deceit: book 1*. Arno, New York.
Harvey, P. H. and Clutton-Brock, T. H. (1985). Life history variation in primates. *Evolution* **39**, 559–81.
Harvey, P. H., Martin, R. D. and Clutton-Brock, T. H. (1987). Life histories in comparative perspective. In *Primate societies* (ed. B. B. Smuts, D. L. Cheney, R. M. Seyfarth, R. W. Wrangham and T. T. Struhsaker), pp. 181–96. University of Chicago Press, Chicago.
Hauser, M. D. (1987). Behavioral ecology of free-ranging vervet monkeys: proximate and ultimate levels of explanation. Ph.D. thesis. University of California, Los Angeles.

Hauser, M. D., Cheney, D. L. and Seyfarth, R. M. (1986). Group extinction and fusion in free-ranging vervet monkeys. *American Journal of Primatology* **11**, 63–77.
Hayes, C. (1951). *The ape in our house.* Harper, New York.
Hayes, K. C. and Hayes, C. (1952). Imitation in a home-raised chimpanzee. *Journal of Comparative Physiological Psychology* **45**, 450–9.
Hayes, K. J. and Nissen, C. H. (1971). Higher mental functions of a home-raised chimpanzee. In *Behavior of non-human primates* (ed. A. M. Schrier and F. Stollnitz), pp. 60–115. Academic Press, New York.
Heider, K. G. (1976). Children's development of competency in social structural concepts. *Ethnology* **15**, 47–62.
Heim, A. W. (1970). *The appraisal of intelligence.* Methuen, London.
Herrigel, E. (1953). *Zen in the art of archery.* Routledge and Kegan Paul, London.
Hewes, G. W. (1973). An explicit formulation of the relationship between tool-using, tool-making and the emergence of language. *Visible Language* **7**, 101–27.
Hewitt, C. (1968). Planner MAC-M-386. M.I.T. Artificial Intelligence Laboratory, Project MAC.
Hill, W. C. O. (1953). *Primates, comparative anatomy and taxonomy.* Edinburgh University Press, Edinburgh.
Hinde, R. A. (1968). Dichotomies in the study of development. In *Genetic and environmental influences on behaviour* (ed. J. M. Thoday and A. S. Parkes). Oliver and Boyd, London.
Hinde, R. A. (1976). Interactions, relationships and social structure. *Man* **11**, 1–17.
Hinde, R. A. (1979). *Towards understanding relationships.* Academic Press, London.
Hinde, R. A. (1981). Animal signals: ethological and games-theory approaches are not incompatible. *Animal Behaviour* **29**, 535–42.
Hinde, R. A. (1983). Triadic interactions and social sophistication. In *Primate social relationships: an integrated approach* (ed. R. A. Hinde), pp. 152–4. Blackwell, Oxford.
Hinde, R. A. and Fisher, J. (1951). Further observations on the opening of milk bottles by birds. *British Birds* **44**, 392–6.
Hinde, R. A. and Stevenson-Hinde, J. (ed.) (1973). *Constraints on learning: limitations and predispositions.* Academic Press, New York and London.
Hladik, C. M. (1967). Surface relative du tractus digestif de quelques primates. *Mammalia* **31**, 120–47.
Hladik, C. M. and Hladik, C. M. (1969). Rapports trophiques entre vegetation et primates dans la Foret de Barro Colorado (Panama). *La Terre et La Vie* **1**, 25–117.
Hofman, M. A. (1983a). Encephalization in hominids: evidence for the model of punctuationalism. *Brain, Behavior and Evolution* **22**, 102–17.
Hofman, M. A. (1983b). Energy metabolism, brain size and longevity in mammals. *Quarterly Review of Biology* **58**, 495–512.
Hoffman, M. L. (1981). Perspectives on the difference between understanding people and understanding things: the role of affect. In *Social cognitive development* (ed. J. H. Flavell and L. Ross), pp. 67–81. Cambridge University Press, Cambridge.

Hold, B. (1976). Attention structures and rank specific behaviour in preschool children. In *The social structure of attention* (ed. M. R. A. Chance and R. R. Larsen). Wiley, London.

Hold-Cavell, B. (1985). Showing-off and aggression in young children. *Aggressive Behavior* **11**, 303–14.

Holland, P. C. and Straub, J. J. (1979). Differential effects of two ways of devaluing the unconditioned stimulus after Pavlovian appetitive conditioning. *Journal of Experimental Psychology: Animal Behaviour* **5**, 65–78.

Holloway, R. L. (1973). Endocranial volumes of early African hominids and the role of the brain in human mosaic evolution. *Human Evolution* **2**, 449–59.

Holloway, R. L. (1975). *The role of human social behaviour in the evolution of the brain*. American Museum of Natural History, New York.

Holloway, R. L. (1981a). Exploring the dorsal surface of hominoid brain endocasts by stereoplotter and discriminant analysis. *Philosophical Transactions of the Royal Society (London)* **B292**, 155–66.

Holloway, R. L. (1981b). Revisiting the South African *Australopithecus* endocasts: results of stereoplotting the lunate sulcas. *American Journal of Physical Anthropology* **56**, 43–58.

Holloway, R. L. (1983). Human brain evolution: a search for units, models and synthesis. *Canadian Journal of Anthropology* **3**, 215–30.

Holloway, R. L. and De la Coste-Lareymondie, M. (1982). Brain endocast asymmetry in pongids and hominids: some preliminary findings on the paleontology of cerebral dominance. *American Journal of Physical Anthropology* **58**, 108–10.

Hood, L. and Bloom, L. (1979). What, when, and how about why. *Monographs of Social Research in Child Development* **44**, 1–41.

Hopkins, J. (ed.) (1981). *The yoga of Tibet: the great exposition of secret mantra by Tsong.ka.pa*. Allen and Unwin, London.

Horrocks, J. and Hunte, W. (1983). Maternal rank and offspring rank in vervet monkeys: an appraisal of the mechanisms of rank acquisition. *Animal Behaviour* **31**, 772–82.

Humphrey, N. K. (1973a). Predispositions to learn. In *Constraints on learning: limitations and predispositions* (ed. R. A. Hinde and J. Stevenson-Hinde). Academic Press, London and New York.

Humphrey, N. K. (1973b). The illusion of beauty. *Perception* **2**, 429–39.

Humphrey, N. K. (1974). Vision in a monkey without striate cortex: a case study. *Perception* **3**, 241–55.

Humphrey, N. K. (1980). Nature's psychologists. In *Consciousness and the physical world* (ed. B. Josephson and V. Ramachandran). Pergamon, London.

Humphrey, N. K. (1983). *Consciousness regained*. Oxford University Press, Oxford.

Humphrey, N. K. (1986). *The inner eye*. Faber and Faber, London.

Humphreys, A. P. and Smith, P. K. (1987). Rough and tumble, friendship and dominance in schoolchildren: evidence for continuity and change with age. *Child Development* **58**, 201–12.

Hutchinson, G. E. (1965). (*The ecological theatre and the evolutionary play*. Yale University Press, New Haven.

Huxley, J. S. (1938). Darwin's theory of sexual selection and the data subsumed by it in the light of recent research. *American Naturalist* **72**, 416–33.
Huxley, J. S. (1942). *Evolution. The modern synthesis.* Allen and Unwin, London.
Inglis, I. (1980). Towards a cognitive theory of exploratory behaviour. In *Exploration in animals and humans* (ed. T. Archer and L. Birke). Van Nostrand, New York.
Isaac, G. L. (1976). Stages of cultural elaboration in the pleistocene: possible archaeological indicators of the development of language capabilities. In *Origins and evolution of language and speech,* Annals of the New York Academy of Science, **280**, (ed. S. Harnad, H. Steklis, and J. Lancaster), pp. 275–88. New York Academy of Sciences, New York.
Isaac, G. (1978). Food sharing and human evolution: archaeological evidence from the Plio-Pleistocene of East Africa. *Journal of Anthropological Research* **34**, 311–25.
Isaac, G. L. (1981). Archaeological tests of alternative models of early hominid behavior: excavation and experiments. In *The emergence of man* (ed. J. Young, E. Jope, and K. Oakley), pp. 177–88. The Royal Society and the British Academy, London.
Itani, J. (1963). Paternal care in the wild Japanese monkey, Macaca fuscata. In *Primate social behavior* (ed. C. H. Southwick), pp. 91–7. Van Nostrand, Princeton, New Jersey.
Itani, J. (1965). In *Japanese monkeys* (ed. S. A. Altmann). University of Alberta, Edmonton.
Itani, J. and Nishimura, A. (1973). The study of infrahuman culture in Japan. A review. In *Precultural primate behavior* (ed. E. W. Menzel), pp. 26–50. Karger, Basel.
Jackson, E., Campos, J. J., and Fischer, K. W. (1978). The question of decalage between object permanence and person permanence. *Developmental Psychology* **14**, 1–10.
Jacobson, J. L. (1980). Cognitive determinants of wariness toward unfamiliar peers. *Developmental Psychology* **16**, 347–54.
Jarman, P. J. (1974). The social organisation of antelope in relation to their ecology. *Behaviour* **48**, 215–67.
Jay, P. C. (1965). Field studies. In *Behavior of non-human primates* (ed. A. M. Schrier, H. F. Harlow, and E. Stollnitz). Academic Press, New York.
Jelinek, J. (1980). European *Homo erectus* and the origin of *Homo sapiens.* In *Current argument on early man* (ed. L. Konigsson), pp. 137–44. Pergamon Press, Oxford.
Jerison, H. J. (1961). Quantitative analysis of evolution of the brain in mammals. *Science* **133**, 1012–4.
Jerison, H. J. (1963). Interpreting the evolution of the brain. *Human Biology* **35**, 263–91.
Jerison, H. J. (1973). *Evolution of the brain and intelligence.* Academic Press, New York.
Jerison, H. J. (1976). Paleoneurology and the evolution of mind. *Scientific American* **234**, 90–101.

Jerison, H. J. (1983). The evolution of the mammalian brain as an information processing system. In *Advances in the study of mammalian behavior* (ed. J. F. Eisenberg and D. G. Kleiman), pp. 113–46. The American Society of Mammalogists, Pittsburgh.
Jerison, H. J. (1985). Animal intelligence as encephalisation. In *Animal intelligence* (ed. L. Weiskrantz), pp. 21–35. Clarendon Press, Oxford.
Johanson, D. and White, T. (1979). A systematic assessment of early African hominids. *Science* **203**, 321–30.
Johnston, T. D. (1981). Contrasting approaches to a theory of learning. *The Behavioral and Brain Sciences* **4**, 125–73.
Jolly, A. (1964). Prosimians' manipulation of simple object problems. *Animal Behavior* **12**, 560–71.
Jolly, A. (1966). *Lemur behavior*. University of Chicago Press, Chicago.
Jolly, A. (1983). Dennett's 'Panglossian Paradigm'. *The Behavioral and Brain Sciences* **6**, 366–7.
Jolly, A. (1985). *The evolution of primate behaviour* (2nd Edn.). Macmillan.
Jolly, A. and Oliver, W. C. .O. (1985). Predatory behavior in captive *Lemur* spp. *Zoo Biology* **4**, 139–46.
Jones, G. and Smith, P. K. (1984). The eyes have it: young children's discrimination of age in masked and unmasked facial photographs. *Journal of Experimental Child Psychology* **38**, 328–37.
Jones, K. C. (1983). Inter-troop transfer of *Lemur catta* males at Berenty, Madagascar. *Folia Primatologia* **40** 145–60.
Jones, P. R. (1981). Experimental implement manufacture and use: a case study from Olduvai Gorge, Tanzania. In *The emergence of man* (ed. J. Young, E. Jope and K. Oakley), pp. 189–95. The Royal Society and the British Academy, London.
Judge, P. G. (1982). Redirection of aggression based on kinship in a captive group of pigtail macaques. *International Journal of Primatology* **3**, 301.
Kahneman, D. (1982). Some remarks on the computer metaphor. Unpublished ms.
Kaplan, J. R. (1977). Patterns of fight-interference in free-ranging rhesus monkeys. *American Journal of Physiological Anthropology* **47**, 279–88.
Kawai, M. (1965). Newly acquired precultural behaviour of the natural troop of Japanese macaques of Koshima Island. *Primates* **6**, 1–10.
Kawamura, S. (1959). The process of sub-culture propagation among Japanese Macaques. *Primates* **2**, 43–60.
Kawamura, S. (1963). The process of sub-culture propagation among Japanese macaques. In *Primate social behaviour* (ed. C. H. Southwick), pp. 82–90. Van Nostrand, New York.
Kaye, K. (1982). *The mental and social life of babies*. Methuen, London.
Kelsang, Gyatso, Geshe (1982). *The clear light of bliss*. Wisdom, London.
Klein, D. J. (1978). The diet and reproductive cycles of a population of vervet monkeys. Unpublished PhD thesis. New York University.
Klein, L. L. and Klein, D. B. (1977). Feeding behaviour of the Colombian Spider Monkey. In *Primate ecology: studies of feeding and ranging behaviour in Lemurs, Monkeys and Apes* (ed. T. H. Clutton-Brock), pp. 153–81. Academic Press, London.

Kluver, H. (1933). *Behavior mechanisms in monkeys*. University of Chicago Press, Chicago.
Kohlberg, L. (1976). Moral stages and moralization: the cognitive-developmental approach. In *Moral development and behavior*, (ed. T. Lickona), pp. 31–53. Holt, Rinehart and Winston, New York.
Kohler, W. (1925). *The mentality of apes*. Routledge and Kegan Paul, London.
Konner, M. J. (1975). Relations among infants and juveniles in comparative perspective. In *Friendship and peer relations* (ed. M. Lewis and L. A. Rosenblum). Wiley, New York.
Krebs, J. R. and Dawkins, R. (1984). Animal signals: mind reading and manipulation. In *Behavioural ecology: an evolutionary approach* (ed. J. R. Krebs and R. Dawkins), pp. 380–401. Blackwell.
Krebs, J., MacRoberts, M. and Cullen, J. M. (1972). Flocking and feeding in great tit *Parus major*—an experimental psychology. *Ibis* **114**, 507–30.
Krout, M. H. (1931). The psychology of children's lies. *Journal of Abnormal and Social Psychology* **26**, 1–27.
Kummer, H. (1957). Soziales verhalten einer mantepaviangruppe. Schweiz *Zeitschrift für Psychologie* **33**, 1–91.
Kummer, H. (1965). A comparison of social behavior in captive and free-living hamadryas baboons. In *The baboon in medical research* (ed. H. Vagtborg), pp. 65–80. University of Texas Press, Austin.
Kummer, H. (1968). *Social organization of hamadryas baboons*. University of Chicago Press, Chicago.
Kummer, H. (1971). *Primate societies*. Aldine Press, Chicago.
Kummer, H. (1982). Social knowledge in free-ranging primates. In *Animal mind—human mind* (ed. D. R. Griffin), pp. 113–30. Springer-Verlag.
Kummer, H. and Goodhall, J. (1985). Conditions of innovative behaviour in primates. In *Animal Intelligence* (ed. L. Weiskrantz), pp. 203–14. Oxford University Press, Oxford.
Kummer, H. and Kurt, F. (1963). Social units of a free-living population of hamadryas baboons. *Folia Primatologica* **1**, 4–19.
Kurland, J. A. (1977). Kin selection in the Japanese monkey. *Contributions to Primatology* **12**. S. Karger, Basel.
LaFreniere, P. and Charlesworth, W. R. (1983). Dominance, attention and affiliation in a preschool group: a nine-month longitudinal study. *Ethology and Sociobiology* **4**, 55–67.
Lancaster, J. B. (1975). *Primate behavior and the emergence of human culture*. Holt, Rinehart and Winston, New York.
Lati Rimpochay, Denma Rimpochay, Zahler, L. and Hopkins, J. (ed.) (1983). *Meditative states in Tibetan Buddhism*. Wisdom, London.
Leach, E. (1973). Concluding address. In *The explanation of culture change: models in pre-history* (ed. C. Renfrew), pp. 762–71. Duckworth, London.
LeDoux, J. E. (1982). Neuroevolutionary mechanisms of cerebral asymmetry in man. *Brain, Behavior and Evolution* **20**, 196–212.
Lee, P. C. (1981). Ecological and social influences on development of vervet monkeys. Unpublished PhD thesis, University of Cambridge.

Lee, P. C. (1983). Context-specific unpredictability in dominance interactions. In *Primate social relationships* (ed. R. A. Hinde), pp. 35–44. Blackwell, Oxford.
Lee, P. C. (1986). Environmental influences on development: play, weaning and social structure. In *Primate ontogeny, cognition, and social behavior* (ed. J. G. Else and P. C. Lee), pp. 227–38. Cambridge University Press, Cambridge.
Leggett, T. (1978). *Zen and the ways*. Routledge and Kegan Paul, London.
Lehrman, D. S. (1953). A critique of Konrad Lorenz's theory of instinctive behaviour. *Quarterly Review of Biology* **28**, 337–63.
Leigh, E. E. Jr. and Smythe, N. (1978). Leaf production, leaf consumption and the regulation of folivory on Barro Colorado Island. In *The ecology of arboreal folivores* (ed. G. G. Montgomery), pp. 33–50. Smithsonian Press, Washington D.C.
Leresche, L. A. (1976). Dyadic play in hamadryas baboons. *Behaviour* **57**, 190–205.
Leslie, A. M. (1987). Pretense and representation in infancy the origins of 'theory of mind'. *Psychological Review* **94**, 412–26.
Leutenegger, W. (1973). Encephalization in australopithecines: a new estimate. *Folia Primatologica* **19**, 9–17.
Levi-Strauss, C. (1962). *The savage mind*. Weidenfeld and Nicolson, London.
Lewis, D. (1974). Radical interpretation. *Synthese* **23**, 331–44.
Lewis, M. and Brooks, J. (1979) The search for the origins of self: implications for social behavior and interaction. *Proceedings of the Symposium on the ecology and education of children under 3*, West Berlin.
Lewis, M. and Brooks-Gunn, J. (1979). *Social cognition and the acquisition of self*, Plenum Press, New York.
Lorenz, K. (1935). Der kumpan in der unwelt des vogels. *Journal für Ornithologie* **83**, 137–203, 289–413.
Lorenz, K. (1957). *King Solomon's ring*. Pan, London.
Lorenz, K. (1975). *Evolution and modification of behavior*. Chicago University Press, Chicago.
MacArthur, R. H. and Pianka. E. R. (1966). On optimal use of a patchy environment. *American Naturalist* **100**, 603–9.
Mackay, D. (1965). A mind's eye view of the brain. In *Progress in brain research* Vol. 17, Cybernetics of the Nervous System, (ed. N. Wiener and J. P. Shade) pp. 321–32. Elsevier, New York.
Mackinnon, J. (1978). *The ape within us*. Collins, London.
Mackintosh, N. (1984). In search of a new theory of conditioning. In *The understanding of animals* (ed. G. Ferry), pp. 22–31. Blackwell, Oxford.
MacNamara, J. (1982). *Names for things*. M.I.T. Press, Cambridge, Massachusetts.
Macphail, E. M. (1982). *Brain and intelligence in vertebrates*. Clarendon Press, Oxford.
Macphail, E. M. (1985). Vertebrate intelligence: the null hypothesis. In *Animal intelligence* (ed. L. Weiskrantz). Clarendon Press, Oxford.
Mace, G. M., Harvey, P. H., and Clutton-Brock, T. H. (1981) Brain size and ecology in small mammals. *Journal of Zoology*, **193**, 333–54.

Main, M. (1980). Avoidance in the service of proximity: the biological origins of detachment and defensive processes. In *Behavioural development: the Bielefeld interdisciplinary project* (ed. K. Immelmann, G. W. Barlow, M. Main, and L. Petrinovitch. Cambridge University Press, Cambridge.

Mainardi, D. (1980). Tradition and the social transmission of behavior in animals. In *Sociobiology: beyond nature/nurture?* (ed. G. W. Barlow and J. Silverbert), pp. 227–56. Westview Press Inc., Colorado.

Marchand, A. (1974). A comparison of children's use of classification and seriation operations in the physical and social domains. *Child Study Journal* **4**, 179–93.

Marcus, D. E. and Overton, W. F. (1978). The development of cognitive gender constancy and sex-role preferences. *Child Development* **49**, 434–44.

Marsden, H. M. (1968). Agonostic behaviour of young rhesus monkeys after changes induced in social rank of their mothers. *Animal Behaviour* **16**, 38–44.

Martin, R. D. (1981). Relative brain size and basal metabolic rate in terrestrial vertebrates. *Nature* **293**, 57–60.

Martin, R. D. (1983). *Human brain evolution in an ecological context*. American Museum of Natural History, New York.

Mason, W. A. (1964). Sociability and social organisation in monkeys and apes. In *Recent Advances in Experimental Psychology* Vol. 1 (ed. L. Berkowitz), pp. 277–305.

Mason, W. A. (1970). Chimpanzee social behaviour. In *The Chimpanzee* (ed. G. Bourne, S. Karger), Basel.

Mason, W. A. (1978). Ontogeny of social systems. In *Recent advances in primatology*, Vol. 1 (ed. D. J. Chivers and J. Herbert), pp. 5–14. Academic Press, London.

Massey, A. (1977). Agonistic aids and kinship in a group of pigtail macaques. *Behavioural Ecology and Sociobiology* **2**, 31–40.

Maynard Smith, J. and Parker, G. A. (1976). The logic of asymmetric contests. *Animal Behaviour* **24**, 159–175.

McFarland, D. (1966). On the causal and functional significance of displacement activities. *Zeitschrift für Tierpsychologie* **23**, 217–35.

McFarland, D. J. (1976). Form and function in the temporal organisation of behaviour. In *Growing points in ethology* (ed. P. P. G. Bateson and R. A. Hinde). Cambridge University Press, Cambridge.

McFarland, D. J. and Houston, A. (1981). *Quantitative ethology: the state space approach*. Pitman, London.

McGonigle, B. O. and Chalmers, M. (1977). Are monkeys logical? *Nature* **267**, 694–6.

McGrew, W. C. (1983). Animal foods in the diet of wild chimpanzees: why cross-cultural variation? *Journal of Ethology* **1**, 46–61.

McGrew, W. C. and Tutin, C. E. G. (1978). Evidence for a social custom in wild chimpanzees. *Man* **13**, 234–51.

McGrew, W. C., Tutin, C. E. G. and Baldwin, P. J. (1979). Chimpanzees, tools, and termites: cross-cultural comparisons of Senegal, Tanzania and Rio Muni. *Man* **14**, 185–214.

McNab, B. (1980). Food habits, energetics and the population biology of mammals. *American Naturalist* **116**, 106–24.

McNamara, J. and Houston, A. (1985). A simple model of information use in the exploitation of patchily distributed food. *Animal Behaviour* **33**, 553–60.

Mead, G. K. (1934). *Mind, self and society*. University of Chicago Press, Chicago.

Meltzoff, A. N. and Moore, M. K. (1983). Newborn infants imitate adult facial gestures. *Child Development* **54**, 702–9.

Menzel, E. (1971). Communication about the environment in a group of young chimpanzees. *Folia Primatologica* **15**, 220–2.

Menzel, E. (1974). A group of young chimpanzees in a one-acre field. In *Behaviour of non-human primates* (ed. A. M. Schrier and F. Stollnitz), Vol. 5. Academic Press, New York.

Menzel, E. W. and Davenport, R. K. (1962). The effects of stimulus presentation variables upon chimpanzees' selection of food by size. *Journal of Comparative and Physiological Psychology* **55**, 235–9.

Menzel, E. W. and Draper, W. A. (1965). Primate selection of food by size: visible versus invisible rewards. *Journal of Comparative Physiological Psychology* **59**, 231–9.

Menzel, E. W. and Johnson, M. K. (1976). Communication and cognitive organisation in humans and other animals. *Annals of the New York Academy of Sciences* **280**, 131–42.

Mertl Millholen, A. (1977). Habituation to territorial scent marks in the field by *Lemur catta*. *Behavioural Biology* **21**, 500–7.

Milgram, S. (1974). *Obedience to authority: an experimental view*. Harper and Row, New York.

Miller, N. E. (1937). Analysis of the form of conflict reaction. *Psychological Bulletin* **34**, 720.

Miller, N. E. (1944). Experimental studies of conflict. In *Personality and the behaviour disorders* (ed. Hunt). Ronald, New York.

Miller, N. E. and Dollard, J. (1941). *Social learning and imitation*. Yale University Press, New Haven.

Miller, W. C. (1950). Communication to a meeting of the Institute for the Study of Animal Behaviour. (Private communication).

Milton, K. (1977a). The foraging strategy of the Howler Monkey in the tropical forest of Barro Colorado Island, Panama. Unpublished PhD thesis, University of New York.

Milton, K. (1977b). Food search strategy in the Barro Colorado Howler Monkey. In *Stability of tropical environments and populations*, IV International Symposium on Tropical Ecology, Panama, 7–11 March, 1977 (ed. H. Wolda) pp. 3–16. La Nacion Press, Panama.

Milton, K. (1977c). Behavioural adaptations to diet by the Barro Colorado Howler Monkey. *American Journal of Physical Anthropology* **47**, 150–1.

Milton, K. (1979). Factors influencing leaf choice by Howler Monkeys: a test of some hypotheses of food selection by generalist herbivores. *American Naturalist* **114**, 362–78.

Milton, K. (1980). *The foraging strategy of Howler Monkeys: a study of primate economics*. Columbia University Press, New York.

Milton, K. (1981). Food choice and digestive strategies of two sympatric primate species. *American Naturalist* **117**, 496–505.

Milton, K. (1985). Mating patterns of woolly spider monkeys, Brachyteles archnoides: implications for female choice. *Behavioural Ecology and Sociobiology* **17**, 53–9.

Milton, K. (1986). Features of digestive physiology in primates. *News in Physiological Sciences* **1**.

Milton, K. (1987). Primate diets and gut morphology: implications for hominid evolution. In *Food and evolution: toward a theory of human food habits* (ed. M. Harris and E. B. Ross), pp. 96–116. Temple University Press, Philadelphia.

Milton, K. and May, M. (1976). Body weight, diet and home range in primates. *Nature* **259**, 459–62.

Milton, K., Van Soest, P. J. and Robertson, J. (1980). Digestive efficiencies of wild Howler Monkeys. *Physiological Zoology* **53**, 402–9.

Minkeka, S., Davidson, M., Cook, M. and Keir, R. (1984). Observational conditioning of snake fear in rhesus monkeys. *Journal of Psychology* **15**, 355–72.

Mitchell, R. W. and Thompson, N. S. (ed.) (1986). *Deception: perspectives on human and nonhuman deceit*. State University of New York Press.

Morris, C. (1955). Foundations of the theory of signs. In *International encylopedia of unified science* (ed. O. Neurath, R. Carnap and C. Morris), pp. 77–137. Chicago University Press, Chicago, Illinois.

Morris, D. (1962). *The biology of art*. Knopf, New York.

Morris, J. G. and Rogers, Q. (1983). Nutritionally related metabolic adaptations of carnivores and ruminants. In *Plant, animal and microbial adaptations to terrestrial environments* (ed. N. S. Margaris, M. Arianoutsou-Faraggitaki, and R. J. Reiter), pp. 165–80. Plenum, New York.

Morton, E. (1977). On the occurrence and significance of motivation-structural rules in some bird and mammal sounds. *American Naturalist* **111**, 855–69.

Munn, C. A. (1986). Birds that 'cry wolf'. *Nature* **319**, 143–5.

Neill, S. R. St. J. (1976). Aggressive and non-aggressive fighting in 12 to 13 year old pre-adolescent boys. *Journal of Child Psychology and Psychiatry* **17**, 213–20.

Neisser, U. (1963). The multiplicity of thought. *British Journal of Psychology* **54**, 1–14.

Netto, W. J. and van Hoof, J. A. R. A. M. (1986). Conflict interference and the development of dominance relationships in immature *Macaca fascicularis*. In *Primate ontogeny, cognition and social behaviour* (ed. J. G. Else and P. C. Lee), pp. 291–300. University of Cambridge Press, Cambridge.

Newell, A. (1982). The knowledge level. 1980 presidential address, American Association for Artificial Intelligence. *Artificial Intelligence* **18**, 87–127.

Nishida, T. (1987). Local traditions and cultural transmission. In *Primate societies* (ed. B. B. Smuts, D. L. Cheney, R. M. Seyfarth, R. W. Wrangham, and T. T. Struhsaker). Chicago University Press, Chicago.

Oakley, K. (1959). *Man the tool-maker*. University of Chicago Press, Chicago.

Oatley, K. (1985). Representations of the physical and social world. In *Brain and Mind* (ed. D. A. Oakley). Methuen, London.

Oden, D. L., Thompson, R. K. R., and Lemack, D. (in press). Spontaneous transfer of matching by input chimpanzees (*Pan troglodytes*). *Journal of Experimental Psychology: Animal Behaviour Processes*.

O'Keefe, J. and Nadel, L. (1978). *The hippocampus as a cognitive map.* Oxford University Press, Oxford.
Olivier, R. C. D. and Laurie, A. (1974). Habitat utilization by hippopotamus in the Mara River. *East African Wildlife Journal* **12**, 249–71.
Omark, D. R. and Edelman, M. S. (1975). A comparison of status hierarchies in young children: an ethological approach. *Social Science Information* **14**, 87–107.
Owens, N. W. (1975). Social play behaviour in free-living baboons, *Papio anubis. Animal Behaviour* **23**, 387–408.
Packer, C. (1977). Reciprocal altruism in olive baboons. *Nature* **265**, 441–3.
Packer, C. and Pusey, A. (1982). Cooperation and competition within coalitions of male lions: kin selection or game theory? *Nature* **296**, 740–2.
Palameta, B. and Lefebvre, L. (1985). The social transmission of a food-finding technique in pigeons: what is learned? *Animal Behaviour* **33**, 892–6.
Paradise, E. and Curcio, F. (1974). Relationship of cognitive and affective behaviors to fear of strangers in male infants. *Developmental Psychology* **10**, 476–83.
Parker, G. A. (1984). Evolutionarily stable strategies. In *Behavioural ecology* (ed. J. R. Krebs and N. B. Davies), pp. 30–61. Blackwell Scientific Publications, Oxford.
Parker, S. and Gibson, K. (1979). A developmental model for the evolution of language and intelligence in early hominids. *The Behavioral and Brain Sciences* **2**, 367–408.
Passingham, R. E. (1981). Primate specialisation in brain and intelligence. *Symposium of the Zoological Society, London* **46**, 361–88.
Passingham, R. E. (1982). *The human primate.* W. H. Freeman, New York.
Payne, R. B. (1985). Behavioural continuity and change in local song populations of village indigobirds, *Vidua chalybeata. Zeitschrift für Tierpsychologie* **70**, 1–47.
Petter, J. J. (1962). Recherches sur l'écologie et l'éthologie des lémuriens Malgaches. *Memoires du Musée National de l'Histoire Naturelle, Serie A* **27**, 1–146.
Petter, J. J. (1965). The lemurs of Madagascar. In *Primate behavior* (ed. I. De Vore). Holt, Rinehart and Winston, New York.
Piaget, J. (1932/1965). *The moral judgment of the child.* The Free Press, New York.
Piaget, J. (1952). *The child's conception of number.* Routledge and Kegan Paul, London.
Piaget, J. (1963a). *The psychology of intelligence.* International Universities Press, New York.
Piaget, J. (1963b). *The origins of intelligence in children.* Norton, New York.
Piaget, J. and Inhelder, B. (1967). *The child's conception of space* (trans. F. Langlon and J. Lunzer). Norton, New York.
Piaget, J. and Inhelder, B. (1969). *The psychology of the child.* Routledge and Kegan Paul, London.
Parker, S. T. and Gibson, K. R. (1975). A developmental model of the evolution of language and intelligence in early hominids. *The Behavioural and Brain Sciences* **2**, 367–408.
Pickert, S. M. and Wall, S. M. (1981). An investigation of children's perspectives of dominance relations. *Perceptual and Motor Skills* **52**, 75–81.

Plotkin, H. C. and Odling-Smee, F. J. (1981). A multiple level model of evolution and its implications for sociobiology. *The Behavioral and Brain Sciences* **4**, 225–68.
Potts, R. (1984). Home base and early hominids. *American Scientist* **72**, 338–48.
Pollock, J. I. (1977). The ecology and sociology of feeding in *Indri indri*. In *Primate ecology: studies of feeding and ranging behaviour in lemurs, monkeys and apes* (ed. T. H. Clutton-Brock), pp. 38–71. Academic Press, London.
Poulson, K., Kintsch, E., Kintsch, W. and Premack, D. (in press). Children's comprehension and memory for stories. *Journal of Experimental Child Psychology*.
Premack, D. (1973). Cognitive principles? In *Contemporary approaches to conditioning and learning* (ed. F. J. McGuigan and D. B. Lumsden), pp. 287–319.
Premack, D. (1976). *Intelligence in ape and man*. Lawrence Erlbaum, Hillsdale.
Premack, D. (1983). The codes of man and beast. *The Behavioral and Brain Sciences* **6**, 125–67.
Premack, D. (1984). Upgrading a mind. In *Talking minds* (ed. T. G. Bever, J. M. Carroll, and L. A. Miller). MIT Press, Cambridge, MA.
Premack, D. (1986). *Gavagai! or the future history of the animal language controversy*. MIT Press, Cambridge MA.
Premack, D. and Premack, A. (1982). *The mind of an ape*. Norton, New York.
Premack, D. and Woodruff, G. (1975). Problem-solving in chimpanzee: test for comprehension. *Science* **202**, 532–5.
Premack, D. and Woodruff, G. (1978). Does the chimpanzee have a theory of mind? *The Behavioral and Brain Sciences* **1**, 515–26.
Pylyshyn, Z. (1980). Computation and cognition: Issues in the foundations of cognitive science. *The Behavioral and Brain Sciences* **3**, 111–32.
Quiatt, D. (1984). Devious intentions of monkeys and apes? In *The meaning of primate signals* (ed. R. Harre and V. Reynolds), pp. 9–40. Cambridge University Press.
Quine, W. V. O. (1960). *Word and object*. MIT Press, Cambridge, MA.
Quirling, D. P. (1950). *Functional anatomy of the vertebrates*. McGraw-Hill, New York.
Reynolds, P. C. (1981). *On the evolution of human behavior*. University of California Press, Berkeley.
Reynolds, V. (1976). *The biology of human action*. Freeman, San Francisco.
Reynolds, V. (1986). Primate social thinking. In *Primate ontogeny: cognition and social behaviour* (ed. J. G. Else and P. C. Lee). Cambridge University Press.
Rheingold, H. L. and Eckerman, C. O. (1969). The infant separates himself from his mother. *Science* **168**, 78–83.
Richard, A. (1977). The feeding behaviour of *Propithecus verreauxi*. In *Primate ecology: studies of feeding and ranging behaviour in lemurs, monkeys and apes* (ed. T. H. Clutton-Brock), pp. 71–96. Academic Press, London.
Richard, A. (1978). *Behavioural variation*. Bucknell University Press, Lewisburg.
Richard, A. (1985). Social boundaries in a Malagasy prosimian, the sifaka (*Propithecus verreauxi*). *International Journal of Primatology* **6**, 553–68.
Rightmire, G. P. (1981). Stasis in the evolution of *Homo erectus*. *Paleobiology* **7**, 200–15.

Rightmire, G. P. (1986). Stasis in *Homo erectus* defended. *Paleobiology* 12, 324–5.
Ristau, C. (in press). Injury feigning and other anti-predator behaviors by plovers: intentional behaviour? In *The minds of other animals* (ed. C. Ristau and P. Marler). Lawrence Erlbaum, Hillsborough, N.J.
Robinson, J. G. (1986) Seasonal variation in the use of time and space by wedge-capped capuchins, *Cebus olivaceus*. *Smithsonial contributions to zoology*.
Rodman, P. S. (1977). Feeding behaviour of orangutans in the Kutai Nature Reserve, East Kalimantan. In *Primate ecology: studies of feeding and ranging behaviour in lemurs, monkeys and apes* (ed. T. H. Clutton-Brock), pp. 384–414. Academic Press, New York.
Roitblat, H. L. (1982). The meaning of representation in animal memory. *The Behavioral and Brain Sciences* 5, 352–406.
Rosa, K. R. (1976). *Autogenic training*. Gollancz, London.
Rozin, P. (1976). The evolution of intelligence and access to the cognitive unconscious. In *Progress in psychology*, 6 (ed. J. N. Sprague and A. N. Epstein), pp. 245–80. Academic Press, New York.
Russell, B. (1905). On denoting. *Mind* 479–93.
Russell, C. and Russell, W. M. S. (1961). *Human behaviour*. Andre Deutsch, London.
Russell, J. K. (1983). Altruism in coati bands; nepotism or reciprocity. In *Social behavior of female vertebrates* (ed. S. K. Wasser), pp. 263–90. Academic Press, New York.
Russell, J. (1984). *Explaining mental life*. Macmillan, London.
Ryle, G. (1949). *The concept of mind*. Hutchinson, London.
Saarni, C. (1979). Children's understanding of display rules for expressive behaviour. *Developmental Psychology* 15, 424–9.
Saarni, C. (1984). An observational study of children's attempts to monitor their expressive behavior. *Child Development* 55, 1504–13.
Sacher, G. A. (1982). The role of brain maturation in the evolution of the primates. In *Primate brain evolution* (ed. E. Armstrong and D. Falk), pp. 97–112. Plenum Press, New York.
Sade, D. S. (1967). Determinants of dominance in a group of free-ranging rhesus monkeys. In *Social communication among primates* (ed. S. A. Altmann), pp. 94–114. University of Chicago Press, Chicago.
Sahlins, M. (1974). *Stone age economics*. Tavistock Publications, London.
Savage-Rumbaugh, E. S. (1986). *Ape language: from conditioned response to symbol*. Colombia University Press, New York.
Savage-Rumbaugh, E. S., Rumbaugh, D. M., Smith, S. T., and Lawson, J. (1980). Reference: the linguistic essential. *Science* 210, 922–5.
Savage-Rumbaugh, E. S., McDonald, K., Sevek, R., Hopkins, W. D., and Rubert, E. (1986). Spontaneous symbol acquisition and communicative use by pygmy chimpanzee (*Pan paniscus*). *Journal of Experimental Psychology: General* 115, 211–35.
Sayre, K. M. (1976). *Cybernetics and the philosophy of mind*. Routledge and Kegan Paul, London.

Scarr, S. and Salapatek, P. (1970). Patterns of fear development during infancy. *Merril-Palmer Quarterly*, **16**, 53–90.

Schaffer, H. R. (1974). Cognitive components of the infant's response to strangers. In *The origins of fear* (ed. M. Lewis and L. A. Rosenblum). Wiley, New York.

Schank, R. C. (1976). Research at Yale in natural language processing. Research report 84, Yale University Department of Computer Science.

Scheurer, J., and Thierry, B. (1985). A further food-washing tradition in Japanese macaques. *Primates* **26**, 491–4.

Schiller, P. H. (1957). Manipulation patterns in the chimpanzee. In *Instinctive behavior* (ed. C. H. Schiller), pp. 264–87. International University Press, New York.

Schoener, T. W. (1971). Theory of feeding strategies. *Annual Review of Systematics and Biology* **2**, 369–403.

Schultz, A. H. (1968). The recent hominoid primates. In *Perspectives in human evolution* (ed. S. Washburn and P. C. Jay), pp. 122–95. Holt, Rinehart and Winston, New York.

Schwartz, R. (1980). How rich a theory of mind? *The Behavioral and Brain Sciences* **3**, 616–8.

Scott, D. K. (1980). Functional aspects of prolonged parental care in Bewick's swans. *Animal Behaviour* **28**, 938–52.

Searle, J. (1980). Mind, brains and programs. *The Behavioral and Brain Sciences* **3**, 417–57.

Sekida, K. (1975). *Zen training, methods and philosophy*. Weatherhill, New York.

Seligman, M. E. P. (1970). On the generality of the laws of learning. *Psychological Review* **77**, 406–18.

Seligman, M. E. P. and Hager, J. J. (ed.) (1972). *Biological boundaries of learning*. Appleton-Century-Crofts, New York.

Selman, R. L. (1981). The child as a friendship philosopher. In *The development of children's friendships* (ed. S. R. Asher and J. M. Gottman), pp. 242–72. Cambridge University Press, Cambridge.

Seyfarth, R. M. (1977). A model of social grooming among adult female monkeys. *Journal of Theoretical Biology* **65**, 671–98.

Seyfarth, R. M. (1980). The distribution of grooming and related behaviours among adult female vervet monkeys. *Animal Behaviour* **28**, 798–813.

Seyfarth, R. M. (1981). Do monkeys rank each other? *The Behavioral and Brain Sciences* **4**, 447–8.

Seyfarth, R. M. (1983). Grooming and social competition in primates. In *Primate social relationships* (ed. R. A. Hinde), pp. 182–90. Blackwell Scientific Publications, Oxford.

Seyfarth, R. M., and Cheney, D. L. (1984). Grooming, alliances and reciprocal altruism in vervet monkeys. *Nature* **308**, 541–3.

Seyfarth, R. M., Cheney, D. L., and Marler, P. (1980a). Vervet monkey alarm calls. *Animal Behaviour* **28**, 1070–94.

Seyfarth, R. M., Cheney, D. L., and Marler, P. (1980b). Monkey responses to three different alarm calls: evidence of predator classification and semantic communication. *Science* **210**, 801–3.

Shallice, T. (1972). Dual functions of consciousness. *Psychological Review*, **79**, 383–93.
Shannon, C. (1949). *The mathematical theory of communication* (with an introductory essay by Warren Weaver). University of Illinois Press.
Sherrod, L. R. (1981). Issues in cognitive-perceptual development: the special case of social stimuli. In *Infant social cognition* (ed. M. E. Lamb and L. R. Sherrod), pp. 11–36; Lawrence Erlbaum Associates, Hillsdale, New Jersey.
Sherry, D. F. and Galef, B. G., Jr (1984). Cultural transmission without imitation: milk bottle opening by birds. *Animal Behaviour* **32**, 937–8.
Shultz, T. R. and Cloughesy, K. (1981). Development of recursive awareness of intention. *Developmental Psychology* **17** 465–71.
Shultz, T. R., Wells, D., and Sarda, M. (1980). Development of the ability to distinguish intended actions from mistakes, reflexes and passive movements. *The British Journal of Social and Clinical Psychology* **19**, 301–10.
Sigg, H. (1980). Differentiation of female positions in hamadryas one-male units. *Zeitschrift für Tierpsychologie* **53**, 265–302.
Sigg, H. and Stolba A. (1981). Home range and daily march in a hamadryas baboon troop. *Folia Primatologica* **36**, 40–75.
Silk, J. B. (1982). Altruism among female *Macaca radiata*: explanations and analysis of grooming and coalition formation. *Behaviour* **79**, 162–88.
Silk, J. B., Samuels, A., and Rodman, P. S. (1981). The influence of kinship, rank and sex on affiliation and aggression between adult female and immature bonnet macaques (*Macaca radiata*). *Behaviour* **78**, 111–37.
Simons, E. L. (1964). The early relatives of man. *Scientific American* **211**, 50.
Simpson, G. G. (1945). The principles of classification and a classification of the mammals. *Bulletin of American Museum of Natural History* **85**, 1.
Slater, P. J. B. (1983). The study of communication. In *Animal behaviour, Vol. 2: communication* (ed. T. R. Halliday and P. J. B. Slater), pp. 9–42. Blackwell, Oxford.
Sluckin, A. M. (1979). Avoiding violence in the playground. *Educational Research* **21**, 82–3.
Sluckin, A. M. (1981). *Growing up in the playground: the social development of children*. Routledge and Kegan Paul, London.
Sluckin, A. M. and Smith, P. K. (1977). Two approaches to the concept of dominance in preschool children. *Child Development* **48**, 917–23.
Smith, P. K. (1979). The ontogeny of fear in children. In *Fear in animals and man* (ed. W. Sluckin). Van Nostrand Reinhold, Wokingham.
Smith, P. K. (1982). Does play matter? Functional and evolutionary aspects of animal and human play. *The Behavioral and Brain Sciences* **5**, 139–84.
Smith, P. K. and Delfosse, P (1980). Accuracy of reporting own and others companions in young children. *British Journal of Social and Clinical Psychology* **19**, 337–8.
Smith, P. K. and Lewis, K. (1985). Rough and tumble play, fighting and chasing in nursery school children. *Ethology and Sociobiology* **6**, 175–81.
Smith, P. K. and Sloboda, J. (1986). Individual consistency in infant-stranger encounters. *British Journal of Developmental Psychology* **4**, 83–91.
Smuts, B. (1985). *Sex and friendship in baboons*. Aldine Publishing Co., Chicago.

Snell, G. D. (ed.) (1941). *Biology of the laboratory mouse*. Blakiston, Philadelphia.
Sokolov, E. N. (1960). Neuronal models and the orienting reflex. In *The central nervous system and behaviour*. Macy, New York.
Sordahl, T. A. (1981). Sleight of wing. *Natural History* **90**, 43–9.
Sroufe, L. A. (1977). Wariness of strangers and the study of infant development. *Child Development* **48**, 731–46.
Stammbach, E. (1978). On social differentiation in groups of captive female hamadryas baboons. *Behaviour*, **67**, 322–38.
Stammbach, E. and Kummer, H. (1982). Individual contributions to a dyadic interaction: an analysis of baboon grooming. *Animal Behaviour* **30** 964–71.
Sternberg, R. J. (1984). A contextualist view of the nature of intelligence. *International Journal of Psychology* **19**, 307–34.
Sternberg, R. J., Conway, B. E., Ketron, J. L., and Bernstein, M. (1981). People's conceptions of intelligence. *Journal of Personality and Social Psychology* **41**, 37–55.
Strayer, F. F., Wareing, S., and Rushton, J. P. (1979). Social constraints on naturally occurring preschool altruism. *Ethology and Sociobiology* **1**, 3–11.
Struhsaker, T. T. (1967a). Auditory communication in vervet monkeys (*Cercopithecus aethiops*). In *Social communication among primates* (ed. S. A. Altmann), pp. 281–324. University of Chicago Press, Chicago.
Struhsaker, T. T. (1967b). Social structure among vervet monkeys (*Cercopithecus aethiops*). *Behaviour* **29**, 83–121.
Strum, S. C. (1981). Processes and products of change: baboon predatory behavior at Gilgil, Kenya. In *Omnivorous primates* (ed. R. Harding and G. Teleki), pp. 255–302. Columbia University Press, New York.
Sullivan, H. S. (1955). *The interpersonal theory of psychiatry*. Tavistock, London.
Sutherland, E., and Smith, M. T. (1980). *The exercise of intelligence: the biosocial preconditions for the operation of intelligence*. Garland Publishing, New York.
Swenson, C. H. (1973). *Introduction to interpersonal relations*. Scott, Foresman, Brighton.
Takasaki, H. (1983). Mahale mountain chimpanzees taste mangoes . . . toward acquisition of a new food item? *Primates* **24**, 273–5.
Taylor, E., Steele, C. and Roberto, K. (1982). Preschool children's discrimination of age. *Perceptual and Motor Skills* **54**, 539–42.
Taylor, J. (1975). *Superminds*. Macmillan, London.
Taylor, L. and Sussman, R. W. (1985). A preliminary study of kinship and social organisation in a semi-free-ranging group of *Lemur catta*. *International Journal of Primatology* **6**, 601–15.
Teleki, G. (1974). Chimpanzee subsistence technology: materials and skills. *Journal of Human Evolution* **3**, 575–94.
Teleki, G. (1981). The omnivorous diet and eclectic feeding habits of chimpanzees in Gombe National Park, Tanzania. In *Omnivorous primates* (ed. R. Harding and G. Teleki), pp. 303–43. Columbia University Press, New York.
Terrace, H., Pettito, L, Saunders, R., and Bever, T. (1979). Can an ape create a sentence? *Science* **206**, 891–902.

Testa, T. J. (1974). Causal relationships and the acquisition of avoidance responses. *Psychological Reviews* **81**, 491–505.
Thompson, S. K. (1975). Gender labels and early sex-role development. *Child Development* **46**, 339–47.
Thorndike, E. L. (1911). *Animal intelligence.* Macmillan, New York.
Thornhill, R. (1979). Adaptive female-mimicking behaviour in a scorpion fly. *Science* **205**, 412–4.
Thorpe, W. H. (1956). *Learning and instinct in animals.* Methuen, London.
Thorpe, W. H. (1958). The learning of song patterns by birds, with especial reference to the song of chaffinch, Fringilla coelebs. *Ibis* **100**, 535–70.
Tinbergen, J. (1951). *The study of instinct.* Clarendon Press, Oxford.
Tinbergen, J. (1981). Foraging decisions in starlings (*Sturnus vulgaris*). *Ardea* **69**, 1–67.
Tobias, P. V. (1981). The emergence of man in Africa and beyond. *Philosophical Transactions of the Royal Society* (London) **B292**, 43–56.
Tobias, P. V. (1983). Hominid evolution in Africa. *Canadian Journal of Anthropology* **3**, 163–85.
Tolman, E. C. (1948). Cognitive maps in rats and men. *Psychological Review* **55**, 189–208.
Trevarthen, C. (1974). Conversations with a two-month old. *New Scientist* **62**, 230–5.
Trevarthen, C. (1982). The primary motives for cooperative understanding. In *Social cognition: studies of the development of understanding* (ed. G. Butterworth and P. Light), pp. 77–109. Harvester Press, Brighton.
Trivers, R. L. (1971). The evolution of reciprocal altruism. *Quarterly Review of Biology* **46** 35–57.
Trivers, R. (1985). *Social evolution.* Benjamin Cummings.
Tversky, A. and Kahneman, D. (1977). Casual schemata in judgments under uncertainty. In *Progress in social psychology* (ed. M. Fishbein). Lawrence Erlbaum Associates, Hillsdale, N.J.
Vandell, D. L. (1980). Sociability with peers and mother during the first year. *Developmental Psychology* **16**, 355–61.
Vasek, M. E. (1986). Lying as a skill: the development of deception in children. In *Deception: perspectives on human and nonhuman deceit* (ed. R. W. Mitchell and N. S. Thompson). State University of New York Press, Albany.
Vaughn, B. E. and Waters. E. (1980). Social organisation among preschool peers: dominance, attention and sociometric correlates. In *Dominance relations: an ethological view of human conflict and social interaction* (ed. D. R. Omark, F. F. Strayer and D. G. Freedman). Garland STPM Press, New York.
Vogel, C. (1985). Helping, cooperation, and altruism in primate societies. *Contributions to Zoology* **31**, 375–89.
Voltaire. (1759). *Candide*, Paris.
Waddington, C. H. (1960). *The ethical animal.* Allen and Unwin, London.
Walters, J. (1980). Interventions and the development of dominance relationships in female baboons. *Folia Primatologica* **34**, 61–89.
Ward, A. A. Jr. (1948a). The cingular gyrus—area 24. *Journal of Neurophysiology* **11**, 13–23.

Ward, A. A., Jr (1948b). The anterior cingular gyrus and personality. *Research Publications of the Association for Nervous and Mental Diseases* **27**, 438–45.

Warren, J. M. (1965). Primate learning in comparative perspective. In *Behavior of non-human primates* (ed. A. M. Schrier, H. F. Harlow, and F. Stollnitz). Academic Press, New York.

Washburn, S. L. (1960). Tools and human evolution. *Scientific American* **203**, 62–75.

Washburn, S. and Hamburg, D. A. (1965). The study of primate behaviour. In *Primate behavior* (ed. I. de Vore). Holt, Rinehart and Winston, New York.

Washburn, S. L., Jay, P. C., and Lancaster, J. B. (1965). Field studies of Old World monkeys and apes. *Science* **150**, 1541.

Wason, P. C. and Johnson-Laird, P. N. (1972). *Psychology of reasoning*. Batsford, London.

Watson, J. S. and Ramey, C. T. (1972). Reactions to response-contingent stimulation in early infancy. *Merrill Palmer Quarterly* **18**, 219–27.

Watts, D. P. (1985). Observations on the ontogeny of feeding behavior in mountain gorillas (*Gorilla gorilla beringei*). *American Journal of Primatology* **8**, 1–10.

Weinstein, B. (1945). The evolution of intelligent behaviour in rhesus monkeys, *Genetical Psychology Monographs* **31**, 3.

Weisbard, C. and Goy, R. (1976). Effect of parturition and group composition on competitive drinking order in stumptail macaques. *Folia Primatologica* **25**, 95–121.

Weiskrantz, L. (1985). Categorisation, cleverness and consciousness. In *Animal intelligence* (ed. L. Weiskrantz), pp. 3–19. Clarendon Press, Oxford.

Wenger, F. (1943). Racial differences in the colon in natives of Bolivia. *American Journal of Physiological Anthropology* **1**, 313–23.

West, M. (1979). Physiological effects of meditation: a longitudinal study. *British Journal of Social and Clinical Pyschology* **18**, 219–26.

Westoby, M. (1974). An analysis of diet selection by large generalist herbivores. *American Naturalist* **108**, 290–304.

Whalely, B. (1969). *Strategem: deception and surprise in war*. MIT Center for International Studies, Cambridge, Mass.

White, R. (1985). Thoughts on social relationships and language in hominid evolution. *Journal of Social and Personal Relationships* **2**, 95–115.

White, S. H. (1965). Evidence for a hierarchical arrangement of learning processes. In *Advances in child development and behaviour* (ed. L. P. Lipsitt and C. .C. Spiker). Academic Press, New York.

Whitehead, J. M. (1986). Development of feeding selectivity in mantled howling monkeys, *Allouatta palliata*. In *Primate ecology and conservation* (ed. J. Else and P. C. Lee). Cambridge University Press, Cambridge.

Whiten, A. and Byrne, R. W. (1986). The St. Andrews catalogue of tactical deception in primates. *St. Andrews Psychological Reports*, no. 10.

Whiten, A. and Byrne, R. W. (1988) Tactical deception in primates. *The Behavioral and Brain Sciences* **11** (2), in press.

Whiten, A. and Byrne, R. W. (in press). Machiavellian monkeys. Cognitive evolution and the social world of primates. In *Cognition and social worlds* (ed. A. R. H. Gellatly, D. R. Rogers, and J. A. Sloboda). Oxford University Press, Oxford.

Wickler, W. (1968). *Mimicry in plants and animals*. McGraw-Hill, New York.
Wiener, N. (1961). *Cybernetics* MIT Press, Cambridge, Massachusetts.
Wiley, R. H. (1983). The evolution of communication: information and manipulation. In *Animal behaviour, Vol. 2: communication* (ed. T. R. Halliday and P. J. B. Slater), pp. 156–89. Blackwell, Oxford.
Wilson, E. O., Durlach, N. I., and Roth, L. M. (1958). Chemical releasers of necrophoric behavior in ants. *Psyche* **65**, 108–14.
Wimmer, H., and Perner, P. (1983). Beliefs about beliefs; representation and constraining function of wronged beliefs in young children's understanding of deception. *Cognition* **13**, 103–28.
Wimmer, H., Hogrefe, J. and Sodian, R. (forthcoming). A second stage in children's conception of mental life: Understanding sources of information.
Winograd, T. (1970) Procedures as a representation for data in a computer program for understanding natural language. TR–84, M.I.T. Artificial Intelligence Laboratory Project MAC.
Wittenberger, J. (1981). *Animal social behavior*. Duxbury Press, Boston.
Wolpoff, M. (1986). Stasis in the intepretation of evolution in *Homo erectus*: a reply to Rightmire. *Paleobiology* **12**, 325–8.
Woodgush, D., Dawkins, M., and Ewbank, R. (ed.). (1981) *Self-awareness in domesticated animals*. University Federation for Animal Welfare, Potters Bar.
Woodruff, G., and Premack, D. (1979). Intentional communication in the chimpanzee: the development of deception. *Cognition* **7**, 333–62.
Woodruff, G. and Premack, D. (1981). Primitive mathematical concepts in the chimpanzee: proportionality and numerosity. *Nature* **293**, 568–70.
Woodruff, G., Premack, D., and Kennell, K. (1978). Conservation of liquid and solid quantity by the chimpanzee. *Science* **202**, 991–4.
Worthy, G. A. J. and Hicks, J. P. (1986). Relative brain size in marine mammals. *American Naturalist* **128**, 445–9.
Wrangham, R. W. (1975). Behavioural ecology of chimpanzees in Gombe National Park, Tanzania. Unpublished PhD thesis, University of Cambridge.
Wrangham, R. W. (1977). Feeding behaviour of chimpanzees in Gombe National Park, Tanzania. In *Primate ecology: studies of feeding and ranging behaviour in lemurs, monkeys and apes* (ed. T. H. Clutton-Brock), pp. 504–38. Academic Press, New York.
Wrangham, R. W. (1980). An ecological model of female-bonded primate groups. *Behaviour* **75**, 262–300.
Wrangham, R. W. (1981). Drinking competition in vervet monkeys. *Animal Behaviour* **29**, 906–10.
Wrangham, R. W. (1982). Mutualism, kinship and social evolution. In *Current problems in sociobiology* (ed. King's College Sociobiology Group), pp. 269–89. Cambridge University Press, Cambridge.
Wrangham, R. W., and Waterman, P. G. (1981). Feeding behaviour of vervet monkeys on *Acacia xanthophloea* and *Acacia tortilis*: with special reference to reproductive strategies and tannin production. *Journal of Animal Ecology* **50**, 715–31.
Wynn, T. (1978). Tool-using and tool-making. *Man* **13**, 137–8.
Wynn, T. (1979). The intelligence of later Acheulean hominids. *Man* **14**, 371–91.

Wynn, T. (1981). The intelligence of Oldowan hominids. *Journal of Human Evolution* **10**, 529–41.
Wynn, T. (1985). Piaget, stone tools, and the evolution of human intelligence. *World Archaeology* **17**, 32–43.
Wynn, T. (1986). Archaeological evidence for the evolution of modern human intelligence. In *The Pleistocene perspective*, vol. 1 (ed. M. Day and R. Foley). Proceedings of the World Archaeological Congress, Allen and Unwin, London.
Wynn, T. (in press). *The evolution of spatial competence*. University of Illinois Press, Urbana.
Wynne-Edwards, V. C. (1962). *Animal dispersion in relation to social behaviour*. Oliver and Boyd, Edinburgh.
Yerkes, R. M. and Elder, J. H. (1936). Oestrus, receptivity and mating in the chimpanzee. *Comparative Psychology Monograph* **13**, 39.
Zajonc, R. C. (1965). Social facilitation. *Science* **149**, 269–74.
Zimmerman, R. R. and Torrey, C. C. (1965). Ontogeny of learning. In *Behavior of non-human primates* (ed. A. M. Schrier, H. F. Harlow, and F. Stollnitz). Academic Press, New York.
Zivin, G. (1977). On becoming subtle: age and social rank changes in the use of a facial gesture. *Child Development* **48**, 1314–21.
Zuckerman, S. (1930). The menstrual cycle of the primates. I. General nature and homology. *Proceedings of the Zoological Society of London* 691–754.
Zuckerman, S. (1932). *The social life of monkeys and apes*. Kegan, Paul, Trench, Trubner, London.
Zuckerman, S. (1933). *Functional affinities of man, apes and monkeys*. Harcourt, New York.

Index

abstract representations, *see* representation
AI, *see* artificial intelligence
aggression, *see* agonistic interactions
agonistic interactions 3–4, 20, 36, 39–46, 57, 114–31, 134–52
  see also approach-avoidance conflict
alarm calls 189, 258
alliances 4, 8, 29, 101–2, 104–6, 126–52, 267, 317–26, 352, 376
altruism, reciprocal 58, 186, 266, 352, 358, 362
analogical reasoning 70, 170
apes, *see* phylogenetic differences
approach-avoidance conflict 35, 38, 43–5, 48
artificial intelligence 52, 58, 63–4, 196–7
attention 32, 214–20, 351, 354–8
attention structure 106–7
Australopithecus 276–7, 302
awareness, *see* self-reflection

behaviourism 61–2, 161, 180–1, 194
belief 160, 163–4, 172–9, 182, 184, 367–71
brain
  proportions 47–8, 271, 274–5
  size 30, 62, 64, 132–3, 271, 273–84, 293–304, 371–3
  see also metabolic rate

children
  deception, *see* developmental change
  social cognition 94–109, 163–4
  theory of mind, *see* mind-reading
  see also developmental psychology
cognitive development, *see* developmental change, in cognition
competition, *see* agonistic interactions
computation, *see* decision-making; information processing
concealment 212–23, 229–32
concepts, *see* social relationships; knowledge
conditioning 55, 168, 192
conflict, social 4–5, 34–7
consciousness 62, 349–62, 363, 365, 377
  see also self-reflection

cooperation 4, 304, 352
  see also alliances
counter-deception 220, 243
creativity 55, 191–2, 281–2, 327–43, 362
  see also intelligence
cultural transmission 18, 64, 284, 327–43, 348, 374
  see also imitation
curiosity 53

deception 58, 123–4, 158–9, 161–3, 167–71, 175, 194–5, 199–200, 205, 224–5, 238–9, 376
  in birds 161, 222, 370–1
  clues v. leakage 244–6, 371
  inhibition v. simulation 235–7, 239, 244–5, 247–9
  see also counter-deception; tactical deception
decision-making 19, 132, 136–44
developmental change
  in cognition 94, 95, 99–101
  in deception 227–8, 238–52, 371
  in social knowledge 76–7, 79–80, 94–109, 163
developmental psychology (as a discipline) 64, 161
distraction 161, 208, 212–23, 232–3
dominance, *see* rank

empathy 80–1, 352–58
encephalization, *see* brain, size
expertise, use and knowledge of other animals' 155–9, 309–26

flexibility of behaviour 52, 56, 65, 122, 143, 287–92
foraging as an intelligent behaviour 57, 269, 278–9, 285–304, 371–6
friendship, *see* alliances

game-playing 6, 19, 61
generation overlap, *see* life-history variables

hierarchy
 of control 62, 351
 social, see rank
Homo erectus 279–82
hunting 266, 272, 302

imaginary objects 208, 228, 232, 234–5
imitation 16, 20, 31, 95, 342
inference, transitive 103–4, 151, 256
information processing 63, 132–3, 136–40, 157, 180, 183, 196, 345
inhibition of behaviour 3, 162–3, 167–70, 213, 216–20, 244–6
inhibition of attention to objects 218–20, 221, 230–1
innovation, see creativity
insight 32, 348
intelligence
 creative 4–5, 15–17
 definitions of 13–15, 50–65
 everyday 50
 technical 2, 5, 7–8, 15–17, 21–2, 28, 31–2, 108, 134, 256–70
intentionality 60–2, 160–202, 238, 284, 351–8, 363–78
 levels of 61, 183–97, 215–23, 240–4
intentional system, see intentionality

kinship 91, 136–41
 see also social relationships
knowledge 51, 182, 255–70, 309–26, 370
 knowing how v. knowing that 51–2
 of social relationships, see social relationships
 see also representation

language
 necessity for cognition 160, 162, 172, 174
 original function 303
leadership 125–31, 155, 209, 213
leakage, see deception, clues v. leakage
learning 4–5, 28, 32, 116, 168–71, 287–92, 348
learning set 54
lemurs, see phylogenetic differences
life-history variables 58
local enhancement 335–6

meaning 71–4, 198–202
 see also representation
memory 58, 269, 288–93, 351, 370

mental map 372, 375
mental states, see theory of mind
metabolic rate 299–302
mind, theory of, see theory of mind
mind-reading 60–2, 157–202, 211, 215–16, 270
monkeys, see phylogenetic differences
mothers and offspring 113–23, 140, 143–4, 292

natural psychology, see mind-reading

object permanence 96–7
observational learning, see imitation; local enhancement; social facilitation

person permanence 96–7
personality, representation of 174
phylogenetic differences 4, 17, 21, 29–33, 46–7, 57, 132–5, 142, 145, 150–2, 158, 161, 164, 173, 175, 179, 201, 221–3, 267–9, 271, 278, 295–9, 368, 375–7
Piagetian theory 94–109, 278–82
planning 19, 357, 376
play 59, 114, 171
predation 30, 258, 263–5
pretence, see imaginary objects
problem-solving 3, 4, 6, 13–14, 23, 28, 54–7, 166, 250, 256
prosimians, see phylogenetic differences
protected threat 29, 113–21

rank, knowledge of 103–4, 107–8, 125–31, 132–52
reasoning, analogical, see analogical reasoning
reciprocation, see altruism, reciprocal
redirected aggression 69–70, 76–7, 79, 150
relationships, see social relationships
representation 62, 69–84, 92, 350
 declarative v. procedural 52
risk 137–9

seeing, comprehension of 163–9, 195, 215, 235
self, knowledge of, see sense of self
self-reflection 62–4, 98, 349, 352–62
sense of self 97–8, 352
seriation, see inference
social complexity 7, 20–1, 29, 57–9, 82–3, 113–52, 304, 347

Index 413

social concepts, *see* social relationships
social facilitation 118, 335–6
social learning, *see* imitation; local enhancement; social facilitation
social relationships 35–7
  known by animals/children 69–93, 101–4, 145–50, 256, 267–70, 368–9
  learned before object relationships 96–7, 250, 270
  used in redirected aggression 69–70, 76–7, 79, 150
social transmission, *see* cultural transmission; imitation; social facilitation
status, *see* rank
stimulus enhancement, *see* local enhancement
suppression of vocalization, *see* inhibition of behaviour
symbol-use in deception 226, 233–4, 236

tactical deception 56, 205–52, 370–1, 377
  categorization of 207–9, 213–4, 251
  defined 206
taxonomic differences, *see* phylogenetic differences
teaching 292
theory of mind 61, 98, 160–79, 215–23
  see also intentionality; mind-reading
tools 266, 270–84, 326–43, 373–4
transitive inference, *see* inference, transitive
triadic interaction 36–7, 39, 43–6, 57, 113–31, 207, 215, 217–8, 222–3, 228–9, 367
  see also alliances
tripartite relations, defined 114
  see also triadic interaction

unreliable signals, animals' reaction to 71–5, 78, 83, 206